ƒP

BATTLE
of WITS

The Complete Story of
Codebreaking in World War II

STEPHEN BUDIANSKY

THE FREE PRESS

New York | *London* | *Toronto* | *Sydney* | *Singapore*

*f*P

THE FREE PRESS
A Division of Simon & Schuster Inc.
1230 Avenue of the Americas
New York, NY 10020

Designed by Susan Hood

Manufactured in the United States of America

10   9   8   7   6   5   4   3   2   1

Library of Congress Cataloging-in-Publication Data

Budiansky, Stephen.
    Battle of wits: the complete story of codebreaking in World War II /
    Stephen Budiansky.
        p.   cm.
    Includes bibliographical references and index.
    1. World War, 1939–1945—Cryptography.   2. World War,
1939–1945—Military intelligence.   I. Title.

D810.C88 B83   2000
940.54'85—dc21
                                                                00-028418

ISBN 0-684-85932-7

In memory of Bernard Budiansky
1925–1999
"One father is more than a hundred schoolmasters"

Knowledge is more than equivalent to force.
—Dr. Johnson

It is not necessary to be crazy to be a cryptanalyst. But it always helps.
—Joseph Rochefort

Contents

List of Maps

Midway

GREEN water breaking across her flat topside, the U.S. carrier *Hornet* plunged through a heavy sea and strong wind. A few hours earlier, as dawn broke on April 18, 1942, lookouts aboard the American ship had spotted a Japanese patrol boat. The *Hornet*'s planes were intending to carry out their raid under cover of darkness, but Admiral William Halsey, whose nickname "Bull" had not been bestowed lightly, gave the order to launch at once—wind, wave, and daylight be damned. The ship swung into the wind, and at 7:25 A.M. the first of the twin-engine bombers groaned off the flight deck.

The Army pilots at the controls of the B-25s had practiced short takeoffs from the comfortably dry land of a Florida airstrip but had never once tried it at sea. No one else ever had, either. Landing a B-25 on a carrier was impossible. Flying a B-25 off a carrier was, by comparison, merely insane. But the medium-weight bombers were the only aircraft in the American arsenal with a prayer of completing the daredevil mission. If all went according to plan, they would fly five hundred miles to Japan, drop their load, then continue another eleven hundred miles to a safe landing in unoccupied China.

Leading the attack was an unflappable test pilot, Lieutenant Colonel James H. Doolittle; his was the first of the lumbering bombers to catapult down the heaving deck. Over the next hour fifteen others followed. One pilot hung on the verge of a stall for so long as he struggled to get airborne that, Halsey later recalled, "we nearly catalogued his effects." Thirteen of the planes headed for Tokyo, roared in over the rooftops from different directions, and dropped their four bombs apiece. The three others hit Nagoya and Osaka. Ever since the attack on Pearl Harbor four months earlier, President Franklin D. Roosevelt had been pressing for just such a morale-boosting coup to bolster some of America's wounded pride. The Doolittle raid had been his pet project, and he was exultant with the news. Asked by reporters where the planes had come from, FDR grinned and said, "Shangri-La."

Doolittle's raiders did essentially no damage—except to the Japanese psyche. On that they scored a direct hit. The Japanese Army claimed it shot down nine of the marauders; the true figure was zero. But the premature launch of the planes had added almost two hundred miles to their planned mission, strong head winds burned up still more fuel as they made their way toward China, and most of the crews ditched or bailed out. Eight who landed in Japanese-occupied territory were taken prisoner and three of them were executed, ostensibly for the crime of bombing civilian targets but in reality in an access of Japanese fury and mortification.

In the months since his lightning strike against the American fleet on December 7, 1941, Admiral Isoroku Yamamoto, Commander in Chief of Japan's Combined Fleet, had grown accustomed to the adulation of a grateful public; each day brought sacks of adoring letters. After the bombs fell on Tokyo he was rattled to find he had become the target of hate mail. He was wracked, too, with anxiety over the Emperor's personal safety.

Where had the bombers come from? Yamamoto pointed to Midway Island, America's westernmost outpost in the Pacific since the Philippines, Guam, and Wake Island had been overrun in Japan's seaborne blitzkrieg. It was a plausible conclusion, even if Shangri-La was actually closer to the mark. Midway was twenty-five hundred miles from Japan and thirteen hundred miles from Honolulu. Thus, Yamamoto argued, as long as it remained in American hands, bombers could fly from Hawaii to Midway, and from Midway to strike Dai Nippon. Japan's defensive perimeter would have to be pushed back farther still.

Two days later, Yamamoto's fleet air officer, Captain Yoshitake Miwa, noted in his diary that if further raids on the mainland were to be prevented, "there would be no other way but to make a landing on Hawaii. This makes landing on Midway a prerequisite. This is the very reason why the Combined Fleet urges a Midway operation." But, truth be told, Yamamoto had had his eyes on Midway for months. Untouched by the "victory fever" that swept through Japan's high command, Yamamoto insisted that unless America could be forced swiftly to accept a negotiated settlement, Japan was ultimately doomed. The grand admiral was a gambler and something of a playboy; he had been to Harvard to learn English, served in Washington as a naval attaché, and his knowledge of America's industrial power made him view war with the United States as folly. But if war was inevitable, he had consistently argued, Japan's only hope was to risk all on a knockout blow. America's industrial might would take months or even years to fully mobilize: thus Yamamoto's bold stroke on December 7. Unfortunately, that had left the job only half done. America's battleships had been caught at anchor at Pearl Harbor, but her aircraft carriers, at sea on the morning of the Japanese attack, had escaped.

Throughout March and early April a bitter fight roiled Japan's high command. Yamamoto pressed his case with mounting impatience: To draw the American carriers into the decisive battle, Japan must seize an objective that the United States would have to defend. If his plan to attack Midway was not approved, he would resign. The Naval General Staff sputtered. Midway might be of strategic value to an America on the defensive, the staff insisted, but it was worthless to Japan. Midway was a rocky atoll hardly larger than the small airstrip that stretched from one end of the island to the other; it could hold no more aircraft than a single carrier. The naval staff preferred a thrust to the south to cut off Australia, or, even more ambitiously, to seize Ceylon and India and link up with the German forces in the Near East. The Japanese Army, its eyes on China and on the threat Russia would pose if it entered the Pacific war, declared it would have nothing to do with Yamamoto's scheme, either.

But as the dust from Doolittle's bombs settled, the Army staff came forward with a new demand: It now insisted that the Army must be included in Yamamoto's forthcoming assault on Midway.

————

Four thousand miles to the east, Midway had become an obsession to another man during that winter and spring of 1942, a man as anonymous as Yamamoto was famous. Commander Joseph J. Rochefort came by his anonymity as much by the nature of his personality as by the necessity of his vocation. Above his desk hung a notice that read: "We can accomplish *anything* provided no one cares who gets the credit." He would later have reason to question the wisdom of that principle. But putting in twenty-hour shifts in a windowless basement was not a calling that appealed to the glory seekers in the United States Navy anyway. "The Dungeon," they called their cheerless command post in the basement of the administration building at Fourteenth Naval District Headquarters at Pearl Harbor; its more formal name was Station Hypo.

Rochefort had enlisted in the Navy in 1918 with vague dreams of becoming a naval aviator. Nothing but the oddest of chances determined that 1942 would find him in charge of breaking JN-25, the Japanese Fleet General Purpose Code—the code that carried the operational orders of the Combined Fleet, the code that would, in short, tell where Japan was going to strike next.

Rochefort was driven but unflamboyant, a conventional career sailor who had pursued a conventional career path: sea duty, engineering school, ensign's commission, more sea duty. Rising from the enlisted ranks, he was an outsider to the elite fraternity of officers who had graduated from the Naval Academy. The coincidence that deflected him out of the ordinary course of duty occurred while he was serving aboard the battleship *Arizona* in 1925. The ship's executive officer, Commander Chester C. Jersey, liked crossword puzzles. Rochefort did too. Jersey remembered that fact when he was posted to Navy Department headquarters in Washington later that year. The Navy needed someone to work on codes, and Jersey recommended Rochefort. The informality of it all would seem fantastic by the standards of the huge and bureaucratic postwar Navy. But in 1925, the Navy's cryptanalytic staff consisted of a grand total of one person, and administrative matters throughout the service were frequently settled through personal contact and word of mouth.

Lieutenant Laurance F. Safford, the Navy's one-man code breaking bureau, had not set out to be a cryptanalyst, either. In 1924 he was assigned the task of developing new codes for the Navy. No one in the Navy was paying much attention to foreign countries' codes at the time, and they certainly weren't trying to break them. But Safford figured that to make a good code he ought to first see what other navies were doing. And so the "research desk" was born in Room 1621 of the old Navy Department Building on the Mall in Washington.

When Rochefort showed up for duty in October 1925, Safford put him through a six-month course in cryptanalysis that basically consisted of tossing him cryptograms to try to solve. When Safford was called to sea duty in February 1926, the "course" ended, and Rochefort, more or less by default, found himself officer in charge of the research desk. Under him was one cryptanalyst and one assistant with "no particular abilities." That was it.

Rochefort's first dose of cryptanalysis left him decidedly disinclined for another. It was not that there was any particular pressure on him to produce results. No one in the Navy had much of an idea what he was up to anyway, and no one would have understood it if he had. But the work had a way of generating its own compulsive pressures. Rochefort would come home every evening at five or six o'clock with his stomach in knots from the tension of the problem he was tackling. It would be eight or nine at night before he could manage to force down his supper. He developed an ulcer and greeted his recall to sea duty in 1927 with unfeigned relief.

But in those two years Rochefort scored America's first victory in a long shadow war with the Japanese Navy. Left over from 1918 was most of a $100,000 secret naval intelligence slush fund. To conceal it from Congress, the money was deposited in a Washington bank in a personal account belonging to the Director of Naval Intelligence. Whenever a new DNI took over, his predecessor just handed the money over to him along with the keys to the office. The money had begun to burn a hole in the pockets of successive DNIs, and in the early 1920s the incumbent decided to get rid of some of it by financing a series of break-ins at the Japanese consulate in New York City. The Japanese Navy's "Red" code book was secretly photographed and, over the course of several years, laboriously translated by linguists hired with more of the DNI's secret funds. (Just how hard it was to use up $100,000 was shown in 1931, when an acting DNI, in a fit of conscience for which his successors never forgave him, returned the money to the Treasury. The balance was $65,000.)

A complete code book was a windfall, but there was still one crucial piece missing. Like almost all of the Japanese Navy codes that Rochefort and his colleagues would encounter over the course of their long battle of wits with their Japanese counterparts, Red was an enciphered code. Every word or syllable likely to be used in a message was assigned a numerical value—that was the "code" part. But such a simple one-for-one substitution would not hold up a team of Boy Scouts, much less a determined military foe, for very long. So before the Japanese Navy sent any coded message over the airwaves, it was given a second disguise. The code clerk opened a second book, which contained page after page of random numbers; starting at the top of a page, he added the first of these random "additives" to the first code group of his message, the second to the second, and so on. An indicator buried in the message would tell what page in the additive book he had used for this "encipherment" of the basic code, so that the recipient could turn to that same page and strip off the additive before looking up the meaning of each code group.

Thanks to the DNI's black-bag jobs, Rochefort had the code book. What he did not have was the additive book. To make matters worse, the Japanese changed the additive book frequently. With nothing to go on but the raw traffic that the Japanese Navy put out over the airwaves, Rochefort's job was to reproduce an additive book that he had never seen.

Breaking a code when one has the underlying code book but no additive book is like finding a way across a strange country without a map or a compass. Breaking a code when one has neither code book nor additive book is like finding a way across a strange country with both eyes closed. Doing the former was what had given Rochefort his ulcer in 1926. His task in 1942 was to do the latter.

America's very success, in September 1940, in breaking the Japanese diplomatic cipher, code named "Purple," had the ironic effect of distracting

attention from where it could have been more profitably focused in the fateful months leading up to Pearl Harbor. The Purple cipher carried the highest-
level diplomatic messages of the Japanese Empire; this was intelligence of
such remarkable value that it was given the code name MAGIC. The Purple cipher was generated by a complex machine. It used a cascade of rotating
switches to encipher every letter of a message in a different key from the
last or the next. In one position of the switches the letter A would become
G; in the next it would become P. The U.S. Army's code breakers had, in
eighteen months of intense effort, deduced the wiring and setup of the machine without ever seeing one, a feat of pure analysis the likes of which had
scarcely before been seen. After hastily soldering together telephone
switches and relays to produce a replica of the machine, they proceeded to
decode the Japanese messages almost as quickly as they arrived.

On the morning of December 3, 1941, a Purple message came through ordering Japan's embassy in Washington to destroy its code books, and even
one of its two vital Purple machines. Frank Rowlett, a senior cryptanalyst of
the Army's Signal Intelligence Service, arrived at his office at noon that day
from a meeting, plucked this latest MAGIC decrypt from his in-box, and proceeded to read its contents with mounting incredulity. With only a single
machine it would obviously be impossible for the embassy to continue its
normal flow of business. Colonel Otis Sadtler, who was in charge of distributing the MAGIC decrypts, showed up in Rowlett's office at that moment and
began to pepper him with questions. Had the Japanese ever sent anything
like this before? Could they be getting ready to change their codes? Perhaps they suspected their current codes had been broken? Then the only
possible meaning of this extraordinary message sank in. Sadtler pulled himself to attention. "Rowlett, do you know what this means? It means Japan is
about to go to war with the United States!" And, decrypt in hand, Sadtler
took off literally running down the corridor of the Munitions Building to
alert the head of Army intelligence.

On the night of December 6, an aide interrupted the President at a White
House dinner to deliver him the latest MAGIC decrypts. These erased all remaining doubt. Japan was preparing to break off diplomatic relations. War
was inevitable.

But diplomatic communications are not the place where military orders
are delivered. America knew that Japan was going to strike; it did not know
where she would strike. To know that would require breaking into the
Japanese naval codes, and there was only one catch: Since mid-1939, America had not read a single message in the main Japanese naval code on the
same day it had been sent. For most of the period from June 1, 1939, to De-

cember 7, 1941, the Navy was working on naval messages that were months, or even over a year, old.

Partly this was a matter of manpower, partly it was a matter of human nature. MAGIC was such a dazzling prize that it blinded its possessors to the smaller but sometimes more valuable gems that lay buried among the dross and slag of supply orders and fleet maneuvers. JN-25 was the most recent descendant of the Japanese Navy's Red code; like its predecessors it was an enciphered code. At the time the new code first appeared on June 1, 1939, the U.S. Navy's Washington code breaking staff had grown to about thirty-six hands. By this time the "research desk" had acquired the official bureaucratic designation of OP-20-G, designating it as part of the Office of Naval Communications, OP-20, and Safford was back in charge after several tours of sea duty. The staff of thirty-six included translators, clerks, radio direction-finding experts, intelligence analysts, and officers responsible for the security of the Navy's own codes; only a handful were trained cryptanalysts, and of these only two or three could be spared to work on the new code, which was initially given the designation AN-1.

Over the course of months they laboriously punched every message on IBM cards and searched for any clues that would give them a toehold on this completely uncharted terrain. But everything about it had the air of scholarly inquiry, far from the heat or urgency of battle. AN-1 was a "research" project, not a "current decryption" job; reconstructing the meaning of thirty thousand code groups and piecing together thirty thousand random additives was not going to be the work of a moment. The Japanese Navy's radio signals were intercepted by U.S. Navy operators in Hawaii, Guam, and the Philippines. The operators transcribed the Morse code signals by hand onto message blanks, bundled them up, and once a week handed them over to the captain of one of the Dollar Line's "President" passenger liners that plied the Pacific. The captains, all members of the Naval Reserve and therefore authorized to act as couriers of confidential documents, dropped the packages in the mail to Washington when their ships reached the West Coast. A very small amount of priority traffic could be entrusted to the Pacific "Clippers" of Pan American Airways; in the hull of each of these airplanes a small steel strongbox had been welded into place just for this purpose, the keys held by naval officers along the route. But delays of weeks from the time a message was transmitted by the Japanese to the time it arrived in Washington were the norm.

Throughout 1939, 1940, and 1941 a slow-motion cat and mouse game ensued. Sorting the IBM cards by number and printing out huge catalogues of every code group in every message, the Navy cryptanalysts began to detect a few ghosts of the underlying code. The numbers in each message that appeared to serve as indicators, telling the recipient what page of the additive book had been used, were not quite as perfectly random as they should have been. The numbers bunched up, meaning that some lazy Japanese code clerks were reusing the same pages of additive over and over. That was a classic error; messages enciphered with the same additive pages could in principle be cracked. It was not until fall of 1940, however, that the first real break came—and it came just in time to be rendered obsolete by a completely new, and much more complex, code book that the Japanese brought into service on December 1, 1940.

By the summer of 1941, as tensions in the Pacific grew, every section of OP-20-G was desperately short of help. Messages in the Purple cipher, which could be read in their totality and almost always the same day they were transmitted, claimed first priority. Meanwhile seven thousand AN-1 messages were pouring in each month by mail to Washington, while only sixteen men could be spared to work on them. As the Navy began calling up Naval Reserve officers throughout the summer and fall, that figure crept up by about one per month. Station Cast, the Navy's intercept station at Cavite in the Philippines, was also trying its hand at compiling the reams of printouts and worksheets needed to tease apart the AN code and its additives; since spring 1941 a highly secret collaboration between Cavite and the British government's code breakers in Singapore had been under way on the project as well. But it simply wasn't enough.

Slowly and laboriously, the new code book was being reconstructed; again, inexorably, on August 1, 1941, the Japanese introduced a new, 50,000-group additive book that sent the code breakers back to the beginning. By November 1941 only 3,800 code groups had been identified, along with only 2,500 additives reconstructed in the current system. It was far less than 10 percent of the total, nowhere near enough to read current traffic.

Conspiracy theorists continue to weave elaborate scenarios "proving" that America had advance warning of the Japanese attack, with one branch of the "FDR knew" theorizers insisting that AN traffic was in fact being read in 1941. Yet month-by-month progress reports, internal histories, war diaries, logs—some declassifed only in 1998—are all in agreement: Not a single AN message had *ever* been read currently by the time of Pearl Harbor, and not a single AN message transmitted at any time during 1941 was read by December 7.

Five years later, with the war safely won, a few of OP-20-G's cryptanalysts were tidying up loose ends and decided to go back and try to crack the unread AN-1 traffic that had piled up in the months just before Pearl Harbor. What they found was enough to break an intelligence officer's heart.

Over and over, the orders to the Japanese fleet during October and November 1941 repeated a single theme: Complete all preparations and be on a total war footing by November 20. Several messages referred to exercises in "ambushing" the "U.S. enemy." And one signal, dispatched November 4, ordered a destroyer to pick up torpedoes that Carrier Divisions 1 and 2 "are to fire against anchored capital ships on the morning in question." None specifically mentioned Pearl Harbor, and indeed many other intelligence indications in those critical months pointed to the Philippines, or even the Panama Canal, as possible targets of Japanese naval action if war broke out. Yet the pre–Pearl Harbor AN traffic, had it been broken at the time, would certainly have conveyed heavy hints of what was to come.

———

In the chaos following the Japanese attack, mail service from the Pacific was thrown into disarray. On December 4, the Japanese had again changed the additive book for AN-1; it was back to square one yet again, and Washington fretted away a month waiting for enough current intercepts to arrive in the mail to renew the attack. But in the meanwhile the decision was made to allow "the field" to begin work without delay. On December 10, Rochefort's Station Hypo, which had been shunted off to work on a dead-end problem before Pearl Harbor, was given the go-ahead by Safford to tackle AN-1 on its own.

The atmosphere throughout Hawaii in the days following the Japanese attack was one of stunned demoralization. During the attack a spent bullet actually ricocheted off the chest of Admiral Husband E. Kimmel, Commander in Chief of the Pacific Fleet, who would shortly become the scapegoat for the worst military disaster in American history. "Too bad it didn't kill me," Kimmel muttered. Two thousand four hundred and three Americans were dead. Two hundred aircraft were destroyed on the ground: Dutifully heeding warnings to be on the alert, the Army commanders had crowded their planes together wingtip to wingtip in midfield, well away from the perimeter fence and the Japanese saboteurs everyone imagined were lurking about the island. The Japanese torpedo planes and bombers caught all but one of the nine American battleships of the Pacific Fleet in port that morning, and all were left damaged or immobilized by the attack. *Arizona,* blown in two when her magazine went up, took eleven hundred of her crew with her to oblivion. *Oklahoma* lay capsized in the mud, never to see action again. The others sank at their moorings, had gaping holes ripped in their sides, or had run aground or were wedged between other crippled ships. Japan's force of ten battleships now had a seemingly insuperable command in the Pacific.

If Pearl Harbor had been asleep, the American forces in the Philippines under General Douglas MacArthur had been comatose. The Philippines had been everyone's bet for where the Japanese blow would fall. When Secretary of the Navy Frank Knox was delivered the news of Pearl Harbor he exclaimed, "My God, this can't be true, this must mean the Philippines!" The Japanese would oblige soon enough. MacArthur received word of the Pearl Harbor attack just as it was ending, at 3:00 A.M. on December 8, Manila time. In the preceding weeks MacArthur had confidently assured Washington that with enough air power he could drive the Japanese back into the sea if they dared to come ashore. On that assurance he had been shipped dozens of top-of-the-line B-17 long-range bombers. When the decisive moment came, MacArthur, apparently frozen in indecision, barricaded himself in his penthouse suite in a downtown Manila hotel and did nothing. Nine hours later Japanese bombers and Zeros appeared over Clark Field; instead of meeting the swarm of enemy fighters they fully expected, the Japanese pilots looked down and rubbed their eyes in disbelief at the sixty neatly parked planes on the field below. That evening, Roosevelt kept a long-scheduled appointment with newsman Edward R. Murrow. FDR pounded his fist on the table in frustration: The American planes had been destroyed "on the ground, by God, on the ground!" he exclaimed.

Three days later, the pride of Britain's Singapore-based Asiatic Fleet, the battleship *Prince of Wales* and the battle cruiser *Repulse,* were sunk by Japanese torpedo bombers flying from Saigon. Not a single Allied battleship or cruiser was left west of Hawaii. Japan, for the moment, was the unchallenged master of the Pacific and Indian oceans.

At Pearl Harbor, the thoroughly shaken American commanders were certain that the Japanese were going to hit them again. Crews were hastily set to work tearing down fences, welding them together, and dangling them into the water around the docked ships as crude antitorpedo barriers. "Of course we had no knowledge whether that kind of net would be any good at all," admitted Rear Admiral Claude Bloch, commander of the Fourteenth Naval District, "but it was the best we had." A hypercautious mentality set in, bordering on paralysis. A carrier task force sent to relieve Wake Island was recalled at the last minute by Vice Admiral William S. Pye, who had been given temporary command of the fleet after Kimmel's ouster; the ships were actually in sight of the besieged atoll when the recall order came, setting off a near mutiny aboard the carrier *Saratoga.* When FDR received the news he said it was a worse blow than Pearl Harbor.

Rochefort, blaming himself for not foreseeing the Pearl Harbor attack, reacted characteristically, driving himself and his men without mercy. Summoned to the Dungeon on December 7 by an 8:00 A.M. telephone call from his deputy—Lieutenant Commander Thomas Dyer had gone outside to see what all the commotion was and had "caught on fast," he later recalled, when he saw a torpedo bomber three hundred yards away emblazoned with the rising sun insignia—Rochefort rushed to the headquarters he would scarcely leave for the next six months. "I can offer a lot of excuses," he would later say, "but we failed in our job. An intelligence officer has one job, one task, one mission—to tell his commander, his superior, today what the Japanese are going to do tomorrow." He was determined not to be caught flat-footed again.

On December 1 Station Hypo had completed its hasty move from the second floor of the Administration Building to the Dungeon. This was partly for security—the basement was sealed off from the rest of the building, with a single steel door at each end leading directly to the outside—and partly to accommodate a growing staff, which had doubled from twenty-three in June to forty-seven in December. Dyer had managed to get funding to rent a few precious IBM machines back in 1938, but the sensitive equipment had to be kept air-conditioned in tropical Hawaii and that was one detail that apparently got lost in the rush to move to a war footing. When the code breakers took up residence in the Dungeon everyone began hacking and coughing constantly. This went on for two months; finally when Rochefort was able to spare someone for a moment he sent a man to check out the air-conditioning system and see if there was something wrong with it. The man came back shortly and reported he had found the trouble: There was no air-conditioning system. When the air-conditioner finally was installed it tended to function erratically and Rochefort took to belting a smoking jacket over his uniform to ward off the chill when it was running full blast. The smoking jacket became part of the Rochefort legend, as did the pair of carpet slippers he wore to ease his sore feet from standing on a hard concrete floor twenty or twenty-two hours a day. But those who knew him said this painted a false image. He was a tall, thin, pale, and driven man, but he was no eccentric. His quiet doggedness inspired a loyalty that his men never forgot: A half-century later Forrest E. Webb, who ran IBM machines at Station Hypo, would say simply, "Rochefort was my ideal of an ideal man. He never raised his voice, but he knew that what he said was law and everybody believed it. He made sure people knew what they were doing and left them to it." He *was* a man with a mission.

But they all joked about being crazy. Dyer hung a sign above his desk that read, "You don't have to be crazy to work here—but it helps." He also kept a bucket of pep pills on his desk and every so often reached in, scooped up a handful, and popped them into his mouth. Dyer would often go for forty-eight hours at a stretch. The Dungeon itself was a large open area, about sixty by a hundred feet. The IBM machines were consuming three million punch cards a month and churning out huge stacks of printouts that were kept in boxes or just piled on the floor. There was no time to file or index anything; that led to more jokes about the crazy code breakers. Someone would run across a code group in one message that would ring a vague bell; he would mention it to the others and someone else would immediately reach down halfway into a stack of printouts and pull out a message from a month before.

Hypo was five thousand miles closer to the Japanese fleet than Washington was; it also had its own intercept station for picking up the Japanese radio traffic, just thirty miles away at Heeia. There was no sitting around waiting for the post office to deliver parcels of intercepted traffic, though there was still an absurd lack of the most basic and obvious conveniences of the communications age. No teletype circuit or radio linked Pearl Harbor and Heeia; a jeep, or sometimes a motorcycle or even a bicycle, was dispatched to pick up the intercepts.

The two years of lagging struggle against AN-1, which was now officially known as JN-25, had not been in vain. Although OP-20-G had never managed to catch up with the latest changes in the code book and additive tables, the code breakers by this time thoroughly understood the principles involved. It was just a matter of manpower, and with America mobilizing for war the manpower problem was being addressed swiftly. When Safford had tapped Rochefort for the command of Hypo in June 1941 he promised him first dibs on any officers who had had Japanese language training. After Pearl Harbor, Rochefort took anyone he could get. He cut a deal with the local personnel officer: As new drafts flooded in from the West Coast they would be lined up with their service records and Rochefort would take anyone who looked promising. The personnel office didn't know what to do with the ship's band from the crippled battleship *California,* left without a job by the Japanese torpedoes; Rochefort said, "I'll take them." The musicians were immediately put to work running the IBM machines. Some would wind up spending the rest of their careers in cryptanalysis.

On March 18, 1942, they caught up at last. The order to begin "current decryption" of JN-25 went out. The war against the Japanese code makers was won. Now the war against the Washington code breakers began.

The odd and ill-defined bureaucratic relationship between Station Hypo and OP-20-G was the stuff that Washington intrigues are made of. Technically Rochefort reported to the commandant of the Fourteenth Naval District in Honolulu, a practical enough man to leave Rochefort to his own devices and let him deal directly with Captain Edwin Layton, the intelligence officer for the Pacific Fleet. But it was OP-20-G that had tapped

Rochefort for the job and that loosely coordinated the division of labor among all the various intercept stations. In late January 1942, Commander Safford, Rochefort's old mentor, had been shunted aside in a Byzantine power play at Navy headquarters. Although Washington and Honolulu had established a close collaboration on JN-25, exchanging additive and code groups via a secure radio link that employed the never-to-be-broken SIGABA/Electric Cipher Machine, Rochefort and his new nominal superiors in Washington were on an inescapable collision course. Washington was now demanding central control over all code breaking and intelligence. Safford had been a firm believer in decentralization, and Rochefort insisted he was answerable only to the new Pacific Fleet commander, Admiral Chester W. Nimitz. He was also tactless enough to make it abundantly clear what he thought of Washington's meddling.

The simmering tension broke out into open warfare almost as soon as current reading of JN-25 began. Rochefort had developed a close working relationship with Layton, and the two men would speak by phone several times a day. Layton had come to respect Rochefort's reliability and caution. Both men were fluent in Japanese and Layton knew that Rochefort personally translated more than a hundred of the five hundred to a thousand messages that were being deciphered each day. Knowing that most commanders tended to dismiss intelligence reports as the work of alarmists, and knowing too that Rochefort was never one to exaggerate, Layton routinely upped Rochefort's estimates to compensate for their subsequent discounting. If Rochefort reported four enemy carriers in an area, Layton would change it to six. So when Layton's phone rang on May 14 and he heard Rochefort at the other end of the line exclaiming, "I've got something so hot here it's burning the top of my desk!" he dropped everything and headed right over to the Dungeon. The hot document proved to be a partial decrypt in which the words *koryaku butai,* invasion force, were followed by the geographical designator AF. *Koryaku butai* had appeared in orders for the invasions of Rabaul, Java, Sumatra, and Bali that Hypo had already read. AF had been tentatively identified as Midway. Rochefort argued that the clincher was an order that air base equipment was to be shipped to Saipan to be in position for the "AF ground crews." AF was obviously an island air base; it was, Rochefort insisted, a matter of simple deduction to see that it had to be Midway.

Nimitz was quickly convinced. On May 17 he ordered his three remaining aircraft carriers to return at once from the South Pacific. The following day he sent an order to Midway, via secure undersea cable, canceling previous orders to one of the U.S. submarines based there. The new instructions: BELIEVE ENEMY WILL ATTACK MIDWAY USING PLANES LAUNCHED FROM A POSI-

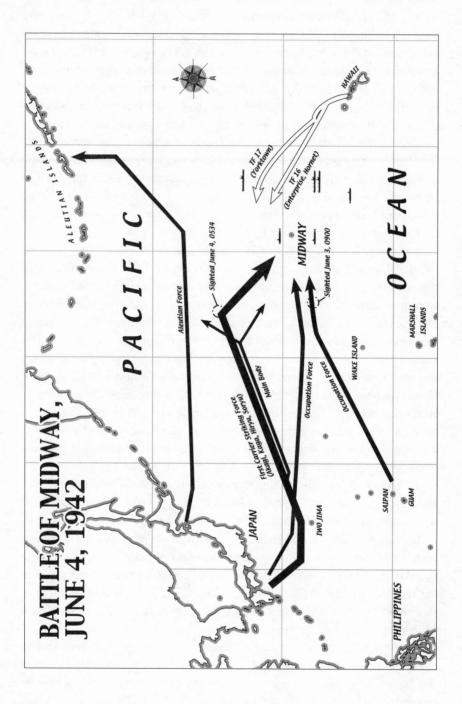

BATTLE OF MIDWAY,
JUNE 4, 1942

PACIFIC

OCEAN

JAPAN

ALEUTIAN ISLANDS

Aleutian Force

Sighted June 4, 0534

First Carrier Striking Force
(Akagi, Kaga, Hiryu, Soryu)

Main Body

MIDWAY

Occupation Force

Occupation Force

Sighted June 3, 0900

IWO JIMA

WAKE ISLAND

SAIPAN

GUAM

MARSHALL
ISLANDS

PHILIPPINES

HAWAII

TF 17
(Yorktown)

TF 16
(Enterprise, Hornet)

TION FIFTY MILES NORTHWEST OF MIDWAY X PATROL THAT AREA UNTIL FURTHER
ORDERS. That same day the Seventh Air Force, at Hawaii, was placed on a
special alert; its B-17 bombers were taken off reconnaissance duty and kept
loaded with demolition bombs to be ready to carry out strikes against enemy ships on a moment's notice. Other B-17s were moved in from the mainland, and plans were set in motion to fly some to Midway itself.

But then Commander John H. Redman, who had maneuvered himself
into the command of OP-20-G in the January reshuffling, began to throw
cold water in copious quantities. Washington, in short, rejected Midway as
the target of the gathering Japanese offensive. Several weeks earlier Redman had scrawled an irritated note on a memo that had identified AF as
Midway: "Comment said AF=Midway, but pointed out that comm. zone
designations not identical with area designations." On May 14, OP-20-G had
sent an intelligence summary to Admiral Ernest King, the Navy Commander in Chief in Washington, concluding that the Japanese were preparing a
"coordinated air and submarine attack on the Hawaiian Islands." Then OP-
20-G insisted that Hypo had bungled its additive tables; the Japanese orders
were not to attack AF but rather AG, Johnston Island. On May 16 and May
17 OP-20-G was warning of a second Japanese offensive, this one directed
against Alaska and the Aleutian Islands. Some intelligence officials in Washington suspected the whole thing was a Japanese deception operation; the
real target might even be the West Coast of the United States. A message intercepted May 19 seemed especially fishy. A Japanese seaplane unit informed the Bureau of Personnel in Tokyo that its "next address" would be
Midway, presumably so its mail could be forwarded. As Army Chief of Staff
George C. Marshall would later note, that seemed to lay it on just "a little bit
too thick."

Yamamoto's plan was the most elaborate seaborne ambush ever conceived
by the mind of man. Five separate forces, a total of two hundred ships and
250 aircraft including eleven battleships, eight carriers, and twenty-three
cruisers, would move across a million square miles of the Pacific in a tightly
orchestrated plan. While the Northern Force staged a diversion in the Aleutians, the Striking Force with four carriers and two battleships would neutralize Midway's fighters and bombers, clearing the way for the dozen
troopships of the Occupation Force to land five thousand men and seize the
island. Meanwhile, a screen of twenty submarines would establish a picket
line between Midway and Hawaii to warn when the American carriers sor-
tied from Pearl Harbor to the rescue. That would be the signal for the final
act to begin. The Main Body, a huge armada built around seven battleships

and a carrier, would hang back hundreds of miles to the west until the American ships committed themselves; it would then spring forward for the kill. Yamamoto would personally command the operation from the battleship *Yamato,* the world's largest, a seventy-two-thousand-ton vessel whose 18.1-inch guns could throw a thirty-two-hundred-pound shell twenty-five miles. An Estimate of Situation issued just before the battle began by Admiral Chuichi Nagumo, commander of the Striking Force, summarized the Japanese plan, and the Japanese confidence in it: "Although the enemy lacks the will to fight, it is likely that he will counterattack if our occupation operations progress satisfactorily. . . . After attacking Midway by air and destroying the enemy's shore based air strength to facilitate our landing operations, we should still be able to destroy any enemy task force which may choose to counterattack."

Nimitz summoned a final staff meeting for Wednesday, May 27, to review his own estimate of situation. Nimitz was prepared to stake everything on Rochefort's analysis, but it was a huge gamble. It would mean leaving Hawaii defenseless as he rushed his carriers to Midway before the Japanese arrived. Attending the Wednesday morning meeting would be General Delos Emmons, the Army commander in Hawaii, and General Robert C. Richardson, whom Marshall had sent from Washington. In the meanwhile, Rochefort and one of his staff hit on a scheme they hoped would get the meddlers back at OP-20-G to shut up. While attending engineering school at the University of Hawaii, Lieutenant Commander Jasper Holmes had spent some time at the Pan Am repair facility on Midway. He recalled that all of the island's water came from a desalination plant. With Nimitz's approval, Rochefort and Layton on May 19 sent instructions via the undersea cable to Midway. The radio operators there were to send an uncoded "flash" message reporting that the distillation plant was broken. Two days later Tokyo Naval Intelligence sent a signal in JN-25 reporting that "AF Air Unit" had sent a message to Hawaii reporting it had only a two weeks' supply of fresh water and asking for an immediate resupply. The Japanese signal was broken both by Station Hypo and by the U.S. Navy intercept unit in Melbourne, Australia; Rochefort shrewdly laid low and said nothing. The next day Melbourne forwarded the intercept to Washington with the comment, "This will confirm identity AF." To keep the Japanese—and Washington—from learning that it had all been a setup, Layton even arranged for Hawaii to send a reply to Midway reporting that supplies were on the way.

Rochefort stayed up all the Tuesday night before Nimitz's staff meeting going over the months of messages. Disheveled and beat, he showed up half an hour late but was able to avert the wrath of a room full of waiting admirals and generals when he reported that Station Hypo had broken the last remaining piece of JN-25, a separate code-within-the-code used for dates. He then flourished the payoff: a message dated May 26 ordering destroyer escorts for the troopships to depart from Saipan on May 28, proceed at eleven knots, and arrive at Midway June 6. An earlier decrypt had revealed that air attacks against Midway would commence from a point to the northwest of the island on day "N – 12." That fixed the likely day for the Japanese air assault at June 3 or 4.

That same day, May 27, the Japanese changed both the code book and the additive tables for JN-25 and imposed radio silence on the Midway and Aleutian forces. The code breakers were blacked out. But Nimitz had everything he needed already. He knew where the Japanese would strike and with what forces, he knew when, and he knew exactly what he had to do to get there first.

The fast carrier task force had been born of necessity in the aftermath of Pearl Harbor. With the Battle Fleet crippled, Nimitz had studied his assets and reassembled them as best he could into a fighting force. Despite the manifest success of naval air power at Pearl Harbor, and despite the manifest vulnerability of even battleships to air attack, many traditionalists in the navies of America and Great Britain continued to insist that nothing could take the place of the heavily armored and heavily gunned behemoths. Nimitz boldly rejected that view, assigned the few surviving battleships of the Pacific Fleet to convoy duty between Hawaii and the West Coast, and began perfecting the high-speed hit-and-run techniques that carriers made possible. Nimitz's plan for Midway was simplicity itself compared to the baroque evolutions of Yamamoto's plan: He would get there first and ambush the ambushers. To carry out the mission, he had two task forces available, TF 16, with *Hornet* and *Enterprise,* and TF 17, which had been composed of *Lexington* and *Yorktown.* But the Battle of the Coral Sea had left the *Lexington* a flaming wreck on May 7. *Yorktown* was damaged in the same fight by a bomb that plunged through her flight deck and exploded below. Nimitz now ordered the ship to be repaired and ready for action in three days, a job that in peacetime would have taken three months. Fifteen hundred men worked around the clock, shoring up bulkheads with wooden timbers and doing more patching than repairing, but she steamed out of Pearl on May 30, ready for action.

Meanwhile the marines and airmen on Midway itself braced as best they could. The local Marine commander, a First World War veteran who firmly believed in the efficacy of barbed wire, strung miles of it around the island. There was so much dynamite stockpiled that it finally began to pose more of a threat to the defenders than the attackers; tons were dumped at sea in preparation for the Japanese attack. Briefed on the enemy plan, the island's commanders were astonished at the level of detail provided. A facetious rumor went around that Tokyo Rose was on the American payroll and sent coded messages in her propaganda broadcasts.

Nimitz's final orders to his task force commanders instructed them to proceed on the principle of "calculated risk." The war would never be won by commanders who never took a chance. Just don't take foolish chances, Nimitz was saying.

———

A little before 6:00 A.M. on June 4 a PBY Catalina float plane droned through the bright morning sky. Lieutenant Howard B. Ady and his crew had been searching a sector northwest of Midway since well before dawn. Then came the electrifying message from Ady's plane: PLANE REPORTS TWO CARRIERS, TWO BATTLESHIPS, BEARING 320 DEGREES, DISTANT 180 MILES, COURSE 135 DEGREES, SPEED 25 KNOTS. Only an hour earlier Nimitz had asked Layton to give him a specific prediction of when and where the Japanese carriers would be first spotted. Layton swallowed hard and hazarded 0600, from the northwest at a bearing of 325 degrees, at a distance 175 miles from Midway. When Nimitz received the PBY's report in his operations room he could not resist tweaking his intelligence officer; turning to Layton he dryly commented, "Well, you were only five minutes, five degrees, and five miles out."

Fifteen minutes later a patrol of six Marine F4F Wildcat fighters from Midway ran headlong into a hornet's nest—an incoming wave of Japanese Zeros and bombers. Captain John F. Carey, leading the Wildcats, went after one of the bombers but was instantly struck in the legs by machine-gun fire from the tail gunner. He was able to make it back to Midway and land with both tires punctured, but with no strength in his legs he was unable to apply the brakes. The plane crashed into a revetment and two ground crewmen pulled him from the wreckage and wrestled him to cover—just as the first bombs began to crash about the airfield. Most of the Midway fighter force consisted of slow and outmoded F2A-3 Buffalos, nicknamed "Flying Coffins," and they were no match for the Zeros. Buffalos were so slow that a Zero flying *level* could outpace a Buffalo in the steepest dive it could safely execute. In all, fifteen of the twenty-six Midway fighters were shot out of the sky; others landed in the midst of the Japanese bombing and were destroyed on the ground. Only two of the planes would ever fly again.

But Midway was better prepared with antiaircraft armament, and that evened the score. Sixty-seven of the 108 Japanese attackers were destroyed or damaged so badly as to be put out of action. At 7:00 A.M. Lieutenant Joichi Tomonaga, leading the attack, urgently radioed Nagumo: another strike was needed. Nagumo agreed, then hesitated. Ninety-three aircraft aboard the carriers *Akagi* and *Kaga* had been held back from the first wave, fitted with torpedoes and armor-piercing bombs to be used against any American ships that might appear. But there were no reports of American ships; surely the U.S. carriers were still back at Hawaii. Nagumo hesitated a few more minutes, then finally gave the order to replace the planes' weapons with land-attack bombs. The process would take an hour.

Reconnaissance was not a strong point of the Japanese Navy. At 7:28 A.M. a float plane reported ten enemy ships; it took the plane another forty minutes to incorrectly identify them as cruisers and destroyers, and it was almost a full hour after his initial report that the pilot almost casually added: ENEMY FORCE ACCOMPANIED BY WHAT APPEARS TO BE AN AIRCRAFT CARRIER. Alarmed, Nagumo ordered the armament changed out once again in preparation for a strike against the American carrier. But at that moment Tomonaga's returning strike force was circling and running dangerously low on fuel, waiting to land. They would have to be recovered first, refueled, and relaunched before the bombers could be brought up to the flight decks— still more maddening delay.

In command of TF 16 was Admiral Raymond Spruance, who had taken Halsey's place at the last minute when Halsey had been packed off to a Honolulu hospital suffering from an odd dermatitis that covered his entire body. Spruance was almost the opposite of the pugnacious Halsey, a cerebral and even cautious commander with cool, steady judgment. But in the Battle of Midway, Spruance stretched "calculated risk" to the limit. When the Japanese Striking Force was located, Spruance quickly determined that it would be several hours before he would be in the best position to launch his planes. He decided not to wait; risking everything, he let loose with an all-out attack at once. Spruance knew that striking immediately would increase the odds of catching the Japanese ships at their point of maximum vulnerability, just as they were recovering the Midway strike force. He also knew it meant that his own torpedo bombers would run out of fuel before they could make it back to their ships. With luck they might be able to land at Midway; more likely, they would have to ditch their planes and make the best of it.

The Japanese had already dodged a series of ineffectual attacks from Midway-based B-17 and B-28 bombers and outmoded SBU Vindicator dive bombers (the pilots of the latter sardonically called them "Wind Indicators" for their habit of spinning around when landing in a crosswind). The American carriers were equipped with more modern aircraft, but these at first seemed destined to the same fate as the Midway force. Three squadrons of TBD Devastator torpedo bombers were cut to pieces by antiaircraft fire and by the swarm of fifty Zeros protecting the Japanese fleet. It was almost a massacre: Of the forty-one planes that attacked, only four made it back. But just as the melee was ending at about 10:20 A.M., forty-nine SBD Dauntless bombers from *Yorktown* and *Enterprise* slipped in unnoticed at fourteen thousand feet. Lieutenant Commander Clarence Wade McClusky, air group commander of the *Enterprise,* had led thirty-two of the Dauntlesses to the Japanese fleet's last reported position only to find empty ocean. Running short of fuel, he at last spotted the wake of a Japanese destroyer and decided to follow it. The decision paid off: A few minutes later, and there below him, in full view, were *Kaga* and *Akagi.* McClusky pushed his nose down, heading straight for the carriers in a seventy-degree dive. *Kaga,* its deck crowded with planes and scattered with armament and fuel lines, took a direct hit. The ship's communications officer hurried toward the bridge to urge the captain to move to safety; then two more bombs struck, and when he looked again the bridge was gone. *Akagi*'s deck went up in a chain reaction of exploding planes and armament. Commander Minoru Genda, air officer of the First Air Fleet, who had been confined to bed with pneumonia and who had dragged himself, feverish, to the bridge to watch his air crews launch the first wave, now surveyed the carnage and uttered a single word of ironic understatement: *Shimatta*—"we goofed." The carrier *Soryu* meanwhile was hit twice by the Dauntless squadron from *Yorktown,* led by Lieutenant Commander Maxwell F. Leslie. The ship erupted in flames so intense that the hangar doors melted. The fourth Japanese carrier, *Hiryu,* shrouded in haze, escaped for the moment, and she at least would have her revenge. *Hiryu* immediately launched an attack against the *Yorktown;* one bomb smashed through the ship's side and sailed through the coffee urn in the ready room before lodging in the stack and exploding, knocking out five of the boilers and slowing the ship from thirty knots to a crawl. A series of torpedo hits finished her off, and the order to abandon ship was given at 2:55 P.M.

But the American forces would have the final word that fateful day. All airworthy dive bombers left on the *Enterprise,* twenty-four planes, were ordered out against *Hiryu.* No fighter escort could be spared; they had to remain to protect the American ships. At 4:45 P.M. the dive bombers spotted the enemy, and four bombs set her ablaze. In the space of a day, four of the six carriers that had launched the attack against Pearl Harbor had been destroyed. Japan lost more than three hundred aircraft and three thousand men. Yamamoto had obtained his decisive confrontation.

Nagumo kept the news from Yamamoto as long as he dared; when the Commander in Chief was finally told that his gamble had failed, he sank into a chair stunned and speechless. Demoralized and hesitant, Yamamoto at first ordered a cruiser bombardment of Midway for the following morning, then countermanded it. Yamamoto's huge battleship force, stripped of its air cover, was now like a short-armed, muscle-bound boxer. It could only land a blow against an opponent who grappled in a close embrace, and Spruance prudently kept his distance, pulling back to the east through the night to where he could still threaten with his aircraft without being threatened by Japanese guns. Yamamoto's huge battle force still outgunned the Americans by orders of magnitude. But with his carriers gone, he was left with no choice but to retire from the battlefield.

———

The most direct result of the Battle of Midway was to halt the tide of Japanese expansion. From the moment Yamamoto's ships steamed to the west in confusion and defeat, Japan was on the defensive, and would remain so throughout three grueling years of island combat. In the three months following Pearl Harbor, Japan had been an unstoppable juggernaut, seizing oil and rubber fields that her war machine so vitally needed in South Asia, throwing out a thousand-mile-deep defensive perimeter in half the time war planners had allotted for the task. Now, in a day, the Imperial Japanese Navy had suffered its first decisive defeat in three centuries, and Japan was in the entirely new position of trying to cling to what she had conquered, rather than looking to new conquests.

But Midway was also one of those moments that concentrate forces of history, that in one intense burst crystallize what might have otherwise taken years to coalesce from the fog of events. Midway decisively announced the end of the age of the battleship: The battleship's brawn was simply no match for the long reach of the carrier. Of even farther-reaching consequence, the American victory at Midway moved code breaking and signals intelligence from an arcane, little-understood, and usually unappreciated specialty to the very center of military operations. Even Nimitz, a rare example of a true intellectual among military commanders, had been doubtful about the value of "radio intelligence," as it was then known; if it had failed at Pearl Harbor, he reasoned, it did not make sense to place much faith in it. Layton had persuaded him otherwise, pointing out that JN-25 had not been cracked in time to warn of Pearl Harbor and that the Japanese Navy had maintained radio silence during the actual operation. But most commanders looked upon intelligence in general with suspicion, if not with the outright contempt that was the characteristic view that men of action of that era held for subtlety or innovation. There was certainly nothing subtle about the huge billboard that Halsey ordered erected on a hillside on one of the Solomon Islands. Visible to passing ships it bore a simple and crude admonition to his troops:

KILL JAPS. KILL JAPS.
KILL MORE JAPS.
You will help to kill the yellow bastards
if you do your job well.

The way to win, in the view of such fighting admirals, was to fight—and think about it later, if at all. The Battle of Midway was won through fighting, to be sure. Bravery, resourcefulness, and a not inconsiderable dose of luck all played their part. But the one indispensable element in the victory was the thinking, and nothing but thinking, that had cracked JN-25.

Three days after the battle ended, Rochefort told everyone at Station Hypo that he "didn't want to see them for three or four days." He expected everyone would just go home and catch some sleep. Instead the code breakers organized a house party on Diamond Head and got their boss up there; it wound up, Rochefort later recalled, as a "straight out-and-out drunken brawl" that lasted the entire three days. Rochefort said he was grateful for one thing: The party's organizers at least had had the sense to stay away from a hotel, which might have left them to the tender mercies of the shore patrol. Then everyone shook off their hangovers and went right back to twenty- and twenty-two-hour shifts to tackle the new code book and additives that the enemy had just introduced into JN-25.

The denouement of the Battle of Midway was not one of the U.S. Navy's finest hours. An outnumbered, outgunned, and battle-stricken force had just changed the course of history through a feat of code breaking and intelligence analysis that laid bare the enemies' intentions; scarcely ever had a military commander known so precisely what the opposing commander was planning and thinking. Nimitz was full of praise for Rochefort and his men. Nimitz enthusiastically forwarded to Admiral King in Washington a recommendation from the Fourteenth Naval District commandant that Rochefort be awarded the Distinguished Service Medal for his part in the victory. Rochefort, with a keener measure of Washington politics than his commander, strongly advised against it: It would only "make trouble." He was right. Redman was continuing his campaign to centralize control of all radio intelligence work under OP-20-G in Washington and Rochefort was continuing to resist; it was quickly degenerating into a fight over who deserved credit for breaking JN-25 and correctly anticipating the Japanese plans for Midway. The honors for breaking JN-25 were properly shared. Of the 110 vital messages broken in advance of Midway, 49 were read simultaneously by both stations, 26 by Hypo only, and 35 by Washington only. But when it came to drawing the correct conclusions from those messages, Hypo won hands down. Had Nimitz been swayed by Washington's analysis, the Japanese ambush would very likely have succeeded.

Yet Redman was now claiming sole credit for the victory at Midway, and in that atmosphere there was no way he could let an award for Rochefort go through without a fight. Just a few weeks after Midway, on June 20, Redman sent a memorandum to the Vice Chief of Naval Operations baldly asserting that "experience has indicated that units in combat areas cannot be relied upon to accomplish more than the business of merely reading enemy messages and performing routine work necessary to keep abreast of minor changes in the cryptographic systems involved." Simultaneously, Redman's older brother, Captain Joseph R. Redman, now Director of Naval Communications, was complaining that Station Hypo was, "by virtue of seniority, in the hands of an ex-Japanese language student" who was "not technically trained in Naval Communications." Rochefort should be replaced with "a senior officer trained in radio intelligence rather than one whose background is in Japanese language," he insisted. The Redman brothers' behind-the-scenes lobbying paid off two days later when Admiral King accepted the advice of his chief of staff and denied Rochefort a medal. The argument, as officially stated, was that Rochefort had "merely efficiently used the tools previously prepared for his use," which had a grain of truth, and that "equal credit is due" to Washington for "correct evaluation of enemy intentions," which was a whopper. King the next day sent "all U.S. Naval radio intelligence activities" a "well done," the naval equivalent of a pat on the head.

A year later, Commander John S. Holtwick, who had run the IBM machines at Station Hypo, called on Joseph Redman, now a rear admiral. In the course of conversation Redman casually remarked that Station Hypo had "missed the boat at the Battle of Midway," but Washington had saved the day. The lie took Holtwick's breath away—especially since Redman had to know that Holtwick *knew* the opposite was true. That was the first real inkling the Hypo crew had of how completely Washington had stolen credit for the victory at Midway. Shortly before his death in 1985, Dyer wrote: "I have given a great deal of thought to the Rochefort affair, and I have been unwillingly forced to the conclusion that Rochefort committed one unforgivable sin. To certain individuals of small mind and overweening ambition, there is no greater insult than to be proved wrong." Two of the Station Hypo team, Dyer and Holmes, finally did receive the Distinguished Service Medal after the war. Rochefort finally did, too—in 1985, nine years after his death.

The Redmans' argument in favor of centralization was not entirely phony. The loose organization of the Navy's intercept units made less sense in a day of secure radio links and rapid exchange of information than it had even a few years before. There was also a pressing need to avoid unnecessary duplication in the huge labor involved in breaking a new code, as well as a need to make sure that *all* relevant intelligence bearing on a given problem came to a central point for correlation. But centralization became a convenient club to beat Rochefort with, and the Redman brothers beat away unmercifully. Finally on September 15, Rochefort, with Nimitz's approval, sent a blistering memo insisting that he in effect was answerable only to Nimitz, and Washington should butt out. Payback came on October 22, 1942: Rochefort was summoned to the Navy Department for "temporary additional duty." When Nimitz protested, he was assured that Washington simply needed Rochefort's expert advice. Rochefort once again read the situation more accurately than his boss; he told everyone that he was not going to be coming back. A month later Nimitz received a letter, which had traveled via surface mail, informing him that Rochefort's "temporary" duty had become "permanent." Nimitz was furious, and for two weeks refused to speak to John Redman—who, in an odd turn of events, had since been promoted to become Nimitz's fleet communications officer. But in the end there was nothing Nimitz could do about it.

Rochefort, meanwhile, proceeded to make "several mistakes in a great big hurry," as he himself put it. Worn out, suffering from bronchitis, and made more prickly than ever by the Redmans' attempt to steal the credit for his work, Rochefort said he would not accept any assignment in radio intelligence unless he was sent back to Honolulu as officer in charge. Failing that, he demanded combat duty. Cryptanalysts were forbidden to enter combat zones: they knew too much that they might give away if captured and tortured by the enemy. But Rochefort pulled every string he could think of and was offered command of a destroyer, only to turn it down because the ship was leaving at once from San Francisco and he had promised his wife they would visit their son at West Point that same weekend—"a very stupid thing for me to have done," Rochefort later said. He ended up in command of a floating dry dock in San Francisco. He never worked on codes again.

CHAPTER I

"No Good, Not Even for Intelligence"

HERBERT Osborn Yardley was a boastful drinker, a boastful womanizer, and a boastful cryptanalyst. He was right about the drinking; the other two claims seem to have been on less solid ground. Joseph Rochefort judged Yardley an "opportunist and a so-so cryptanalyst." Others had far less polite words for the garrulous, poker-playing Indianan. One naval intelligence officer opined that Yardley should be "hanged at the yardarm." In the summer of 1931 there were few within the small community of American code breakers who would have disagreed.

To call that community "small" was actually an understatement; "skeletal" was closer to the mark. A combination of Depression-era austerity and high-minded optimism had just about done them in altogether; then came Yardley's blow, from which it seemed they might never recover.

Yardley may have been a so-so cryptanalyst but he was a born salesman, the kind of man's man who knew how to sweep aside doubts and misgivings over a bottle of whiskey or a round of golf. He had been president of his high-school class, captain of the football team, precociously self-confident and wise in the ways of the world. Born in 1889 in the small midwestern town of Worthington, Indiana, Yardley spent much of his time from age sixteen on hanging out in saloons and mastering the mathematical and psychological subtleties of poker; finally at age twenty-three he landed his first steady job, as a State Department code clerk. When America entered the First World War in April 1917, the twenty-seven-year-old clerk, in a feat of salesmanship extraordinary even by his standards, talked the War Department into creating a code breaking bureau and placing him in charge. He received a lieutenant's commission and a hefty budget and proceeded to assemble an impressive staff of linguists, philologists, and assorted other scholars who occupied a series of anonymous office quarters around downtown Washington.

The bureau was officially designated MI-8, a branch of the Military Intelligence Division. One of its most dramatic feats was deciphering a note that had been found on a suspected German secret agent who was arrested in Texas in January 1918. Yardley's chief assistant, John M. Manly, who in civilian life had been chairman of the University of Chicago's Department of English, spent three days working on the captured note. It proved to be a transposition cipher, a complex anagram in which the 424 letters of text had been rearranged in a prescribed pattern. At the end of the three days Manly had grasped the pattern and extracted the original text, which was damning in the extreme—addressed to Germany's ambassador in Mexico City, it actually identified the bearer as a secret agent. That was quite sufficient for a military court to sentence him to death. (The spy, Lothar Witzke, whose improbable alias was Pablo Waberski, had his sentence commuted by President Woodrow Wilson and he was eventually released, in 1923.)

When the war ended, Yardley exercised his considerable powers of persuasion once again and in May 1919 secured the approval of the State and War departments in a scheme to secretly keep MI-8 afloat as a permanent peacetime operation. The two departments would jointly fund Yardley's outfit; to ensure secrecy, and to comply with congressional restrictions that limited what the State Department could spend for its headquarters operations within the District of Columbia, Yardley set up a front company in New York City whose ostensible business was compiling commercial codes, a convenient if unimaginative cover story. He began with a staff of sixteen and a personnel budget of $36,000 a year; the departments also covered Yardley's overhead for rent and supplies. Yardley's own salary was a princely $6,000, a substantial increase over the $900 a year he had started at as a code clerk just a half dozen years earlier. In 1921 he received a raise to $6,900. With his instinctive flair for self-dramatics, Yardley would later call his new organization the "Black Chamber"; at the time it went under the more prosaic name of the Cipher Bureau.

One of its first tasks was to break the Japanese diplomatic codes. In a typical bit of Yardley bravura, he informed his masters he would resign if he did not have the problem solved within a year. The Black Chamber was able to redeem that promise, which would lead to its greatest intelligence coup in November 1921. The Washington Naval Conference had convened to set limits on the tonnage of capital ships in the navies of Great Britain, the United States, France, Japan, and Italy. Japan's opening bid was for a ratio of ten for Britain and America to seven for Japan. On November 28, the Japanese Foreign Office cabled instructions to its ambassador in Washington to hold out for 10:7 but "in case of inevitable necessity" to offer 10:6.5; if pressed to the limit, he was authorized to accept even 10:6. The Black Chamber promptly relayed the contents of this message to Secretary of State Charles Evans Hughes. It did not take a genius to figure out how to put this information to use at the negotiating table. Hughes held out for 10:6. On December 10, Japan gave in.

In 1924 the Black Chamber's budget was cut in half, part of a government-wide austerity measure. Then, in early May 1929, there landed on the desk of the new Secretary of State, Henry L. Stimson, a series of decrypts from the Black Chamber, of whose existence he had not previously known. Stimson was shocked; he considered it "highly unethical" and ordered the operation shut down at once.

It was a decision that would become famous as an example of America's babe-in-the-woods innocence in foreign affairs, forever encapsulated in the words Stimson used two decades later in his autobiography to explain his actions: "Gentlemen do not read each other's mail." But in the context of the times, this was not as naive a sentiment as it would later sound to those who had witnessed the barbarity of Nazism. In 1929, not only American diplomats but the general populace of most of the Western world believed that the horrors of the Great War had made another such conflict unthinkable. A series of solemn international agreements had been sealed setting limits on arms and advancing the principle of arbitration. Stimson, a Yale-educated lawyer, was hardly alone in his commitment to the emerging framework of international law. Any system of law depended on certain basic rules of conduct; flouting those rules would breed contempt for and ultimately undermine the rule of law itself. It was particularly unscrupulous, in Stimson's view, to be spying on the very people—foreign ambassadors—with whom one was conducting negotiations. Stealing opposing counsel's brief was the work of shysters, not honorable officers of the court, and certainly not gentlemen.

Of course this rather overlooked the fact that no other government in the world appeared to be constrained by the same scruples. And it might have been a more robust policy had the United States government taken at least a few steps to make its own mail a bit more difficult to read. Even among its own code clerks the State Department was a laughingstock for its simplistic codes. The American humorist James Thurber, who had a brief career as a code clerk during and after the First World War, said that a standard rumor had it that the Germans from time to time sent taunting messages mocking Washington for its childish ciphers; on one occasion they supposedly offered the helpful suggestion that the Americans' transparent device of combining two ciphers might have been more effective had they used two different ciphers, which they named. The story was apocryphal, no doubt. But it wholly captured the underlying truth of the matter. Thurber said the American codes were "as easy to see through as the passing attack of a grammar-school football team." The symbol for a quotation mark in one code was something like ZOXIL; as a short-cut, UNZOXIL was permitted to be used to indicate the end of a quotation. This, Thurber noted, left the clerks with the depressing sense they were engaged in nothing more than an "exercise in block lettering." The extra layer of protection that President Woodrow Wilson devised for use in the telegrams he sent to his adviser Colonel House in Europe must have only added to the amusement of code breakers in foreign capitals. It is hard to imagine that Wilson's device of referring to the U.S. Secretary of War as MARS and the Secretary of the Navy as NEPTUNE posed a terrible challenge. There is something slightly touching about the innocence of an age in which the President of the United States personally spent hours late into the night encoding and decoding such messages.

Throughout the 1920s and 1930s, in fact up until America's entry into the Second World War, the British Foreign Office read American diplomatic codes with regularity and apparent ease. At war's end in 1918, the United States had introduced a new diplomatic code with additive tables that changed quarterly. It took the British a year to solve the first one, but progress was rapid after that, and during the Washington Naval Conference the British were reading the American codes as readily as the Americans were reading the Japanese. When Colonel Alfred McCormack, a U.S. military intelligence officer, visited the British radio intercept station at Beaumanor in 1943, the station's commander casually remarked that he used to read the U.S. State Department ciphers and that it had been "lots of fun." The Japanese were not far behind. They at first resorted to the direct means of rummaging through the trash of the American embassy in Tokyo for copies of the plain text of transmitted messages; with such "cribs" it was not too hard to figure out which transmitted code groups stood for which words. By the end of 1932 Japan's code breakers had recovered five thousand code groups in the State Department's "Gray" code. In a bit of overreaching, they then broke into the office of the American naval attaché and bungled the job, which led to a general tightening of security and the introduction of new codes. Nonetheless, by 1941 at least Japan had once again mastered most of the State Department codes. In August of that year Joseph C. Grew, the American ambassador in Tokyo, was startled when a pro-American official of the Japanese Foreign Ministry asked that the information he was secretly providing to Grew be transmitted to Washington in one specific American code, which he named: All the other U.S. diplomatic codes, he explained, had been broken.

Though nominally still part of MI-8, the Black Chamber was receiving 70 percent of its funding from the State Department, so Stimson's decision meant the Black Chamber was finished. Yardley, whose salary by this time had grown to $7,500 a year, supplemented by several thousand more from a real estate business he ran on the side, was offered a job in Washington at a salary he was expected to refuse, and he did. He was out of work, just as the Depression took hold. A year later he was flat broke. Code breaking was not exactly a marketable skill, but Yardley had always had other talents, not the least of which was spinning stories, not least when the hero of the story was Herbert Osborn Yardley. On January 31, 1931, he cut his last official government connection, resigning his commission as a major in the Army Reserve Corps. On April 4 he dropped his bombshell. An article in the *Saturday Evening Post,* the best-read magazine in America, appeared under the byline of Herbert O. Yardley; it described in great detail the study of secret inks. Two weeks later a second installment followed, this one on codes; two weeks later, it was ciphers. The culmination was *The American Black Chamber,* a book that became an immediate bestseller upon its publication in the summer of 1931. In the book and articles Yardley revealed the whole story of the Black Chamber, described how America had read Japan's codes during the Washington Naval Conference, and heaped unflattering criticism on the State Department for its shortsightedness in shutting down the operation.

The reaction in Japan was stunning. The Osaka *Mainichi* and Tokyo *Nichi-Nichi* newspapers sent their New York correspondents to interview Yardley and their stories appeared under a wave of indignant headlines:

BETRAYAL OF INTERNATIONAL TRUST
TREACHERY AT THE WASHINGTON CONFERENCE
DISGRACE TO THE CONVENER OF THE CONFERENCE

Translated into Japanese, *The American Black Chamber* promptly sold thirty thousand copies. Edwin Layton, the future intelligence officer of the Pacific Fleet, was in Japan that summer as a language student; he remembered the sudden cloud of suspicion that Yardley's book cast over him and other Americans. Officials who had been cordial immediately began treating them as common spies.

The small number of officials at War and State who knew the truth about the Black Chamber seethed. They were furious over Yardley's betrayal of one of America's most closely held secrets; they were furious, too, over his grandiose claims and the way he blithely assumed credit for the work of others.

The betrayal may have been even deeper than they suspected. On June 10, 1931, the chief of the Telegraph Section of Japan's Foreign Ministry set down in a memorandum his thoughts on how to counter the impact of Yardley's revelations. One was obvious: tighten up Japan's codes. During the London Naval conference in early 1930 Japanese diplomats had tried out a new cipher machine built by the Navy; such a machine, the memorandum noted, was extremely expensive, but was "the only means which can absolutely withstand scientific attacks." There is little doubt that Yardley's book became one of the most powerful arguments in favor of the adoption of cipher machines by the Japanese Foreign Ministry just a few years later, a step that would greatly complicate the work of Yardley's successors.

The other step the chief of the Telegraph Section proposed was to raise doubts "as to Yardley's sense of responsibleness":

> Subject Yardley in the above is assumed to be the same Yardley who earlier approached our Embassy in the U.S. with the proposition that he would sell the secrets of how the Japanese government's cryptography was solved. The upshot of the proposition was that U.S. $7,000 was passed to him for which we received many copies of decrypted messages together with a set of papers on how to [defeat] cryptography. If, in fact, these two Yardleys are one and the same, his recounting of how Japanese codes and ciphers were broken in his book is a flagrant act of bad faith on his part since we had stressed the need for utmost secrecy as a condition for our paying him the aforementioned U.S. $7,000. He was asked to guarantee that there would be no revelation to others.

The memorandum regretfully concluded that attempting to discredit Yardley by revealing him to be a traitor to his own country would backfire against Japan; only if Yardley confessed publicly should the Foreign Ministry confirm that the payoff took place. According to the memorandum, the deal with Yardley was sealed in June 1930, which was when he was almost broke and before he had started writing his book. It is not beyond the realm of possibility that the author of this memorandum—who was, after all, responsible for the security of Japan's codes and had blown it—was trying to save face by concocting a story of having paid off Yardley. But it seems equally plausible that Yardley was laughing up his sleeve at the ploy of getting the Japanese to pay $7,000 for "secrets" that, a year later, they could purchase along with the rest of the world for $2 at any bookstore.

———

The closing of the Black Chamber left the War Department with a cryptanalytic staff of one. He was William F. Friedman, thirty-seven years old, an intense, reserved, personally fastidious man, the son of Russian Jewish immigrants. Friedman himself had actually been born in Russia, in Kishinev, in 1891, and had arrived in America at the age of one.

The path that took him to cryptology was as roundabout as such paths always are. He studied agriculture at Michigan Agricultural College, a choice dictated principally by the fact that the tuition was free. Agriculture led to genetics and genetics led to his employment by one of those fantastic and uniquely American eccentrics who seemed to crop up in abundance at the turn of the last century. George Fabyan was an uneducated man, but a very rich man. Having made a fortune in textiles, he decided to create his own research institute on his five-hundred-acre estate, Riverbank, in Illinois. Friedman was hired to work in Riverbank's genetics laboratory, but was very soon drawn to the work of the "Department of Ciphers," whose sole and highly improbable task was to prove, by searching for a hidden cipher signature in Shakespeare's printed plays, that their true author was Francis Bacon. However absurd the assigned object of its research was, the Department of Ciphers managed to carry out groundbreaking work on the statistics of ciphers, and during the First World War Friedman and others at Riverbank were called on to help train Army cryptographers. In 1921, unable to take his boss's crackpot theories and dictatorial manner any longer, Friedman quit and went to work for the Army Signal Corps.

The Black Chamber, under the Military Intelligence Division, broke codes, but it was the Signal Corps' job to make codes, and that was what Friedman did throughout the 1920s. It was not the sanest bureaucratic division of labor, and a full month before Stimson acted to shut down the Black Chamber, the Army had itself decided to pull the plug on Yardley and move all work on codes to the Signal Corps. A growing dissatisfaction with Yardley's dwindling productivity may have added impetus to the decision: For the last few years the Black Chamber had produced almost no results at all.

The Signal Corps knew how to operate radios, and it had William Friedman. It had no radio intercept stations to speak of, not much money, and little institutional knowledge or understanding of what the point of radio intelligence was. But like the Military Intelligence Division, the Signal Corps was obsessed with secrecy to a degree that was suffocating. Elaborate steps were taken to keep the new operation under wraps. To get around the manifest legal, or at least political, difficulties of continuing the eavesdropping that had just got the Black Chamber shut down, the official instructions to the Chief Signal Officer dated May 10, 1929, explained that in peacetime, the interception of foreign radio signals and their decipherment would be restricted to "training" and "research." In April 1930 Friedman was authorized to hire three "junior cryptanalysts" at $2,000 a year. They were three schoolteachers—Frank Rowlett, who came from the small town of Rocky Mount in southern Virginia, an area principally known for its moonshine whiskey production, and Abraham Sinkov and Solomon Kullback, both from New York City. All had taught mathematics, all were in their early twenties, and none had a clue about cryptanalysis. All were told to deny that they were working on anything related to codes; if anyone asked what they did, Friedman told them, they could say they were conducting a statistical analysis of War Department communications. Within the confines of the office of the Chief Signal Officer, the bureau was known as the Signal Intelligence Service. Outside it did not officially exist at all.

It was striking that, with Friedman's new hires, three-quarters of the professional staff of the U.S. Army's code breaking bureau was Jewish. Striking, but not entirely coincidental: In America in the 1920s and 1930s the federal government was practically the only major employer of college-educated professionals that would hire Jews. Jews were graduating from college in ever greater numbers, but the only doors open to them were in family businesses, as independent professionals such as doctors and dentists—and in government. Anti-Semitism in America before the Second World War was overt, widespread, and unapologetic. In major cities not only private clubs but major hotels and the more desirable neighborhoods barred Jews. Large corporations basically did not hire Jews, period. Nor did universities: In 1927 Yale, Princeton, Johns Hopkins, Chicago, Georgia, and Texas each had a single Jewish faculty member. There were two Jews on the faculties of Berkeley and Columbia, three at Harvard, and four at City College of New York. Those select few had received letters of recommendation that read, for example, "[He] is a Jew though not the kind to which one takes exception" or "He is one of the few men of Jewish descent who does not get on your nerves and really behaves like a gentile to a satisfactory degree." The United States military was certainly no exception to such sentiments. The few Jews who enrolled at the U.S. Naval Academy were harassed, shunned, and ostracized. In 1922, a Jewish midshipman finished second in his class at Annapolis; the editor of the school's yearbook placed his picture on a perforated page that could be conveniently torn out and thrown away.

But Civil Service reforms, designed to eliminate political patronage and cronyism, had set precise qualifications for civilian government positions; applicants took a test and if they passed were placed on a list of qualified individuals, and government managers could hire only from the list. The intention was not to eliminate discrimination, but to a considerable degree that was its practical effect. With doors firmly shut elsewhere, Jews sought positions in the federal government in disproportionate numbers. Kullback and Sinkov were both teaching in the New York Public Schools and hoping for something better to do with their mathematics degrees from City College when Sinkov showed Kullback a Civil Service notice announcing an examination for the position of government "mathematician." Both eagerly took the test.

It was not just for purposes of cover that Friedman's office was assigned responsibilities that extended well beyond code breaking. He and his staff of three untrained cryptanalysts were also supposed to overhaul the War Department's codes, study secret inks, oversee the printing and distribution of codes to field units, train regular officers of the Signal Corps in the use of codes, ensure the security of U.S. military communications, and even help out the overworked and undermanned War Department Code Room by taking shifts coding and decoding routine traffic. The three new recruits spent months learning the basics from mimeographed sheets of a course Friedman had drafted, and for many more months at a stretch were set to work on tasks of pure drudgery, such as tabulating, completely by hand, the frequency distribution of all five-letter groups that appeared in fifty thousand letters of telegraphic text. To compile a new code book for the War Department, they entered sixty thousand words in alphabetical order on three-by-five index cards, also by hand. The cards then had to be placed in a random order so they could each be assigned a five-letter code group; the solution was to push the furniture aside to clear a space in the secure vault where they worked, then throw the cards into a stream of air from a fan. When the cards landed on the floor they were given a final stirring, then picked up.

With the loss of the Black Chamber, the American government's major source of intercepted traffic had been lost as well. Yardley had maintained an informal understanding with the commercial cable companies in New York by which they handed over hard copies of any interesting cable traffic. What had been legal under wartime censorship regulations, though, had become manifestly illegal in peacetime, and the companies had been growing increasingly uneasy about the practice. With Yardley gone it came to an abrupt halt.

Much of the transatlantic and transpacific traffic handled by the commercial cable firms actually traveled by radio as opposed to undersea cable—they were "radiograms" rather than "cablegrams"—so in theory it was possible for anyone with the right sort of radio receiver and a strategically stationed antenna to intercept it. Embassies were not in general allowed to have their own direct radio links to their capitals, so coded diplomatic telegrams passing into and out of America all were carried by these commercial firms. In addition, the militaries of foreign nations maintained their own networks of radio stations; those channels used for long-range communication—such as between Tokyo and the Mandate Islands that it controlled in the Pacific—traveled via short-wave signals that propagated by skipping off the ionosphere and could, under the right conditions, be picked up far beyond their point of origin. So the first step was obviously to identify the frequencies used by countries of interest and start copying down any likely traffic.

Less than a decade before the outbreak of the Second World War, the U.S. Army's capability for doing these things was essentially zero. In 1931 Friedman was able to get approval for an "experimental" monitoring station to be built at Battery Cove, Virginia, so that his trainees could "secure actual intercept material," but its reach was woefully inadequate. An official decision had been made to concentrate on both military and diplomatic communications of Japan, Mexico, and Russia, but little in the way of foreign military traffic could be picked up at all from the East Coast of the United States. The situation with diplomatic traffic was more promising, but it was clear that intercept stations would have to be set up closer to the source if anything useful was to be obtained.

Paralyzed by a cheapness of almost comical proportions and by the ever-present dread of discovery by higher authority, the Signal Corps made slow progress. From his post as Signal Officer of the Army's Ninth Corps Area, based at San Francisco's Presidio, Lieutenant Colonel Joseph O. Mauborgne sent a series of plaintive appeals to Washington for help in setting up and operating a listening post. He had no personnel available to assist him. Indeed he had been specifically instructed that "to obtain the greatest secrecy" he had to handle the monitoring of foreign stations entirely on his own. The handbook supplied by the intelligence staff in Washington was several years out of date in its listings of Japanese military radio stations. In January 1933 he was reduced to sending a personal plea to the Chief Signal Officer in Washington for $71.38 so he could buy four 72-volt batteries; that way he wouldn't have to shut down his radio every time the batteries needed recharging. To maintain security, the radio itself had to be hidden in the basement of Mauborgne's private quarters, and this caused problems, too: Mauborgne was having to pay the electric bill out of his own pocket. This time he had to go personally to the Commanding General of the Ninth

Corps Area for approval so that a second-hand electric meter, rummaged from the Signal Corps inventory, could be installed in his quarters to measure how much power the equipment used, so he could thus be reimbursed by the government. Mauborgne did not know the Japanese syllabic Morse code, so he was unable to copy the traffic himself; Washington ultimately sent him a Dictaphone recorder so he could record the signals and forward them to Washington for transcription. It wasn't until 1937 that the station at the Presidio was up to what was considered full staffing: a sergeant, a corporal, and three privates to operate the radios, plus several other enlisted men to keep records and repair the equipment—when they weren't also doubling as drivers.

By the time war broke out in Europe the Signal Corps had small intercept units operating in Hawaii, the Philippines, the Panama Canal, Texas, California, and New Jersey. But it wasn't until October 17, 1939, that Signal Corps headquarters decided that "in the near future" it would be a good idea to obtain a large volume of German diplomatic and military attaché traffic. By this time Stimson was long gone and the State Department was once again back among the satisfied consumers of other gentlemen's mail, now supplied by the Army and the Navy rather than its own Black Chamber. But there was, in the place of Secretary Stimson, a new reason to fear what would happen if other parts of the government learned what the Signal Corps was up to. The Communications Act of 1934 made it a crime, punishable by fine and imprisonment, for anyone to monitor and divulge to a third party the contents of any radio message. In 1935, Mauborgne, soon to become Chief Signal Officer, issued a less than completely reassuring decision that "these penalties apply, of course, to the work done by the entire Signal Intelligence Service, but they may be ignored, as this service operates in compliance with existing directions of the Secretary of War." As a legal theory, the notion that an Army regulation took precedence over an Act of Congress was a bit thin.

———

The United States Navy was only slightly better off. The Navy had a few years' head start in both intercept and cryptanalysis, and it had for many years recognized that the difficulty of the Japanese language posed a unique challenge, not only to code breaking but to learning the most basic facts about a country that commanded the world's third-largest navy. One naval officer was sent to Tokyo each year to learn Japanese; starting in 1927 the number was increased to two or three a year. Navies in the 1920s and 1930s tended to be more technically minded than armies. Ships, with all their complex engineering, were their lifeblood, while armies still could scarcely bring themselves to abandon the horse, and the fact that navies routinely had to think in terms of lines of communications and battle fronts thousands of miles long made them naturally quicker to adopt the new technology of wireless communications. The Director of Naval Communications in 1930, Captain Stanford C. Hooper, was a great believer in radio, and more generally in the need for the military to learn from industry and academia in order to stay technologically up to date, and he did what he could to drag the Navy along.

Because the navies of other nations had to communicate over long distances, their signals were easier to pick up from afar than were army communications, and the United States Navy, at a fairly early date, was alert to the possibilities this presented. The Navy set up its first listening post at the American consulate in Shanghai in 1924; a temporary station was established in Hawaii, east of Honolulu, in 1925; and in 1927 permanent stations were set up in Guam and the Philippines. The U.S. Navy also experimented with placing intercept units directly aboard ships.

Another early act of foresight was Commander Safford's recognition that the Japanese Morse code posed a unique challenge to intercepting Japanese signals. The rest of the world used international Morse code for radiotelegraphy, in which each letter is represented by a pattern of dashes and dots, that is long and short beeps. Transmitting a series of beeps is done by simply pulsing a radio transmitter on and off, a simple and robust means of communication that is far less susceptible to atmospheric interference than voice transmission. It is also far more precise and less liable to misinterpretation. Morse code itself was thus not "code" in the sense of a security measure, but rather simply a means of translating letters into a form suitable for dispatch by radio. But the uniqueness of Japanese Morse code certainly added to the difficulties of penetrating Japanese communications, much as did the Japanese language itself. Written Japanese relies mainly on some two thousand Chinese ideograms that stand for entire words. Words can also be written out using a set of forty-eight kana, characters that each stand for a syllable, and this was the logical basis for rendering the Japanese language in Morse code. But before radio operators could copy down Japanese Morse signals they had to learn a whole new set of long and short patterns, which involved actually unlearning the familiar ones. The International Morse code symbol for the letter C ($-\cdot-\cdot$) is, for example, the same as the kana syllable NI in Japanese Morse.

In 1924 the Navy had only two operators who knew Japanese Morse. Safford contracted with the Underwood Typewriter Company to build special typewriters that printed kana. The keys were cleverly arranged so that, at least for the lowercase character set, the kana keys were located in the same place as the corresponding International Morse letter would be on a regular typewriter. When an operator heard – · – ·, he would just hit the key where "C" would normally be, and the machine would print the kana for NI. The machines cost a whopping $160 apiece, and only about forty were acquired, paid for out of the Office of Naval Intelligence slush fund. The Navy also, in July 1928, began a secret training program that would eventually teach 176 enlisted men Japanese Morse. The three-month course took place in a twenty-by-twenty-foot concrete blockhouse that perched atop Wing 6 of the Navy Building in Washington. To get to class each morning, the students had to climb a ladder up to the roof. They quickly became known among themselves as the "On-the-Roof Gang."

Counterbalancing these few steps into the twentieth century were a set of naval traditions and a rigid career structure that dated from the age of sail. A strict rotation rule limited shore duty to two years unless an officer was given an engineer's rating. Continuity at OP-20-G was repeatedly interrupted by the departure of officers for mandatory sea duty. In 1931, six years after Safford had established the Research Desk, he and Joseph Rochefort were still the only two officers qualified to instruct new cryptanalysts. Only five other naval officers had received any cryptanalysis training at all; only one, Lieutenant Joseph N. Wenger, had had more than a six months' course.

OP-20-G's only other chief asset was a formidable force that went by the name of Agnes Meyer Driscoll—"Miss Aggie." A photograph taken probably when she was in her fifties shows a tall, thin woman with sharp eyes and an imperious expression and bearing, and one gets the feeling that she was a commanding presence even as a much younger woman. A civilian and a woman in a key position was an unusual sight in the U.S. Navy of the 1920s. She was thirty-one years old, a low-level code clerk in the Office of Naval Communications, when in 1920 George Fabyan offered to pay her salary and expenses if the Navy would give her leave to come work in the Department of Ciphers at his Riverbank Laboratory. She proved to have a remarkable flair for the work and, upon returning to the Navy the following year, easily cracked a sample cipher message produced on a supposedly "unbreakable" cipher machine that inventor Edward Hugh Hebern was trying to sell to the Navy. In 1923 Hebern hired Driscoll to help perfect an improved machine, but when the company went bust the following year (Hebern had sold $1 million worth of stock on fantastic promises of quick riches) she again returned to the Navy Department, in time to become Safford's first instructor in the mysteries of cryptology.

Like the Army, the Navy, too, was hamstrung by a policy of secrecy bordering on paranoia. Worse, at times it behaved as if the real enemy was not such theoretical opponents as Germany or Japan but the much more tangi-

ble one of the United States Army. When Solomon Kullback was sent to
Hawaii in 1937, he spent a year working at the Army's intercept station at
Fort Shafter without once learning that the Navy was operating its own in-
tercept and code breaking unit just a few miles away.

Meanwhile, generals and admirals in both services showed almost no interest in the work of OP-20-G or the Signal Intelligence Service; indeed
they showed little interest in intelligence of any kind. Technical specialists in
the military of the 1920s and 1930s were looked on with contempt or antag-
onism among career officers, and intelligence specialists were no exception
to this rule. Some of the contempt was fully justified by the poor performance turned in by intelligence officers, but there was a large component of
anti-intellectualism and snobbery and simple careerism at work, too. The
U.S. Army between the wars was more a private social club than the bureaucracy of technocrats it would become by war's end. Promotion up to the
rank of colonel was by seniority within each branch of arms, and line officers
tended to view new technologies as bearing the dangerous potential for creating new career paths that might let upstarts in the back door. Infantry,
cavalry, artillery, signals, and engineering still tended to see themselves as
separate and sovereign entities, a legacy of the long-standing refusal of Congress to permit the Army to have a general staff (that smacked too much of
Prussian militarism). Even when a general staff was authorized in 1903, the
old habits remained and the staff found it difficult to exercise much real
control. Adding to the Army's resistance to innovation was the glacial pace
of promotion. By 1930 it took a captain in the shrinking peacetime Army an
average of twenty-two years to rise to the rank of major. One of those majors, Dwight D. Eisenhower, had so little to do during the 1930s—this despite the fact that he was sole aide to Army Chief of Staff General Douglas
MacArthur—that he almost resigned his commission just to escape the
sheer boredom of it all. Instead, he took to reading Zane Grey novels to
pass the long Washington afternoons away. Major George S. Patton, Jr., stationed at nearby Fort Meyer, Virginia, whiled away his idle hours playing
polo and riding in steeplechases. He had been a major since 1919, which had
given him the time to collect four hundred ribbons and two hundred trophies for his equestrian exploits.

The somnolence of the U.S. Army was mirrored in the creaky wooden
headquarters it occupied on the Mall in Washington. The Munitions Building, and the companion Navy Department building next door, had been
hastily erected in the world war as "temporary" office space to cope with
America's sudden mobilization. Assistant Secretary of the Navy Franklin D.
Roosevelt, who fancied himself an amateur architect, had initially ordered
the Navy's building to be positioned in direct line of sight from the south

portico of the White House. That way it would be so ugly, he reasoned, that "it would just *have* to be taken down at the end of the war." But President Wilson objected that the sawing and hammering would disturb his concentration, so FDR reluctantly told the contractor to shift the site a half-mile down the Mall. The buildings were comb-shaped monstrosities whose backs fronted on Constitution Avenue beginning just west of the Washington Monument and stretching all the way to the Lincoln Memorial, with the teeth projecting south to the very edge of the Reflecting Pool. Decades later they were still there. It was a crime, lamented FDR, "for which I should be kept out of heaven, for having desecrated the whole plan of the loveliest city in the world." (The last of the "temporary" buildings finally came down in 1970.) The Navy Department and Munitions Buildings had no air-conditioning; they did have water-stained ceilings, creaking doors, and walls and floors that would shake whenever someone walked down the corridor, setting the bottles in the Coke machine rattling.

The organizations the buildings represented were relics, too. In 1932 the United States had the sixteenth-largest army in the world; it boasted a single mechanized regiment, a cavalry forbidden by Act of Congress to acquire tanks, and a stockpile of equipment left over from wars of earlier decades, even earlier centuries. Company commanders were personally responsible for the equipment in their charge and nothing could be disposed of; whenever a company moved, it took everything with it. In the 1930s a supply sergeant at Fort Jackson, South Carolina, found on his shelves a dozen objects made of black-painted tin and glass; no one, not even the oldest officer in the regiment, knew what they were for. They seemed to be signal lanterns of some kind. When the sergeant turned one over, he found stamped on the bottom the legend "Mfg USA 1863."

Ambitious plans for a postwar American Tank Corps were scrapped, and its two battalion commanders, Eisenhower and Patton, returned to the infantry and cavalry, respectively. In the 1930s the War Department rejected a design for a medium tank in part because, at fifteen tons, it weighed too much for the Army's bridging equipment. Its American inventor then sold a chassis to the Russians, who used it as the prototype for what would become the most effective medium tank of the Second World War, the T-34. From 1925 to 1940, the U.S. Army had all of about $20 million a year to modernize the equipment of its ground forces. Some far-sighted officers lamented the neglect of mechanization but many did not; an officer riding in a tank felt he looked about as dashing as "a construction worker on his way to excavate a basement," as one observer noted, hardly the spit-and-polish figure he wished to cut. British officers were even worse in this regard. American boys at least grew up tinkering with machines. In Britain anything involving grease and gears was for machinists and chauffeurs to deal with, not gentle-men. The British Army, said one officer in the 1930s, would never take tanks seriously until they could eat hay and whinny.

About the only thing the U.S. Army was good for in 1932 was patrolling the border with Mexico on horseback and rousting a rag-tag assemblage of world war veterans from the nation's capital. The "Bonus Army" had en-camped in Washington to demand immediate payment of the bonus that Congress had promised the veterans, but was not due to be paid until 1945. The twenty thousand men, women, and children were generally orderly and well-behaved but an eyesore, and after a policeman lost control in a scuffle and fired at an unarmed group of veterans, President Herbert Hoover or-dered the Army to clear their encampment off the Mall. MacArthur donned his uniform, ordered up cavalry and tanks, and, with tear gas and bayonets, forced the veterans out of downtown Washington and across the Anacostia River to where the main Bonus Army encampment lay. Hoover sent MacArthur written orders—twice—that he was not to cross the Anacostia bridge in pursuit. MacArthur gruffly told Eisenhower that he was "too busy and did not want either himself or his staff bothered by people coming down and pretending to bring orders," and proceeded to clean out the Ana-costia camp. A seven-year-old boy who had gone back for his pet rabbit was stabbed in the leg by a trooper who shouted, "Get out of here, you little son of a bitch!" MacArthur called a press conference to declare that revolution had been averted. Hoover said nothing about his orders having been defied. MacArthur's lifelong habit of insubordination received much encourage-ment that summer's day.

Naval officers had a somewhat more forward-looking attitude, certainly a more modern appreciation of technology, but they had an equally con-strained budget and an equally bored, or at least discouraged, outlook. The American Navy between the wars was perpetually undermanned, and al-though the treaty limitations set by the Washington and London confer-ences had clearly been to America's advantage—in general the United States accepted limitations on ships it never intended to build in exchange for real cuts in the forces of Britain and Japan, and it was left with a far more modern fleet as well—the psychological impact of scrapping fifteen capital ships, however outmoded they were, was immense. In 1929, the annual Grid-iron Club dinner, to which Washington's high and mighty were invited by the press corps to be skewered by satirical songs and skits, featured a chorus that sang (to the tune of "Little Buttercup" from *H.M.S. Pinafore*):

I'm off to that Conference
That London Conference

Though I can scarcely tell why;
Sadder and wiser, of
Diplomats I'm very shy
Our ships they are slighting,
They say "No more fighting."
We scarcely dare think what it means;
The Navy they're sinking
The Army they're shrinking—
Thank God we still have the Marines.

In such an atmosphere it was hardly surprising that the arcane science of intercepting and decoding enemy radio messages did not excite much interest. Both the U.S. Army and the U.S. Navy had established intelligence divisions, the Office of Naval Intelligence in 1882, the Military Information (later Intelligence) Division in 1885. But these offices were mostly a dumping ground for hacks and has-beens. The post of Director of Naval Intelligence was in 1919 given to an over-the-hill admiral, Albert Parker, who was remembered chiefly for his habit of asking incoherent questions during meetings with subordinates. In January 1920 Parker decreed that the best plan he could come up with for naval intelligence was to return to "the old organization and the old system which has worked so well for us." What that meant was ONI would grind out monographs at a leisurely pace based on reports filed from naval attachés who had skimmed a technical journal or had a few drinks with some foreign official at a party. Most of these papers were elementary profiles of foreign countries and their navies. Most were years out of date. A young naval officer slated for assignment as assistant attaché in Tokyo in the 1930s stopped by to read the ONI files on Japan and found they reflected none of the recent academic studies that he was personally familiar with; "I could have done as well reading books I had in my own library," he said.

It was a vicious circle. Intelligence was viewed with contempt, so capable and ambitious officers stuck tightly to a traditional career path and sought regiments or destroyers to command; the lightweights and rejects that no one knew what else to do with wound up in intelligence, thereby confirming the dim view with which intelligence was generally regarded. The attitude was that intelligence simply provided nothing of immediate importance to the commander in the field: As Eisenhower remarked years later, the Army General Staff before the war treated intelligence as a "stepchild." Those few capable officers who did take the job of intelligence-gathering seriously were generally met with indifference. One of Parker's more able successors lamented, in 1922, "The fact that we have never had brought home to us the need of a real intelligence service operating against a first-class naval power has given us in this respect a false sense of security," but no one paid much attention. When Layton and Rochefort were assigned to Japan in 1929 as language officers, both were given more or less the same lecture by the bored U.S. naval attaché in Tokyo: "I don't know anything about this language, but I understand it's difficult. You have only two duties to perform: One, study and master the Japanese language; two, stay out of trouble. If you fail in either, I'll send you home in the next ship. Payday is once a month. Other than that, I don't want to see you. Good-bye."

In 1936, Layton and Rochefort crossed paths again. Both were serving on the battleship *Pennsylvania*. So were Thomas Dyer and Wesley A. "Ham" Wright, both members of the select fraternity of officers who had been trained as cryptanalysts at OP-20-G. It was a remarkable assemblage of Japanese language and cryptanalytical talent, all the more remarkable for the fact that the ship's commander, Captain Russel Willson, had been Director of Naval Communications during the world war and had himself worked on radio interception and code breaking. One day Rochefort—he was serving as the Pacific Fleet's intelligence officer at this time—suggested to Wright that he try his hand at breaking the radio code that was being used by the ubiquitous Japanese fishing boats that plied the Pacific. The messages turned out to contain information about where fish were. But Wright became so absorbed in the task that he missed a Saturday morning inspection, for which Willson thoroughly chewed him out. Thinking that he had an excuse his commander would sympathize with, Wright explained why he had been late, at which Willson erupted in renewed fury at Wright for "fooling around with that stuff." Understandably perplexed, Wright screwed up his courage and ventured to ask whether Willson hadn't worked on codes himself. "Yes," the captain barked back, "but I had the good sense to get out of it!"

———

Starting from scratch at the beginning of the First World War, Britain had by war's end built a large and professional code breaking establishment, undoubtedly the best in the world. The Admiralty's "Room 40," as it came to be called, performed feats of cryptanalysis that have since become legendary, reading almost all of Germany's naval and diplomatic traffic during the war, including the famous Zimmermann Telegram that revealed Germany's proposed alliance with Mexico against the United States and whose disclosure helped to propel America into the war.

But the British code breakers if anything faced even greater hostility from line officers than did their American counterparts. The exigencies of war had forced the British armed forces to open their officers' ranks to all sorts of people who would never have been allowed in under normal circumstances. Being allowed in was one thing, though; being accepted was quite another. The poet Robert Graves, who had joined the Royal Welch Fusiliers as a "Special Reserve" officer and who was severely wounded in the fighting in France, noted that during peacetime "a candidate for a commission not only had to distinguish himself in the passing-out examination at the Royal Military College, Sandhurst, and be strongly recommended by two officers, but to possess a guaranteed private income that would enable him to play polo and hunt and keep up the social reputation of the Regiment. These requirements were waived in our case; but we were to understand that we did not belong to the 'Regiment' in the special sense." Just in case they missed the signals, Graves and others of his ilk were referred to by the regular senior officers as "warts." They were also informed that they could not expect to receive a medal for any feats on the battlefield; medals were for regular officers only. Even amid the worst of the war in the trenches, officers of "regular" army regiments managed to find time to look down with withering condescension upon newly commissioned officers whose not-quite-right color of khaki showed that they had not been to the right tailor—and thus were not real gentlemen. Such officers as had to be endured for the duration were, without any apparent irony, referred to as "temporary gentlemen."

At the Admiralty, the "professors" of Room 40 were similarly viewed with predictable contempt for their unseamanlike bearing and ignorance of naval tradition. Though commissioned as members of the Royal Navy Volunteer Reserves, they neglected to wear their caps in regulation fashion, forgot to salute senior officers whom they passed in the corridors of Whitehall, and sent solid Navy men into fits with their ignorance of naval terminology. (Admiral Jackie Fisher, the First Sea Lord, on one occasion personally informed the staff of Room 40 that warships did not "run in" and begged them to use the word "proceed" in their translations of German messages.) To many naval officers of small imagination and great veneration for naval tradition, that was enough to seal their fate.

So backward was the Admiralty in its views of intelligence that it began the war still sending ships out to mount standing patrols and peek into enemy harbors, a Napoleonic-War-era concept of reconnaissance that minefields and submarines had made dangerously obsolete; it might have kept on doing so had not an alarmed First Lord of the Admiralty, Winston Churchill, issued an order on September 18, 1914, putting a stop to such tactics as an unjustifiable risk hardly worth the price. His order had not yet been implemented four days later when a German submarine torpedoed three cruisers on slow patrol at Dogger Bank in the middle of the North Sea, killing fourteen hundred sailors.

The Admiralty's Director of Operations during the war, Rear Admiral Thomas Jackson, epitomized the anti-intellectual, "blue water" fighting admiral of yore who had no time for technical innovation or outsiders who presumed to advise him. William F. "Nobby" Clarke, a barrister by profession who had joined the Navy code breaking team in 1916, later wrote that Jackson "displayed supreme contempt for the work of Room 40"; only on two or three occasions did the admiral so much as set foot within its precincts, once to complain that the locked box he had been sent containing decrypted German messages had cut his hand, another time to express with great satisfaction, upon hearing that the Germans had introduced a new code book, "Thank God, I shan't have any more of that damned stuff!"

Room 40 had come into being in a typically casual and typically British fashion; indeed it was almost an accident that it had come into being at all. At the outbreak of war on August 4, 1914, the Marconi Company, the General Post Office, and British naval radio stations intercepted a number of radio signals "of which the only thing that could be said was that they were not British." The signals were copied down and forwarded to the Admiralty, where no one quite knew what to do with them. In due course they were passed on to the Director of Naval Intelligence, Admiral Henry "Dummy" Oliver, who didn't quite know what to do with them either. But a few days later Oliver was on his way to lunch at his club when it suddenly and conveniently occurred to him that he was lunching with "the very man I wanted." Sir Alfred Ewing, the Director of Naval Education, "seemed to be out of a job" now that war had broken out; he was mathematically knowledgeable, having been a professor of applied mechanics at Cambridge University when, in 1902, the forward-looking faction within the Royal Navy had decided that all naval cadets ought to receive a grounding in the basics of engineering and that Ewing was the man to make it happen. Moreover, Ewing had actually dabbled a bit with ciphers, having devised "a rather futile machine" to automate the task of enciphering and deciphering dispatches. Oliver asked Ewing if he would have a look at the intercepts and see if he "could make anything" of them. "Of course I said I should try," Ewing later recalled; "it was a moment when one grasped at even the most unpromising chance of being useful."

Ewing's start was certainly unpromising enough. Knowing next to nothing about codes, he spent several days perusing commercial code books at Lloyd's, the General Post Office, and the British Museum, hoping to get some basic idea of how codes "were or might be built up." Somewhat more promising was his contact with an amateur radio enthusiast, a barrister named Russell Clarke, who said he was picking up German signals at longer wavelengths than the ones intercepted by the Admiralty and could provide a steady supply of intercepts if the Admiralty gave him the facilities do so. Ewing gave Clarke permission to go ahead; eventually Clarke and another amateur enthusiast, a Somersetshire squire named Colonel Hippisley, were running eight stations, all staffed with radio operators supplied by the GPO, and all connected by direct telegraph lines to the Admiralty. Ewing had the Marconi Company set up additional stations.

Another source of intercepts came via a less direct route. On the second day of the war, the British cable-laying ship *Telconia* began pulling up and cutting German undersea telegraph cables. The original intention was a rather simple-minded one of disrupting enemy communications, but it would pay huge and unexpected dividends by forcing German overseas traffic onto radio or onto transatlantic cables controlled by Sweden and America—cables that passed through Britain, where the authorities had not the least compunction about tapping into them. This added to the flood of intercepts converging on the Admiralty, two thousand or more on some days. The messages would arrive addressed simply, "Ewing, Admiralty."

For assistance, Ewing was shipped four language teachers from the naval colleges at Dartmouth and Osborne, readily available because it was their summer vacation; their sole qualifications were a knowledge of German and "a reputation for discretion," as one of the four, A. G. Denniston, later wrote. "Cryptographers did not exist, so far as one knew," he added. The entire staff of would-be code breakers crammed into Ewing's office. Every once in a while someone would come in to see Sir Alfred on a naval education matter, whereupon the cryptographers would have to "pack up their papers as innocently as possible and scuttle into the small boxlike room" occupied by Ewing's secretary next door. The situation became even more strained in mid-October 1914 when the fledgling cryptographers were repeatedly evicted by the arrival of the Russian naval attaché coming to call on their boss; at the same time they began to be "almost rudely discouraged" from entering the secretary's room, now in the possession of "an unknown naval officer" who arrived early and stayed late each day. At last the cryptographers penetrated the veil of secrecy: The Russian had come to turn over a "priceless gift," a complete copy of the German naval code book captured on August 20, 1914, from the German cruiser *Magdeburg*. Churchill would forever after delight in the dramatic tale of how the sea-stained book was found clasped in the arms of a dead German signalman when his body washed ashore. The truth was more prosaic; the code book, now in the Public Record Office, is pristine and shows no trace of water damage, and had no doubt simply been taken from the ship's code room. But in any case the

Russians realized they lacked the means to exploit their windfall, and so turned it over to their allies.

The "unknown naval officer" in the boxlike room was Fleet Paymaster Charles J. E. Rotter, the Naval Intelligence Division's principal German expert. His initial investigations determined that the Germans were broadcasting weather reports using the straight code. In other signals, however, the code groups as transmitted did not make any sense when their meanings were looked up in the book; they were obviously being disguised by some additional encipherment. That was a fine point that seemed to have totally escaped Ewing, who ordered Rotter just to stick to the code book. Years later, Nobby Clarke found page after page of meaningless scribbles that Rotter had obediently produced, "testimony to the incompetent direction of Sir Alfred and the uncomplaining efforts doing a job at the direction of his chief which he must have realised quite early were entirely misdirected." This, no doubt, explained Rotter's long hours. At night he worked out the correct solution, discovering that the encipherment was a simple alphabetic substitution. To prepare a message, the German code clerk would look up a word or phrase (say, "February"), write down the corresponding four-letter code word listed in the book (say, "BIHU"), and then apply a letter-for-letter substitution key (say, B = F, I = A, H = P, U = E, etc.) to produce the enciphered code group ("FAPE"), which would be transmitted in the actual radio signal.

With the code book in hand and the encipherment solved, Room 40 was galvanized into action, and in early November began producing translations of broken German signals. The growing staff also overflowed the capacity of Ewing's office and was assigned the quarters that would give it its name (even after it later moved again to still larger premises): Room 40 of the Admiralty's Old Building.

By all accounts, Ewing never really did catch on to the cryptanalytic work he ostensibly commanded. But by the time he left Room 40 in May 1917 to take up the post of vice chancellor of Edinburgh University, he had managed to recruit a remarkable band of talented workers who more than compensated for his shortcomings. Ewing used his connections with King's College, Cambridge, which he had attended, to recruit many of the scholars who swelled the ranks of Room 40. A few had mathematical knowledge, but there were no actual mathematicians among them; they were mostly historians, linguists, classicists. The best, however, all shared a certain single-mindedness of focus and the scholar's aptitude for piecing together bits of a puzzle into a coherent whole.

One of the most brilliant—and surely one of the most unseamanlike— was Alfred Dillwyn Knox, known as "Dilly." Born in 1883, the second of four sons of the Bishop of Manchester, Knox had a typical upper-middle-class Victorian childhood growing up in a typical "Victorian vicarage." His three brothers all became notable figures in Britain (one was editor of *Punch*). While at Eton, Knox was sent off to take the examination for a Cambridge scholarship; it was wholly characteristic of his intense focus only on what interested him that he wrote two brilliant papers, one in mathematics, one in Greek verse, and left the rest of the examination unfinished. At Cambridge he showed an equally blazing and focused intelligence. He became a ferocious atheist, not merely ignoring God but positively "reproving God for not existing," in the words of his biographer and niece Penelope Fitzgerald. (At Cambridge he also found himself at one point the unintended object of the writer Lytton Strachey's homosexual ardor. The fires of Strachey's passion, however, were abruptly extinguished when Knox acquired a pair of glasses; Strachey wrote to a friend in horror about "the dreadful news about Knox" and couldn't bear to go see him on his next visit to King's.)

Knox eventually took on as his scholarly life's work the task of piecing together the shattered fragments of a papyrus scroll, the satiric dialogues or "mimes" of Herodas, unearthed in an archaeological excavation in 1889. From 1907 to 1914 Knox shuttled between Cambridge and the British Museum where the papyri were housed, ruining his eyesight and digestion with the strain of the work. Yet in many ways it was the perfect preparation for the task of the cryptanalyst.

Knox's work habits at the Admiralty must have done little to reassure the regular naval officers who were already prepared to take a dim view of the goings-on in Room 40. Mainly to avoid the musical parties that his housemate Frank Birch held (and thus spare the guests his "observations about music"), Knox would often work all night at the office and come "doddering home" in the dawn. His office, Room 53, was mostly taken up by a huge table that barely left enough room to squeeze around. But it was also the only room with a bathtub, and Knox made copious use of it, believing that soap and hot water were conducive to the "perception of analogies." After the war Frank Birch produced a privately printed book, "Alice in I.D. 25" (I.D. 25 was Room 40's official title from 1917 on, designating it a branch of the Navy's Intelligence division); the book was read at a celebratory concert the staff held on December 11, 1918, and it included "verses by A. D. Knox," among them this autobiographical one:

The sailor in Room 53
Has never, it's true, been to sea.
But though not in a boat,

He has yet served afloat—
In a bath at the Admiralty.

———

The other person at the Admiralty who became famous for thinking in the bath was the man at the top. Winston Churchill was soaking in his tub at 8:30 A.M. on December 16, 1914, when an aide burst in with a signal reporting that the Germans were bombarding Hartlepool in the northeast of England. Room 40 had correctly forecast the raid and knew from its reading of the decrypts that the German battleship fleet was being ordered out to support the mission. But the British Naval Chief of Staff had already decided that the battleships were not coming out and did not bother to consult Room 40. Ordering a trap laid for the anticipated raiders, he dispatched only a portion of the British fleet from its base at Scapa Flow in Scotland. When word of the German attack arrived, Churchill, grasping the signal "with dripping hand," jumped from his tub, pulled on his clothes, and ran to the Admiralty's War Room where he ordered the positioned British ships to move in for the kill. But the German battle fleet escaped—from "the very jaws of death," as Admiral Fisher, a personal ally of Churchill's in his war against hidebound Admiralty thinking, fumed.

The incident nonetheless helped to put Room 40 on the map. The British fleet's war plans had envisioned constant patrolling of the North Sea and a constant state of high alert at all other times to guard against a breakout of the German fleet from its safe harborage in the well-protected river estuaries of the Heligoland Bight. But Room 40 had proved that this was extravagant and simply unnecessary: Every German sortie was unfailingly preceded by a series of signals ordering minefields swept, aerial reconnaissance mounted, barrier booms opened; on occasion orders from the German Commander in Chief naming the squadrons and flotillas that would take part and their time of departure could even be intercepted. Churchill at once appreciated the huge significance of this weapon. Signals intelligence was, in Churchill's words, "news of the kind that never fails"; it meant the fleet could cease its relentless patrolling; it meant that the Admiralty could rest assured it would never be caught by surprise.

So began a lifelong fascination with secret intelligence. One of Churchill's first acts upon returning to office as Chancellor of the Exchequer in 1924 was to write on November 21 to the Foreign Secretary, Austen Chamberlain, asking "to see the Intercepts as they emerge." He followed up the next day with a second importuning letter, asserting that "I have studied this information over a long period and more attentively than probably any other Minister has done . . . and I attach more importance to them as a means of forming a true judgment of public policy in these spheres than to any other source of knowledge at the disposal of the State."

Churchill was also keenly aware of the extreme importance of keeping this new source of information an absolute secret lest the Germans catch wind of it and change their codes. In this, Ewing noted, Room 40 was amply aided by "the British reputation for stupidity." During the war the Germans never did suspect that their codes were being read, and took remarkably few precautions to prevent the possibility. Later in the war the alphabetic encipherment of the main naval code was changed every night at midnight, but the basic code book remained in use. A second code book, used mainly by submarines and zeppelins, did change in 1916, but copies of the book were salvaged from a zeppelin shot down outside of London and by divers who combed through a grisly tableau of corpses being fed upon by conger eels and dogfish in sunken U-boats. Room 40 was also keenly aware that what they could do the Germans might be doing as well, and strict orders were given that intelligence derived from enemy signals must be sent by courier or telegraph, never by radio. Such intelligence reports were always marked "Notbywit"—not by wireless telegraphy.

But Churchill's enthusiasm remained far and away the exception within the Admiralty; old attitudes die hard, and even as the old-line officers grudgingly conceded that Room 40 could provide some useful information, they insisted that the job of interpreting that information lay in the hands of experienced naval officers. The cryptanalysts were "crossword puzzle solvers," not intelligence analysts who understood naval tactics and strategy. The commanders were set, too, in their traditional view of intelligence as defensive; its chief purpose was to give a warning when the enemy was preparing to move. Little appreciation seems to have been given to the possibility that it could be used offensively, to plan one's own operations. Room 40 was actually forbidden to keep a flagged chart of the whereabouts of German ships on the grounds that the operations divisions took care of that job; it was not until May 1917, when Room 40 officially became part of the Naval Intelligence Division, that this edict was eased. Before that date the cryptanalysts had also been basically forbidden to communicate directly with the Naval Intelligence Division's German Section or Enemy Submarines Section, which resulted in the absurd situation that the authorities responsible for tracking German submarines relied exclusively on sighting reports from British ships.

The worst missed opportunity of the war followed directly from this limited view of signals intelligence and of the capabilities of the men (and, increasingly as the war went on, women) who managed it. On May 30, 1916, Room 40 broke a signal indicating that the entire German High Seas Fleet was preparing to come out the following day for what its new Commander in Chief, Admiral Reinhard von Scheer, intended to be the decisive confrontation of the war. The German commander planned to lure the British force on with a decoy force of battle cruisers; then von Scheer would pounce with his battleships, held at the ready to the rear. It was priceless information, and correctly used it could have given the British an insuperable advantage. But at this decisive moment Admiral Jackson made the unfortunate decision to become his own signals-intelligence expert. On the morning of May 31 he appeared in Room 40 to ask where the German station with the call sign DK was located. He was told that it was at Wilhelmshaven. Jackson left. Had he deigned to tell the code breakers why he was asking, they would have further explained that while DK was normally the call sign of von Scheer's flagship (the one fact, it later became clear, that Jackson did know), DK was always transferred to the fixed radio station at the port in Wilhelmshaven whenever von Scheer put to sea. Confidently acting on his incomplete knowledge, Jackson radioed the British fleet at midday that "the German fleet flagship" was still in port, adding, "Consider it probable that lack of air reconnaissance may have delayed their start." In fact, von Scheer was already well out to sea. Admiral Sir John Jellicoe, commander of the Grand Fleet at Scapa Flow, accordingly bided his time and was still seventy miles away from the action when the commander of the British battle cruiser squadron, Vice Admiral Sir David Beatty, ran into the Germans. There were plenty of other mistakes made that day in what came to be known as the Battle of Jutland, but Jackson's error was a crucial one. The battle, which should have been a victory of annihilation, ended in an indecisive draw. A year later Room 40 was still "simmering with resentment" over the way it had been bungled.

———

Room 40 continued to make itself distinctly unpopular with the regular Navy after the war ended. One of Nobby Clarke's "none too pleasant" tasks was rejecting claims from destroyer commanders for prize bounties. To boost morale during the battle against the U-boats, the Admiralty had passed out medals in abundance for the sinking of submarines. The commanders who had received medals naturally thought they had a pretty good case for receiving the bounty paid for sunken enemy ships. Clarke, who together with Frank Birch had been assigned the job of writing a historical review of intercepted signals, got to be the one to tell them otherwise, when time after time intercepts proved that the sunken U-boats had not been sunk at all. But though the Admiralty readily accepted Clarke's findings when it saved them from having to hand out cash, his historical report contained too many criticisms of the mistakes top officials had made in employing signals intelligence, and it was quietly tucked away on a shelf.

In early 1919, the British Cabinet decided that the secret work of Room 40 and its War Office counterpart (known as MI1b) should continue, and with equal secrecy, in peacetime. But when it came to paying for the operation, no department wanted to get stuck holding the bag. The Admiralty and War Office rightly pointed out that with the war over and Germany disarmed under the Versailles Treaty, there was no more naval and military traffic to intercept, so the bulk of the work for the new "Government Code and Cypher School" would be decrypting diplomatic communications, which was not really their department. (The not terribly deep cover name for the new department was dreamed up by the head of the Foreign Office's Communication Department, later ambassador to Peru, Courtenay Forbes. Like the post–Black Chamber Signal Corps, GC&CS was publicly responsible for developing codes in addition to being secretly responsible for breaking them.) The Foreign Secretary, Lord Curzon, who had little interest in decryption and little faith in its importance, countered that the Admiralty owned most of the intercept stations; logically they should be the ones to take charge. In the end that was what happened. Then the Treasury sent a puzzled rebuke to the Admiralty: Why did it want to spend £65,000 a year to maintain four wireless stations? Not only was this three times as much as budgeted, but it "had also understood that wireless messages were of little value for the purposes for which the Code and Cyphers School had been created" —that is, to intercept diplomatic traffic, which went largely via cable. The Treasury suggested shutting down the radio stations and, since that would mean fewer intercepts coming in, the cryptanalytic staff could also be cut commensurably. The argument had gone completely around a circle. The Admiralty wound up keeping two stations, at Scarborough and Pembroke, while the War Office retained one at Chatham, which did in fact shift their focus from naval and military traffic to intercepting the diplomatic radiograms sent by foreign commercial transmitters.

The final act in this somewhat farcical bureaucratic drama took place three years later. One day during March 1922 Lord Curzon had a meeting with the French ambassador during which he indiscreetly disparaged the views of his Cabinet colleagues. The French ambassador duly reported this back to Paris. GC&CS duly intercepted the ambassador's report and duly circulated the decrypt among its usual list of recipients in the Cabinet. Curzon was apoplectic. He also promptly revised his views of the effectiveness of signals intelligence. Completely reversing his previous arguments, Curzon indignantly pointed out that, since GC&CS was doing no naval or military interception to speak of, it ought to come under his control. The

Admiralty was happy to shed the expense, and on April 1 GC&CS was transferred to the Foreign Office.

GC&CS had been given approval to take on a professional staff of twenty-four "assistants" plus about thirty typists, clerks, and traffic sorters. All of the initial recruits were drawn from the ranks of Room 40 and MI1b. One key early hire was Dilly Knox. In 1920 he had married Olive Roddam, one of the young women who had gone to work for the Admiralty during the war, and in 1921 he sold most of his Great Western Railway shares and used the £1,900 to buy a house and forty acres of woods in a far-out London suburb near High Wycombe where he and his bride settled down. The daily commute by train gave him the time to finish his work on Herodas, which was published in 1922. It also allowed him to escape a suburban life that he found to be one of largely mind-numbing tediousness; Knox fit in with sub-urbanites about as well as he had fit in with the admirals and captains at the Admiralty. Although he did go to the inevitable neighborhood tennis par-ties (where he acquired a reputation for serving with a vicious "unreturn-able" spin), he secretly vented his irritation with local idiocy by jotting down "alarming" verses about his neighbors. A visit from the High Wycombe doc-tor meant death within the month, noted one; the stories told by a retired admiral in the neighborhood gave a new terror to the concept of infinity; the prevalence of adultery was surprising given that all the wives looked the same.

Another of the talented staff of Room 40 to sign on to the new organiza-tion was Oliver Strachey—Lytton Strachey's brother—who initially worked with Knox on American diplomatic codes and who would later, during the Second World War, take charge of breaking the hand ciphers used by Ger-man agents. Frank Birch stayed on briefly as well; in the ensuing years he juggled careers as a historian and Fellow of King's College and a comic ac-tor and theatrical director on the London stage. When he returned to GC&CS to head its German naval subsection in 1939 he could safely claim to have been the only Fellow of King's (and probably the only cryptanalyst, too) to have appeared in the traditional female-impersonator's role of the Widow Twankey in the annual Christmas Pantomime performance of *Alad-din* at London's Palladium theater.

Budgets in the new organization remained extremely tight. In August 1919 Nobby Clarke was invited to join with a promise of salary of £500 plus an annual bonus of £200. He accepted, only to be told that the bonus had been eliminated. When the subsequent discussion took place over what the new department should be called, Clarke proposed "Public Benefactors."

Another source of discontent among the former staff of Room 40 and MI1b was the choice of the man to head GC&CS. "Alice in I.D. 25" had concluded with a song, "Oh, if a time should ever come when we're demobilized," whose final stanza went:

> *We shall not wear mysterious airs nor boast what we could blab,*
> *Nor chase the solitary soap across the slippery slab.*
> *No more delights like these for us:—But* Denniston *will* never
> *Desert his solitary post:—He will go on for ever!*

It was a prescient commentary. Denniston was appointed head of GC&CS in 1919 and would stay at that "solitary post" for the next two decades. Clarke acidly observed that "many of the most capable of those who had worked in Room 40 flatly declined to serve under Denniston whom one of them described as possibly fit to manage a small sweet shop in the East End." But Denniston defied such easy categorization. Though a physically small and at first sight unimpressive man—indeed he was often referred to by his staff as "the little man"—he was also an accomplished and extremely competitive athlete who played golf and a "violent" game of tennis. He had no particular ability for (or liking of) administrative tasks, yet he understood the work of cryptanalysis, and perhaps more to the point he understood cryptanalysts; he knew when to leave them alone and how to fend off impatient politicians and military officials when things were not going well, and he was equally certain to instantly report their successes to higher-ups when things did go well.

Denniston was the operational head of GC&CS but officially carried the title of "assistant director." The director was a retired rear admiral, Hugh "Quex" Sinclair, a former Director of Naval Intelligence who had been brought into the Foreign Office to head both GC&CS and the Secret Intelligence Service, the cloak-and-dagger branch also known as MI6 about which so much, true and otherwise, would be written over the years. Sinclair—always referred to in official correspondence only as "C" (which had been the actual initial of his predecessor as Secret Service head, Mansfield George Smith Cumming)—was clearly the model for Ian Fleming's "M" in the James Bond novels, down to the green light over his office door that indicated when a visitor might enter the inner sanctum. Sinclair was a charismatic commander and a rather notorious man about town; his nickname came from the title character in the play *The Gay Lord Quex*, "the wickedest man in London." The real-life Quex traveled about town in a huge open Lancia touring car, and his personal accouterments included a bowler hat one size too small that was always rammed firmly on his head and a crocodile-skin case that was always stocked with one hundred huge cigars.

While the American code breakers operated between the wars in the shadows of the law, the British, with a nicer sense of bureaucratic form and fewer constitutional barriers to the legal suspension of civil liberties, simply enacted legislation giving the government the power to continue reading cables. Cable censorship had ended with the conclusion of the Versailles Treaty, but under the Official Secrets Act, passed in 1920, the Secretary of State was empowered in peacetime to serve cable companies with a warrant requiring them to hand over all cable messages "for the purpose of scrutiny." The Secretary of State merely had to determine that it was "expedient in the public interest" to do so. He immediately determined that it was, and for the next twenty years every cable sent to or from the United Kingdom passed through GC&CS's hands first. "Between us and the companies there has never been any question as to why we wanted the traffic and what we did with it," Denniston later wrote. "The warrant merely said scrutiny, and the traffic arrived back apparently untouched within a few hours. I have no doubt that the managers and senior officials [of the cable companies] must have guessed the true answer, but I have never heard of any indiscretions. . . . I believe we never failed to return all the traffic, though many million telegrams must have passed through our hands." A slightly less official, backdoor channel was arranged to obtain copies of cables that went through Malta, particularly traffic passing between Tokyo and Europe; even cables sent on special private lines that foreign embassies installed to communicate with their capitals during major diplomatic conferences in London found their way to GC&CS. Sinclair meanwhile quietly arranged with the Police Commissioner for the "loan" of the police radio station on Denmark Hill in northwest London, which was well located for intercepting diplomatic radiogram traffic carried by commercial companies outside the United Kingdom. The Foreign Office paid for the station's upkeep and the salary of the police constables who operated it, and the commissioner "agreed to ask no questions about their work."

It proved easier to spy on former allies than former enemies. GC&CS broke the French and American diplomatic codes in short order; the Italian Navy, which produced the only substantial flow of military radio traffic that could be readily intercepted, had the "delightful habit" of encoding long political editorials from the daily press, which allowed GC&CS to reconstruct the entire naval code book "in the early days." Japanese naval traffic was beyond the reach of the radio receivers but a quantity of Japanese diplomatic and attaché traffic was intercepted and broken, with the help of a single Japanese interpreter supplied by the Admiralty. (It was not until the 1930s that a small intercept unit was established at Hong Kong—this was the unit that moved to Singapore in 1939—to begin work on Japanese naval codes.)

The comparative cryptographic innocence of Britain's First World War allies was not, however, shared by the Germans. The Zimmermann Telegram had exposed the weakness of the German codes, and when the German delegation arrived in Paris after the war it was equipped with a set of codes that GC&CS deemed completely unbreakable. The code used a basic book of 100,000 five-digit numerical code groups, each of which stood for a word or phrase. During the war the Germans had used such code books without any further disguise, or sometimes with only a simple substitution that applied across the board to every code group in the message. The new system was entirely different. In preparing a message, the code groups were enciphered by adding each in turn to a series of numbers drawn from a book of 60,000 random five-digit numbers. Each code group in each message was thus enciphered according to a *different* substitution scheme. To make it even more impenetrable, the message was then subjected to a second such treatment: A sequence of additives drawn from a different starting point in the 60,000-group additive book was added to the message again. This doubly enciphered code, which GC&CS dubbed "Floradora," would remain unbroken until 1942; even then it was mastered only with the help of incredible luck—and a stolen scrap of paper. For their most secret communications, the Germans chose an even more secure method of encipherment. Pages of additives were supplied in booklets; when each page had been used for a single message it was torn out and thrown away. Such a "one-time pad," if constructed properly with a truly random sequence of numbers, is impregnable. The attitude would long prevail at GC&CS that "all German codes were unbreakable," as Frank Birch would recall being told in the early days of the Second World War.

There were many indiscreet leaks and boasts of Britain's wartime accomplishments that had put the Germans on notice, just as Yardley's tell-all had alerted the Japanese. Churchill himself told the story of Room 40 in his 1923 memoir, *The World Crisis*. More details on the Zimmermann Telegram and how it was broken followed. Ewing presented a lecture to the Edinburgh Philosophical Institution on December 13, 1927, describing the history of Room 40 and discussing in considerable detail specific codes and how they were broken. One of Ewing's gravest indiscretions was to reveal that radio direction finding had been used as a deliberate camouflage during the war to conceal the fact that information was being derived from code breaking. The idea of direction finding was simple: An antenna on a receiver can be mechanically or electrically "steered" to determine the direction a signal is coming from. Two receivers at different locations can then triangulate on the source of the signal. The Marconi Company conducted experiments in direction finding in the spring of 1915 and immediately informed the Admiralty of their success; the Admiralty then built up a large network of D/F stations, which proved invaluable in pinpointing the location of enemy ships at sea. Because direction finding was not a secret, it also proved invaluable in diverting suspicion from the real secret of what Room 40 had pulled off. In the Second World War the British would repeatedly use this same ruse, which, remarkably, worked again in allaying German fears that their codes had been broken. But it was a dangerous lapse to have revealed the ploy in 1927.

Reading French and American diplomatic cables was amusing, no doubt, but it offered little in the way of hard intelligence. The one real operational success that GC&CS scored in the 1920s, however, led to the most serious security breach yet, with drastic repercussions. Following Britain's misadventurous intervention in the Russian Revolution, an embargo had been placed on trade with Soviet Russia. In January 1920 British Prime Minister Lloyd George sought a thaw in relations; "commerce," he argued, "has a sobering influence" that would be far more effective in combating Communism than troops or blockades had been. On May 31 the Soviet Commissar for Foreign Trade, Leonid Krasin, was received by Lloyd George at No. 10 Downing Street and negotiations on a trade agreement began.

The Bolshevik government had inherited the czarist codes but realized they had to be thoroughly compromised and so began using a fairly simple transposition code that scrambled the letters of the plain Russian text of a message. GC&CS was well equipped to deal with Russian traffic. After the Revolution, the Czar's chief cryptologist, Ernst Fetterlein, had narrowly managed to flee the country; safely reaching Western Europe, he made contact with the British and French intelligence organizations. A meeting was arranged, and Fetterlein's only question was, "Well, gentlemen, which will pay me the most?" Apparently the British would. Fetterlein was always seen sporting a huge ruby ring on his right index finger, a token of recognition of his services to the Czar, but other than this one bit of ostentation he was a reserved, silent, and withdrawn man, short, with thick-lensed glasses and an equally thick Russian accent. (A later colleague found the only way to draw him into conversation was to say something patently foolish: "And the Czar, Mr. Fetterlein, I believe he was a strong man, with good physique?" Thus provoked Fetterlein would erupt into speech: "The Czar was a weakling who had no mind of his own, sickly and generally the subject of scorn.")

Fetterlein had two young female assistants, also Russian émigrés, and within a few weeks of Krasin's arrival in England, GC&CS was able to hand

Lloyd George a decrypt of the instructions Krasin had been cabled directly from Lenin himself:

> That swine Lloyd George has no scruples or shame in the way he deceives. Don't believe a word he says and gull him three times as much.

More seriously, intercepted telegrams revealed that the Soviets were using their trade mission to slip tens of thousands of pounds to British political organizations and left-wing publications, including the "Hands Off Russia" Committee, the Communist party of Great Britain, and the *Daily Herald* newspaper. Since one of Lloyd George's proposed terms for a trade agreement was a halt to Soviet propaganda and subversion within the British empire, this was at the very least a breach of faith.

Curzon and Sinclair, both rabid anti-Bolshevists, demanded that this evidence of Soviet perfidy be made public at once; Sinclair went so far as to make the extraordinary statement that "even if the publication of the telegrams was to result in not another message being decoded, then the present situation would fully justify it." Even the usually discreet Churchill agreed: The need to expose the attempted subversion "outweighs all other considerations," he said. The Cabinet so ordered. Eight intercepts were handed over to all the national newspapers. Three years later Curzon obtained Cabinet approval for another deliberate indiscretion, and released verbatim texts of traffic between Soviet diplomatic officials in Moscow and Kabul ordering anti-British subversion in India. On that occasion the Soviets backed down, recalling their envoy in Kabul and promising to mend their ways. A few months later they also changed their codes. When the final break in British–Soviet relations came on May 26, 1927, it was also accompanied by the release of an intercepted telegram, this one documenting Soviet espionage activities in Britain. Denniston lamented that the government "found it necessary to compromise our work beyond any question," and indeed that was exactly what it had done. The Soviets reacted swiftly; following the Germans' example of how to deal with British code breakers, they adopted a one-time pad for all diplomatic and trade communications with foreign missions. Like the German diplomatic codes, these new Soviet codes would remain unbroken until well into the Second World War.

The proper use of signals intelligence would involve a delicate balancing act throughout the war years. There was little point of going to the trouble to intercept enemy signals and break their codes if the information was not going to be put to use on the battlefield. On the other hand, careless use endangered the whole enterprise. GC&CS clearly thought that in the case of the Russian decrypts far too much had been sacrificed for far too little.

The compromise of the code breakers' work was only one legacy of this episode. The other enduring legacy was an obsession with domestic subversion within the intelligence agencies that would persist up until the very outbreak of war. The extreme anti-Communism of the British and American intelligence chiefs of the 1920s and 1930s was part of the explanation for their preoccupation with Soviet Bolshevism, but there was more to it than that. The idea of an enemy within touches something deep in human nature, and indeed there was ample and recent precedent for the intelligence services' hunt for subversives at home. In the years preceding the First World War, Britain was swept with a series of Fifth Column panics. It all seems farcical now, but it was deadly earnest at the time, both to the public and to British officialdom. During one debate in the House of Lords it was solemnly alleged that eighty thousand German waiters in Britain were members of a secret army awaiting the signal to seize railway junctions and the forts on the Thames. One of the few people who saw anything funny in this was P. G. Wodehouse, who produced a tale, *The Swoop! Or How Clarence Saved England,* describing how a lone Boy Scout uncovers a German plot to take over the country. (Clarence's keen eye spots an item amid the cricket scores and racing results in the late bulletins in the local newspaper: "Fry not out, 104. Surrey 147 for 8. A German army landed in Essex this afternoon. Loamshire Handicap: Spring Chicken, 1; Salome, 2; Yip-i-addy, 3. Seven ran.") The public was not amused, and the book soon went out of print, a flop.

In contrast to the mythical prewar army of German waiters, Communists in Britain between the wars really were receiving money and marching orders from abroad. Anti-Communists might exaggerate the significance and effectiveness of Soviet schemes to undermine Western governments, and did, but the basic facts were indisputable. And one did not have to be a fanatical anti-Bolshevist to believe that, with the Kaiser defeated, Lenin seemed a far more plausible threat to world order and peace than did a vanquished and crippled Germany, stripped of its air force and U-boats and limited by the Versailles Treaty to a tiny professional army of one hundred thousand men.

———

Yet even long after the threat from a rearmed Nazi Germany should have been glaringly obvious, old attitudes persisted. As late as February 1939, Quex Sinclair was insisting that "alarmist rumours" of new aggressive moves by Germany were being "put forward by Jews and Bolshevists for their own ends." The following month Hitler invaded Czechoslovakia. Throughout the 1920s and 1930s a series of American and British Directors of Naval Intelligence were equally oblivious to the threat from abroad as they focused on what they were convinced was a much more imminent threat at home. Admiral Sir Barry Domville, who had been the British DNI from 1927 to 1930, grew obsessed with discovering what "Masons, Jews, and other secret forces at work in our Society were up to." Captain Hayne Ellis, appointed DNI of the United States Navy in 1931, ordered the Office of Naval Intelligence to step up its surveillance of radical and pacifist groups, including the Women's League for Peace and Freedom and the National Federation of Churches. Ellis spent much of his time corresponding with vigilante and "patriotic" groups and recruiting professors of naval science at Harvard, Yale, Georgia Tech, and Northwestern to spy on "ultra pacifist" students and faculty. By 1936, the monitoring of Communists, labor unions, and opponents of naval expansion had become the primary task of ONI.

By contrast, Nazi Germany was for the most part simply ignored. The obsession with domestic subversion, the neverending turf wars between the intelligence departments, the separation of code breaking from intelligence, the generally low regard in which intelligence was held in the first place, the flip-flopping between indiscreet publicity and suffocating secrecy—all conspired to create a situation in which little information about the growing threat from Germany and Italy was obtained from signals intelligence, and what little was obtained was not permitted to disturb preconceived notions or bureaucratic routine. When GC&CS broke the Italian naval codes, the Admiralty refused, on secrecy grounds, to allow intelligence derived from the decrypts to be distributed to naval commanders. As a result, Nobby Clarke furiously recalled, "demonstrably false" information on Italian forts and vessels obtained from unreliable sources was circulated in its stead. The few times intelligence of immediate operational importance was obtained in the 1930s, the system was simply incapable of handling it. In 1935 a message was read giving details on the Italian submarine patrol line from Malta to Alexandria. An urgent cable to the British Commander in Chief in the Mediterranean was drafted. The next morning the head of the GC&CS naval section was huffily informed by the DNI that the cable had *not* been sent, and would he please bring a copy of the original intercept. "This," Clarke dryly noted, "was carefully examined by the DNI who knew no Italian." Admiral John Godfrey, the extremely capable DNI who assumed command in 1938, commented that in 1939 none of the assistant chiefs of staff or directors of operational divisions of the Admiralty "knew anything about intelligence. I myself knew precious little. There is no particular reason why we should have done, because the subject was swept out of sight during the twenty years of peace."

The British War Office had gone so far as to abolish its Military Intelligence Division in 1928; it was not revived until 1939. The Air Ministry was so committed to its doctrine of strategic bombing that it ignored its own intelligence staff, which was correctly insisting that all indications showed the Luftwaffe was being organized as a shock force to support ground troops. The Foreign Office meanwhile remained notoriously obtuse about GC&CS's work altogether; on one occasion in the 1930s, having partially broken a code used by the Rumanian government, GC&CS intercepted and read a cable from Rumania's minister in London reporting on a meeting with a top Foreign Office official named Orme Sargent. One portion of the decrypt read, "As I was leaving, Mr Sargent said wittily . . ." but there then followed a dozen code groups that the cryptanalysts had been unable to make out. Hoping to learn the meanings of these indecipherable groups, a GC&CS official called on Sargent and showed him the intercept. Sargent was outraged. "I made no witticism to that bloody little man!" he retorted. It was only after the German invasion of Czechoslovakia in March 1939 that the military chiefs of staff agreed to set up a "Situation Report Centre" under Foreign Office chairmanship to collect and distribute intelligence that appeared to require immediate attention.

A few attachés and diplomats accurately reported on Hitler's rearmament plans and the growing might and aggressiveness of the German armed forces, but an equal number blithely ignored it, and the intelligence chiefs, their minds elsewhere, threw in their lot on the side of blissful ignorance almost to a man. Captain William D. Puleston, the U.S. Navy's DNI from 1934 to 1937, concluded in late 1936: "Indications are that [Germany's] relative strength will never be great enough to encourage her to initiate hostilities without powerful allies." His conclusion was based largely on reports from America's special naval attaché in Berlin, who found Hitler to be a "good natured" leader.

In fairness, these were widely held views among Britons loath to fight another European war and Americans content in their isolationism. Underrating their future enemies was par for the course; the only thing surprising was that the men charged specifically with anticipating the moves of potential enemies were equally blinded by their preconceptions. As late as 1941 Puleston was writing that Japanese naval aviators were just not up to American standards when it came to landing and taking off from carriers. (Other articles by naval experts published in the American popular press concurred that the Japanese were "daring but incompetent aviators," suffering from myopia, defects in the inner ear that affected their sense of balance, and a lack of American mechanical know-how and individualism.) In 1938, an American Army captain, Albert C. Wedemeyer, returned to Washington after two years as an exchange trainee at Germany's war college. This was a standard courtesy among the militaries of the world, but Wedemeyer was shocked by what he had witnessed. Not only was the German Army becoming a huge and powerful force but its officers were being inculcated with an aggressive militarism. Yet nobody back at the creaky Munitions Building in Washington seemed interested. A few scoffed that Wedemeyer was easily impressed by Germans because he himself was of German ancestry. Most of his fellow officers were interested only in learning what perquisites German officers of various rank were entitled to, and how that compared to the U.S. Army.

Even in Britain in the spring of 1940, in the midst of war, little had yet occurred to force a change in the deep-seated attitudes of a generation of officers when it came to intelligence. It would take two more years and repeated, incontrovertible proof of what code breaking could do before the military establishment would begin to take intelligence seriously. That spring of 1940, as the phony war ended and German tanks thundered through Holland and Belgium, Peter Calvocoressi—London lawyer, graduate of Eton and Balliol College, Oxford—presented himself at the War Office to volunteer his services. Calvocoressi later wound up as a senior intelligence officer in the air section of GC&CS. But that spring day he was shuttled from one office to another, interviewed, given tests, and finally at the end of the afternoon shunted into a room where an officer sat at a desk before him. Calvocoressi took a surreptitious glance at a sheet of paper that lay on the desk and could see, reading upside down, that it was a summary of his test results. At the bottom was the verdict: "No good, not even for Intelligence."

CHAPTER 2

Nature of the Beast

S ECRET writing is almost as old as writing. But the science of breaking an unknown code arose only with the flowering of mathematics and lexicography in the medieval Arab world. By the fourteenth century, Arab philosophers had established the basic principle for discovering the identity of letters in a simple cipher; it is a principle that has scarcely been improved upon in the centuries since. Substituting one set of letters or symbols for another cannot disguise the fact that some letters occur more frequently in common language than do others. The Arabs noted that alif is the most frequent letter found in Arabic writing, za the least frequent; that certain combinations of letters often occur together in words while others are rare or impossible; and thus simply counting how many times each symbol occurs in a cipher message can go a long way to establishing its true identity. A medieval Arab treatise advises making a frequency count as the first step toward decrypting a cipher message; so does William Friedman in *Elements of Cryptanalysis*, the tutorial he began preparing in the 1920s.

In English, a random sample of prose contains *e* as the most common letter (12 percent), followed by *t* (9 percent), *a* and *o* (8 percent each), *i, n,* and *s* (7 percent each), and *r* (6 percent); bringing up the rear are *j, k,* and *x* (0.5 percent each) and *q* and *z* (0.3 percent each). The longer a message is, the more likely its letter frequency is to approach this average, but even short messages are unlikely to be too anomalous; the most frequently occurring character in a cipher text may not always stand for *e* but it is extremely unlikely to stand for *q* or *z*. Combinations of letters are more revealing still. There are 26 × 26 different possible two-letter combinations in the English alphabet, and tabulating the frequency with which each of these 676 bigrams occurs in a cipher message can swiftly narrow the field of possible solutions. Certain letters almost never occur doubled in English text *(hh, ii, jj, kk, qq, uu, ww, xx, yy)* and so can be ruled out as the identities of any symbols in a cipher message that do show up doubled; the vowels *a, i,* and *o* appear far more frequently adjacent to other letters than they do to one another, a highly significant clue in establishing their identities; the letter *n* is far more likely to be preceded by a vowel than by a consonant; certain pairs of letters occur frequently in one order but rarely or never in the reverse *(ea* versus *ae, lm* versus *ml, rn* versus *nr).* Once the identity of a handful of the cipher-text characters has been established this way, recovering the remaining letters is a skill on a par with solving a crossword puzzle. The blanks are filled in to form likely words; incorrect answers can swiftly be eliminated when they yield contradictions elsewhere in the message. If, for example, the cipher text character X has been identified as *a* and G as *u* in the following message:

$$\text{cipher text} \quad Q \ G \ T \ L \ G \ X \ A \ L \ Z \ L \ F \ P \ F$$
$$\text{plain text} \quad - \ u \ - \ - \ u \ a \ - \ - \ - \ - \ - \ - \ -$$

one could rule out "kumquat" as the first word because, by assigning L = *q,* it would imply an extremely unlikely ensuing sequence of letters:

$$Q \ G \ T \ L \ G \ X \ A \ L \ Z \ L \ F \ P \ F$$
$$k \ u \ m \ q \ u \ a \ t \ q \ - \ q \ - \ - \ -$$

On the other hand, "fun Guam" (which implies L = *g)* seems more promising:

$$Q \ G \ T \ L \ G \ X \ A \ L \ Z \ L \ F \ P \ F$$
$$f \ u \ n \ g \ u \ a \ m \ g \ - \ g \ - \ - \ -$$

Filling in the remaining blanks to form the word "gigolo" would produce additional letter assignments (Z = *i,* F = *o,* P = *l)* that could in turn be cross-checked elsewhere in the message.

Such simple, or monalphabetic, substitution ciphers were widely used in the ancient and medieval worlds. They would scarcely have stumped an expert (or even an amateur) code breaker by the time Herbert Yardley began his career as a cryptanalyst, but the discovery of a systematic technique for unraveling even those simple ciphers was a code breaking insight of such fundamental power that all the innovations by code makers in subsequent centuries could never completely defeat it. The cat was out of the bag. No matter how elaborate a scheme was used to scramble and disguise the original text, its ghost always shone through; the uneven letter frequencies that are a characteristic of all languages leave their distinctive trace, a smile of the Cheshire Cat that will never completely disappear.

The Arab lexicographers recognized several other enduring truisms of ciphers. As early as the eighth century, the Basra grammarian Al-Khalil noted that the stereotypical fashion in which most messages begin constitutes a grave weakness that can be exploited to unravel a cipher. Later cryptanalysts would call this process "cribbing"—making an educated guess of a few key words and then seeing if the resulting letter assignments begin to yield readable text elsewhere in the message. Al-Khalil reported that he had solved a Greek cipher message sent him by the Byzantine Emperor by assuming that it began, "In the name of God." Military and diplomatic messages are especially prone to stereotypical phraseology, and one of the key principles of cipher security that would be emphasized again and again to code clerks over the centuries—and again and again ignored, to their downfall—was the need to avoid reusing stock phrases. Another principle of cipher security that followed directly from the Arabs' discoveries was that the longer the message, the more statistics it would provide to a would-be code breaker on letter frequencies, and the easier it would be to solve.

The advent of the telegraph in 1844 gave urgency to several competing trends in code development that had been slowly evolving since the Renaissance. The telegraph for the first time made rapid communication possible, even as it made communications far less inherently secure. A letter carried by a messenger might be intercepted, but a telegraph message was practically a public announcement. Many people were sure to see any telegraph message as it was handed to a clerk and tapped out over the wires; the lines themselves could be tapped, too, by someone possessing even a small amount of technical know-how. During the American Civil War the Union armies sent more than 6 million telegrams, of which many were intercepted (though none apparently solved) by Confederate forces. Such a volume of traffic placed an unprecedented burden on any cryptographic system. It had to be simple and quick enough to use in the field, otherwise the time spent coding and decoding would undo the speed that telegraphy made possible. Yet it had to be secure enough to withstand cryptanalysis even when an enemy was likely to have thousands of coded messages to work with.

Most of the schemes that were concocted to meet these admittedly conflicting demands were variations on what would come to be known as polyalphabetic ciphers. Instead of a single alphabetic substitution, a series of different cipher alphabets can be used in sequence for each letter of the message in turn. In other words, the letter *a* might be enciphered as X when it appears as the first letter of a message; when it reappears a few letters later it might be enciphered as G. One such scheme, which actually dated back to the fifteenth century, only to be forgotten and reinvented in the late nineteenth century, was the Vigenère tableau. This consisted of a chart containing a series of twenty-six cipher alphabets, each shifted one letter over from the last. Plain text letters were listed across the top of the chart; down the side were the "key" letters that determined which cipher alphabet was to be used:

plain text

		a b c d e f g h i j k l m n o p q r s t u v w x y z
	A	a b c d e f g h i j k l m n o p q r s t u v w x y z
	B	b c d e f g h i j k l m n o p q r s t u v w x y z a
	C	c d e f g h i j k l m n o p q r s t u v w x y z a b
key	D	d e f g h i j k l m n o p q r s t u v w x y z a b c
letter	E	e f g h i j k l m n o p q r s t u v w x y z a b c d
	F	f g h i j k l m n o p q r s t u v w x y z a b c d e
	G	g h i j k l m n o p q r s t u v w x y z a b c d e f
	H	h i j k l m n o p q r s t u v w x y z a b c d e f g

etc.

Thus to encipher the plain text *p* with the key letter B, one copied out the letter found in column p and row B, which is *q*. The key letters were often specified by means of an easily remembered word or phrase written out over the plain text:

key	S W O R D F I S H S W O R D F I S H S W
plain text	m y h o v e r c r a f t n e e d s o i l
cipher text	E U V F Y J Z U Y S B H E H J L K V A H

The virtues of the polyalphabetic encipherment are immediately apparent. In this example, the letter *r* has been enciphered as Z in one occurrence, Y in another. The cipher text H stands for *t* in one place, *e* in another, *l* in yet another. Any straightforward attempt at analysis of letter frequencies would thus be thwarted by these multiple identities. In a well-constructed polyalphabetic cipher, each cipher letter appears with almost equal frequency. The letter frequencies, in other words, become far more random. A monalphabetic cipher follows the skewed distribution of plain language itself, ranging from 12 percent for whatever cipher symbol stands for *e* to 0.3 percent for the symbol for *z*. An ideal polyalphabetic cipher has a dead flat distribution of letter frequencies; every symbol has the same 1 in 26 chance, about 3.8 percent, of showing up in the cipher text.

Another advantage of the scheme is that it is relatively easy to use, and even if the encipherment method is revealed to an enemy, he still cannot read any intercepted messages without the key word—which can be changed willy-nilly, every week or every day or whenever security demands. The key word can be committed to memory with no written evidence to betray it. Finally, polyalphabetic ciphers lend themselves readily to generation by mechanical devices. One of the oldest is the familiar cipher cylinder popularized by Thomas Jefferson, in which a series of disks, each with a different jumbled alphabet printed on its circumference, is mounted on an axle; the disks are rotated so that one row of letters spells out the plain text, and the cipher text is then read off from any other row.

Still, even the best polyalphabetic ciphers bear the ghost of the plain text that lies buried beneath. In 1935, Solomon Kullback produced what is today still regarded as the definitive mathematical analysis of polyalphabetic ciphers. At the time, Kullback had only five years' experience as a cryptanalyst, but he had completed a Ph.D. in statistics at George Washington University in the meanwhile. Expanding on a pioneering series of papers by his boss William Friedman, Kullback showed that even the most complex polyalphabetic cipher can be successfully attacked by frequency analysis— so long as one has enough cipher text to work with.

Friedman's key insight, which Kullback extended, was one of those flashes of brilliant simplicity that time and again characterize cryptology. It had long been recognized that the basic weakness of polyalphabetic encipherment is that at some point the key must repeat itself. If a key is ten letters long, then the first, eleventh, twenty-first, thirty-first (and so forth) letters of the message will all have been enciphered using the same cipher alphabet. Likewise the second, twelfth, twenty-second, and thirty-second letters will all have been enciphered with a second alphabet—and so on. Thus any polyalphabetic cipher with a key of length n can in principle be reduced to a problem of breaking n separate monalphabetic substitution ciphers.

The trouble is that a single message would have to be of enormous length to provide enough letters for a valid frequency count for each of the multiple monalphabetic ciphers that constitute it. To get good statistics, about fifty letters of text are required. The longer the key, the greater the number of separate cipher alphabets and the smaller the sample of cipher text in each. And usually a cryptanalyst has no particular way of even knowing the length of the key to start with. What he really needs to do is to amass a large number of separate messages that all have been enciphered using the same key, and place them one below another such that all the letters in one column will have been enciphered with the same key letter. The cipher text in each vertical column then represents one monalphabetic substitution problem.

But aligning multiple messages this way is also easier said than done. No code clerk would be dimwitted enough to start every message with the first letter of the key. Rather, he would follow some scheme to shift the key with each message; some messages would start the key with SWORDFISH, others with WORDFISHS, others with ORDFISHSW, and so on.

Friedman realized, however, that without solving a single bit of the cipher, without knowing the length of key, without knowing the system used to shift the key from message to message, it is nonetheless possible to place messages in alignment so that every letter in a given column will have been enciphered with the same key letter. The principle involves what he termed the index of coincidences, and its brilliance lies in recognizing the ghostlike way that the underlying unevenness of language can never be completely obscured. "Coincidence" in this case means how often two cipher texts placed one above the other will have the same letter occurring in the same column. If two random strings of polyalphabetic cipher text are compared this way, the number of such coincidences is dictated by the straight laws of chance. At any given spot in a message, the letter A has a 1 in 26 chance of showing up; the chances of an A showing up at that same spot in the second message are likewise 1 in 26; thus the chances of A showing up in both messages at the same spot simultaneously are $1/26 \times 1/26$, which equals 0.15 percent. So the chances that *any* coincidence (A and A, or B and B, or C and C, and so on) will occur at a given spot are 26 times as great: $26 \times 1/26 \times 1/26$, or 3.8 percent.

But when two strings of cipher text that have been enciphered with the same key are properly aligned, something dramatically different occurs. In that case, each vertical pair has been enciphered with the same monalphabetic substitution. In any given column, an *e* in one message will appear as the same cipher text letter as an *e* in the other message. And those high-frequency letters now skew the odds of a coincidence occurring. Whichever cipher-text letter happens to stand for *e* in any given column now has a 12 percent chance of occurring; its chances of occurring in both messages simultaneously are $12/100 \times 12/100$, or 1.4 percent—almost ten times the odds of a coincidence for any given letter in the purely random case. The cipher letter in each column that stands for a low-frequency letter like *z* will, to be sure, have a much smaller chance of appearing simultaneously in both messages (for *z* it is less than .001 percent). But the high-frequency letters so skew the odds that they overwhelm the effect of the low-frequency letters. When the odds of all twenty-six possible coincidences are summed, the to-

tal chance of any coincidence occurring between two properly aligned messages in English works out to about 6.7 percent, nearly double the coincidence rate for a randomly aligned cipher text.

To place two messages in alignment thus becomes a purely mechanical
problem of arithmetic. The two are placed in a trial alignment, the number
of coincidences are counted, and that tally is divided by the total number of
letters. If the messages are properly aligned the result will be near 6.7 percent; if improperly aligned it will be closer to 3.8 percent, and the two messages can then be shifted one letter over relative to each other and the test
run again.

Placing messages in alignment—or "in depth"—is a cryptanalytic technique of enormous power, and it would prove to be the key to breaking
some of the most difficult ciphers that American and British code breakers
would encounter during the Second World War. All, that is, except for the
most difficult of all: the Enigma machine cipher.

———

In its military dictatorship, implacable hostility toward Russia, and persecution of its Jewish minority, Poland in the 1930s had much in common with
Germany's new rulers. Marshal Józef Pilsudski, who had seized power in
Warsaw in a military coup in 1926, efficiently dismantled the country's fledgling parliamentary democracy and, less than a year after Hitler's ascent to
power in Germany in 1933, sealed a ten-year nonaggression pact with the
Nazi government. When Hitler annexed Czechoslovakia's Sudetenland in
the summer of 1938, Poland joined in the pillage, grabbing the Czechoslovak
frontier district of Teschen. "Glorious in revolt and ruin; squalid and shame-
ful in triumph," said Churchill of the Polish rapine; "The bravest of the brave,
too often led by the vilest of the vile!"

It was Russia that was Poland's historical oppressor, and even as war
clouds gathered unmistakably in the west in 1939, Poland ruled off the table
any talk of a defensive alliance that so much as hinted of common cause
with the Soviet Union. As the historian James Stokesbury observed, the
Poles hated the Russians more than they feared the Germans. Yet Poland
was not blind to the covetous eye that Germany cast to the east, nor to the
simple truth that Poland's borders were basically indefensible. Poland's history was one of repeated dismemberment by powerful neighbors on all
sides; the saying went that Poland in fact had no history at all, only neighbors. In the late eighteenth century Germany and Austria had joined Russia
in carving up Poland, and for more than a century the country had ceased to
exist as a sovereign state. The Treaty of Versailles restored Poland's exis-

tence. But German governments from 1919 on made no secret of their desire for a "revision" of the western border of Poland. Germany signed the 1925 Locarno Pact, solemnizing her postwar borders with France and Belgium, but pointedly refused to consider an Eastern Locarno Pact that would do the same for Poland. Another sore spot was the Corridor that Poland had been granted at Versailles. The Corridor linked Poland to the Baltic Sea, with the port of Danzig a free city under international control; it also cut off Eastern Prussia from the fatherland. Prussia was the ancestral seat of German militarism; the Teutonic Knights had colonized the territory in the Middle Ages, and it was still the home of the powerful landed dynasties known as the Junkers who dominated Germany's officer corps. To such men the Corridor was almost a personal affront to their honor. Yet the truth was that it was not just Poland's borders but Poland itself that Germany found intolerable. As early as 1922 General Hans von Seeckt, the commander of the German Army, was insisting to his government that Poland's very existence was "incompatible with the essential conditions of Germany's life." His conclusion: "Poland must and will go." Hitler's designs on Poland were not a feverish aberration of Nazism, but a chronic syndrome of German nationalism.

Poland had a million men under arms, including twelve brigades of cavalry; on paper it was an impressive force, all the more impressive in the eyes of those fatally romantic Polish generals who were convinced that with bravery and enough horses one could do anything. But there were a few farsighted officers who early on grasped a simple fact that had eluded their British and American counterparts. Intelligence, and code breaking in particular, was, in the military terminology of a later era, a "force multiplier." The knowledge of how a potential enemy was developing and deploying his military might made one's own limited forces that much more effective. It was not a weapon that a newly reborn nation, especially one as vulnerable as Poland, could afford to neglect. And so from the start Polish intelligence kept at least one eye firmly fixed on the German armed forces. Poland was obviously in an excellent position to pick up even short-range German Army and Navy radio communications, and by the early 1920s the Biuro Szyfrow, or Cipher Bureau—part of Section II (Intelligence) of Warsaw's Army General Staff—was producing a steady stream of decrypts of German traffic collected by its listening stations.

But quite abruptly success came to a halt. In February 1926 German Navy messages became unreadable; in July 1928 the Army messages followed suit. The Poles suspected that the new German codes were ciphers generated by a machine, but were at an impasse.

The idea of using a machine to generate a cipher had been pushed to the fore by the advent of radio. Marconi's demonstration of transatlantic wireless telegraphy in 1901 had opened the way to a jump forward in the volume and speed of military, naval, and diplomatic traffic, as great an advance over telegraphy as telegraphy had been over written dispatches. Even elaborate polyalphabetic ciphers using long key phrases and frequent changes of key collapsed under the weight of traffic they now were expected to carry. If communication by telegraph had been a public announcement, communication by radio was something approaching a public spectacle. Anyone could listen in, and, as the First World War had amply proved, did.

The weakness of hand ciphers under such circumstances was twofold. First, the key inevitably repeated itself; that allowed messages to be placed in depth. Second, even when the key was changed, the encipherment was based on a limited set of possibilities. The Vigenère tableau, for example, employed only twenty-six different cipher alphabets. To expand that set would create a system that would quickly become too cumbersome to use in the field. It would require complex keying procedures to keep track of and indicate which cipher alphabet to use for each letter of the message; it would require elaborate charts for encoding and decoding.

The various cipher machines that were invented in the 1910s and 1920s attempted to overcome these limitations by using a mechanical system to automatically generate a key of such extraordinary length that hundreds or even thousands of messages could be sent without ever using the same stretch of key twice. Thus messages could never be placed in depth. Moreover, the machines were not limited to twenty-six different cipher alphabets; they could draw upon literally millions of different scrambled alphabets; and the number of different possible key sequences that resulted from mechanically stringing together these millions of alphabets in various permutations could easily reach the millions of millions in even the simplest of these early cipher devices. Changing keys was as simple as resetting a dial or a switch or shifting a plug, and the result was as dramatic as inventing an entirely new cipher. Best of all, machines eliminated much of the drudgery of encoding and decoding messages. The message text was typed into a keyboard and the cipher text emerged automatically, and vice versa.

The cipher machine known as the Enigma had had an unpromising beginning. A German engineer by the name of Arthur Scherbius had devised the name and taken out patents, but both seemed the kiss of death to a string of companies that acquired the rights to them. An early model of the Enigma machine was exhibited at the 1923 Congress of the International Postal Union in Bern, Switzerland, where it was pitched to businessmen as a way to keep the contents of their telegrams secret from the prying eyes of competitors. Mechanically and mathematically it was brilliant. Commercially it was a flop.

As with several other cipher machines patented almost simultaneously in Sweden and the United States, the heart of Scherbius's design was a series of rotating disks that acted to scramble the cipher alphabet. Each of these rotors had a ring of twenty-six electrical contacts on each face. A typewriter keyboard activated a bank of electrical switches that started the encipherment process. Pressing the letter A closed one of these switches, sending an electrical current to one of twenty-six contacts on a fixed ring that abutted the twenty-six contacts on the right face of the rightmost rotor. Inside each rotor a thicket of wires connected the right face to the left face in a fixed but randomly scrambled pattern. Contact number 1 on the right face might be wired to contact 22 on the left face; number 2 might be wired to number 5; number 3 to 16; and so on. As the current emerged at the electrical contacts on the left face of the rotor, it flowed into the contacts on the right face of the abutting middle rotor, where another scrambling took place. The output of the middle rotor was in turn fed into the final, left-hand rotor.

The especially clever part of the design was what happened next. When the current exited the left face of the third rotor it entered a "reflector": This was a half-width disk that had contacts on its right face only. The internal wiring of the reflector joined these right-face contacts to one another in pairs. Thus, upon reaching the end of the line, the current was "reflected" back into the left-hand rotor, but now at a different contact on its left face. The current then traveled through all three rotors again, this time from left to right, undergoing a further scrambling. Upon reaching the right face of the right-hand rotor, the current again passed through the contacts of the fixed entry ring and from there was carried by wires to one of twenty-six lamps, each marked with a letter of the alphabet, indicating the encipherment of the letter whose key had been depressed.

The reflector had the effect of ensuring that a letter could never be enciphered by itself. A could not stand for A. Furthermore, it ensured that for any given setting of the three rotors, the contacts on the fixed ring would be uniquely joined in pairs. If pressing A caused E to light up, then pressing E would cause A to light up. That was a huge advantage in practical operation, for it meant the same machine, set up the same way, could be used both to encode a message and to decode it.

Like all rotor machines, the Enigma was designed to change the scrambling for every successive letter of text typed in. Every time a key was pressed, the first thing that would happen was that a mechanical linkage caused the right-hand rotor to advance one position, one twenty-sixth of a turn, thus shifting all of its contacts over one. A system of notches and pawls

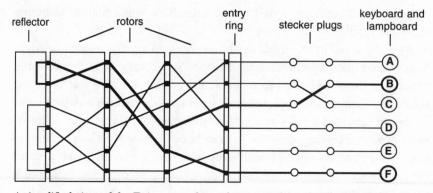

A simplified view of the Enigma machine, showing only six letters. When a key is depressed, current flows through the connections causing a corresponding lamp to illuminate. The bold line traces the path of one such connection, linking B and F. As each key is depressed the rotors move to a new position, creating a new scrambling pattern.

caused the middle rotor to advance to its next position after the right-hand rotor had completed a full circle; when the middle rotor had made a full revolution the left-hand rotor would advance one position. Each time one or more rotors advanced, the interconnections between them would change, sending the current on a new path through the scramblers. Each letter of a message would thus be enciphered with a completely different cipher alphabet; only after all three rotors had gone through the full cycle would the key repeat itself. That meant that about seventeen thousand letters could be typed in without ever reusing the same stretch of key.

Each of the rotors of the Enigma machine was labeled around its circumference, on a movable ring, with the digits 1 to 26 or the letters A to Z; a message could be started at any point in the key cycle by spinning the three rotors on the machine to the desired position. Following either a predetermined schedule list or an indicator transmitted in the message itself, the sender and receiver would set their rotors so that the chosen three-letter indicator (AZG or YSD or whatever) showed through a window in the top panel of the machine. So long as these starting points were spaced well apart from one message to another, each message would be enciphered with a totally unique stretch of key. Placing messages in depth would be simply impossible.

To increase the number of different key sequences available, many sources of extra permutation were built into the machine's design. The rotors, each of which had a different internal wiring, could be taken out of their slots and reinserted in a different left-to-right order. That created six permutations: Any of the three rotors could be placed in the left-hand slot, leaving two choices for the middle slot; whichever rotor was left then went in the right slot, yielding $3 \times 2 \times 1 = 6$ different possible left-to-right rotor combinations (or "wheel orders," as they would more commonly come to be called by the Allied code breakers). To generate even more permutations, additional rotors could be supplied and swapped for one or more of the three. With five rotors to choose from, there were $5 \times 4 \times 3 = 60$ different possible wheel orders.

Additionally, the notch on each rotor that engaged the "turnover" mechanism could be shifted to a new position. The notch was located on the movable ring on which the letters A to Z were printed; by releasing a clip, the ring could be rotated independently of the rotor inside, like sliding a tire around on a wheel. Resetting the rings thus shifted the point during the revolution of each wired rotor at which the rotor to its left would advance, and that broke up each key sequence into a further 26×26, or 676, possible combinations. Being able to change the ring setting had the added bonus of disguising the actual starting position of the wired rotors for any given indicator; telling a recipient to set his rotors to starting position AFQ would not be enough information for him to successfully decode a message unless he also knew how the rings had been set relative to the wired rotors—and there were $26 \times 26 \times 26$ possible ways to do that.

Finally, in later models of the Enigma machine a series of plugs, or *steckers,* added yet another layer of scrambling, one that generated billions of additional permutations in key sequences. Instead of connecting the keyboard and lamps directly to the fixed entry ring that contacted the right face of the right-hand rotor, the steckered Enigma machines interposed a board with twenty-six pairs of jacks. Plugged cables, rather like those found in an old-fashioned telephone switchboard, were connected between chosen letters. Plugging one end of a cable into the F jack and the other end into the W jack would swap those two letters' identities in whatever cipher substitution the rotors generated. With ten cables, 150 million million such permutations were possible.

If this did not afford sufficient security it was hard to imagine what could. Even if a machine fell into enemy hands, it would give a code breaker almost nothing to go on. The number of unique key sequences that the Enigma machine could generate; the ease with which those keys could be changed every day, or even more often; the length of each key, which prevented two messages from ever having to be enciphered with the same stretch of key—all of these features rendered the conventional mathematical tools of cryptanalysis impotent.

If businesses fretting over their trade secrets were not particularly impressed by all this, one customer finally was. Smarting from the 1920s revelations of British success in reading its codes during the world war, the

German Navy and Army seized upon machine-generated ciphers as a way of ensuring that such fiascos as the *Magdeburg* did not happen again. When the Biuro Szyfrow began picking up indecipherable German messages in the late 1920s their first thought was that this was dummy traffic put on the air simply to spoof any would-be code breakers who happened to be listening in. The text was *too* perfectly random; it bore not even a ghost of the skewed letter-frequencies that real ciphers always betray. But as the traffic became more and more common and eventually replaced almost all other cipher traffic of the German Navy and Army, the Polish code breakers realized what they were up against.

Like just about all of the world's cipher bureaus, the Poles were well aware of the Enigma as a result of its unsuccessful commercial marketing. In 1928 the Poles obtained more direct confirmation of German interest in the Enigma: A package sent from Germany arrived at the Warsaw Customs Office for routine inspection. Immediately a protest came from Germany that the package had been misdirected and should be returned at once, unopened. That was of course enough to arouse anyone's curiosity, and technicians from Section II promptly arrived to take a look. It proved to contain a commercial Enigma machine. Using a false address as cover, the Biuro Szyfrow then placed an order for its own machine from the German manufacturer. But when the Polish code breakers tried it out on the intercepted traffic it produced gibberish. Not surprisingly, for in their military version of the machine the Germans had changed the internal wirings of the rotors.

Although the ultimate security of the Enigma cipher did not depend on keeping those wiring patterns a secret, it certainly didn't hurt to add another source of bafflement to anyone who tried to break the cipher from scratch. The number of different ways the three rotors could be wired up inside was nearly 10^{80}—that is, the number one followed by eighty zeroes. Such numbers defy comprehension in everyday terms. If a million million computers on each of a million million planets in each of a million million galaxies spent a million million years trying every possible combination of wiring three Enigma rotors, and took a millionth of a millionth of a second to test each one, they would still have less than one chance in a million million of finding the correct one in that time.

Faced with a task of those daunting dimensions, the Poles gamely decided the time had come to call in the mathematicians. In an act of foresight unparalleled for the times—for the Poles surely had no hopes that it would lead to quick results—two officers from Section II arrived in Poznan University in January 1929 to explain to twenty mystified third- and fourth-year mathematics students that they had been selected by their professor to attend a special course in cryptology. The class would meet twice a week in the evening; the students would be expected to keep up with their regular course work, and the very existence of the cryptology class was to be kept a strict secret.

Poznan University was selected for this experiment not only because of the renown its Mathematics Institute enjoyed, but also because of its location. Poznan was in the western part of the country, the part that had been controlled by Germany until 1919, and its students had almost all attended German-language schools in their youth. In a fitting twist of history, the Mathematics Institute at Poznan was housed in a pseudo–Teutonic Knights castle that had been built at the turn of the century by Kaiser Wilhelm II as the official residence of Germany's Crown Prince, part of the effort to Germanize the Polish territories.

As the course progressed, one by one the students dropped out. Eventually only two of the original twenty survived the ordeal: Jerzy Rozycki and Henryk Zygalski. A third student, Marian Rejewski, had left Poznan in March 1929 before completing the course to enroll in a graduate program in actuarial mathematics at Göttingen in Germany. But in those two months Rejewski had already demonstrated a phenomenal grasp of the mathematics of cryptanalysis, taking only a few hours to break a complex German double transposition cipher the instructor had given the class to work on. When Rejewski returned to Poznan in the summer of 1930 to take up a teaching assistantship, he learned that Rozycki and Zygalski, still students, were now working for the General Staff twelve hours a week in an underground vault at the Army's regional Command Post building, right next to the university on St. Martin's Street. Rejewski was quickly invited to join the group. Intercepted messages were delivered by courier from Warsaw and from a monitoring station that was set up in a nearby suburb of Poznan; the mathematicians were to concentrate on breaking the new cipher keys in various German hand ciphers whenever the keys were changed and dispatch the solutions back to Warsaw. The three jocularly called their vault the "Black Chamber." By September 1932 Zygalski and Rozycki had graduated and the General Staff shut down its cryptological outpost in Poznan and brought all three men to Warsaw. And one morning, probably in October, Major Maksymilian Ciezki, head of the Biuro Szyfrow's German section, called Rejewski into his office and casually asked him if he had some extra time in the afternoons. Rejewski said he did. "In that case," Ciezki said, "I would like you to come in in the afternoons as well. Only don't tell your colleagues anything about this." Rejewski was to come in for two hours a day after his coworkers had gone home. Waiting for him that first afternoon were the Biuro Szyfrow's files on the Enigma. The military cryptanalysts who ran the Biuro had tried and failed to crack the problem. And so they were turning it over to one twenty-seven-year-old mathematician, who was to see what he could do in his spare time. They clearly felt they had nothing to lose by trying. More than four years of work had gotten nowhere.

———

Machine ciphers were a creature of the communications revolution. So too were a very different class of codes that came to the fore in the years between the wars. It had always been recognized that using thousands of different symbols to stand for words or phrases—which is, strictly speaking, the definition of a "code"—is inherently more secure than using twenty-six different symbols to stand for the letters of the alphabet. By sheer weight of numbers a simple code is much more difficult to crack than a monalphabetic cipher. But codes require large code books, which are laborious to compile, cumbersome to carry, and liable to capture. And though they present difficulties, they are ultimately no more impenetrable than are ordinary ciphers. The words that make up language are as patterned as the letters that make up words, if anything more so. Reconstructing a code book is more a linguistic art and less a mathematical exercise than is breaking a cipher, but the process of "book breaking" is perfectly straightforward. Code groups that represent stereotyped or frequently used words or phrases ("to," "from," place names, months of the year) give themselves away by showing up repeatedly, often in the same location, in multiple messages. Once the code group for "stop" is found, texts are delineated into grammatical units that can be more easily parsed and whose meanings may be guessed at, especially by someone familiar with the rules of word order and grammar of the underlying language. The neat habits of message clerks in numbering messages sequentially can readily reveal the code groups that stand for numerals; if a series of intercepted messages begins:

message 1:	ZOXIL	GEZOR	. . .
message 2:	ZOXIL	KUMQT	. . .
message 3:	ZOXIL	ORANG	. . .
message 4:	ARDVK	FABOL	. . .

it is not hard to figure out that GEZOR is seven, KUMQT is eight, ORANG is nine, FABOL is zero, and ZOXIL and ARDVK are two other sequential numbers.

Whether code groups are represented by letters or digits makes no difference in the plan of attack. But codes in which meanings are assigned to code groups in alphabetical order offer another foothold; if one discovers that 52980 means "tunnel" and 52976 means "tuna" then 52978 must fall alphabetically between those two words in such a "one-part" code (so called because encoding and decoding can be done with the same book).

Navies and diplomats tended to favor codes over ciphers for both practical and customary reasons. They had less reason to fear losing code books to capture by hostile forces than did armies; just to make sure, however, it was traditional to bind navy code books between sheets of lead or to print them in water-soluble ink so they could be quickly jettisoned and rendered useless if capture threatened. The codes that came into use after the First World War by the world's major navies and diplomatic services were almost exclusively two-part codes (which the British referred to as "hatted" codes); code groups were assigned in nonalphabetical, random order, which meant that two separate code books were needed for encoding and decoding. The encoding book listed the meanings in alphabetical order, the decoding book listed the code groups in numerical order.

The simple sort of enciphered code used by the German Navy in the First World War, in which a monalphabetic substitution was applied to all code groups across the entire code book, was by the 1920s almost universally replaced among the world's code users by much more complex forms. Some used a complex pattern of transpositions to shuffle the identities of the code groups. Others used encipherment by a running key of additives supplied by a separate key, or additive, book. Just as the Enigma machine ensured that every letter of text even in a lengthy message was disguised with a different substitution cipher, the use of an additive book in an enciphered code ensured that every code group was disguised with a different encipherment. The additive books issued in the 1920s typically contained tens of thousands of random numbers. The process of encipherment was cumbersome but theoretically extremely effective in ensuring security. The code clerk would first write out the text of his message and look up the code group equivalents for each word or phrase. Then underneath the code groups he would copy out a string of random additives beginning at an arbitrarily chosen page and line of the additive book. For example:

plain text: FROM KAGA ESTIMATED TIME OF ARRIVAL 2130

	from	*uppercase follows*	*Kaga*	*stop*	*ETA*	*2130*	*stop*
code text:	21936	48322	01905	38832	87039	11520	38832
additive:	02923	41338	00989	15861	28959	23693	18229
enciphered message:	23859	89650	01884	43693	05988	34113	46051

To simplify matters, and to keep each resulting sum of enciphered text a five-digit number, the addition was done digit by digit, without carrying (e.g., 9 plus 4 equals 3). When the clerk was done, he had a string of five-digit numbers that, to any casual or even not so casual observer, would seem random and meaningless. Although the code group "stop" occurs twice in this message, it appears in the final signal as 43693 in one place and 46051 in another. In another message it would appear as an entirely different number.

The system seems truly impenetrable. No pattern of the underlying language shows through, nor is there any way even to guess which two numbers were summed to produce a given enciphered group. Any enciphered group might stand for *any* code group.

It might seem the longest of long shots, yet there was one tried and true way of teasing out meaning from an enciphered code. If two messages had been enciphered with the same stretch of additive, they could be placed in depth—aligned such that the groups in each vertical column had been enciphered using the same additive. And that could begin a slow and painstaking process of extraction that would begin to lay bare the code groups beneath. Key to this process was the recognition that a *pair* of code groups falling in the same column always retained one characteristic that no encipherment could disguise. Regardless of the additive applied to them, the numerical *difference* between two given code groups in the same column is constant. Searching for recurring differences between columns of messages in depth was the wedge that could split an enciphered code apart. For instance, if three messages such as the following were known to be properly placed in depth:

	1	2	3	4	5
message 1:	0256	7892	8835	7923	0470
message 2:	0003	6286	5684	9901	5236
message 3:	8263	0017	5431	4329	5565

the recurring difference in column 1 (message 1 minus message 2 equals 0253) and column 3 (message 2 minus message 3 equals 0253) is a strong indication that the same pair of code groups appears in each location. That discovery in turn immediately supplies relative values not only for the two code groups responsible for the recurrence (which we can arbitrarily assign the relative values of A = 0000 and B = 0253) but also for the additives for those two columns *and* for any other code groups that fall in those columns:

	1	2	3	4	5
additive:	0003	....	5431	....	....
message 1:	0256	7892	8835	7923	0470
code text:	**0253 (B)**	....	3404 (C)	....	....
message 2:	0003	6286	5684	9901	5236
code text:	**0000 (A)**	**....**	**0253 (B)**	**....**	**....**
message 3:	8263	0017	5431	4329	5565
code text:	8260 (D)	....	**0000 (A)**	....	....

The meanings of the code groups are as yet unknown. But as more and more code groups and additives are recovered the process snowballs. The recovery of the code groups 8260 and 3404 above provides additional pairwise differences among the growing set of known code groups that can be hunted for in other columns: $8260 - 0253 = 8017$; $3404 - 2503 = 1901$; $3404 - 8260 = 5244$. (Again, all addition and subtraction is done digit by digit, without carrying.) This last difference shows up in column 5 (message 1 minus message 2), allowing more recoveries to be filled in:

	1	2	3	4	5
additive:	0003	....	5431	....	7076
message 1:	0256	7892	8835	7923	0470
code text:	0253	....	3404	....	**3404**
message 2:	0003	6286	5684	9901	5236
code text:	0000	**....**	0253	**....**	**8260**
message 3:	8263	0017	5431	4329	5565
code text:	8260	....	0000	....	8599

Thus message 1 reads: B ... C ... C; message 2: A ... B ... D; message 3: D ... A ... E. Once the code group for very common words or punctuation (such as the ubiquitous "stop") is identified, it is possible to speed up the process by subtracting its value from every cipher group in a message to generate a string of "hypothetical additive," then see if stripping this run of additive from other aligned messages yields any other already identified code groups. Whenever such a hit occurs, the "stop" has probably been correctly placed in the message.

When enough additive has been recovered to allow most or all of a series of messages to be "stripped," laying bare their underlying code groups, the same linguistic process of book breaking that works for ordinary unenciphered codes can be brought to bear to assign meanings.

The catch is achieving depth to begin with. If a typical message is fifteen groups long and if the additive book contains thirty thousand groups, as the Japanese naval codes often did, the statistical chance of two randomly picked messages overlapping at all, by even as much as a single group, is a little shy of 1 in 1,000. To minimize the chance of overlap, code clerks were usually under strict orders to pick a different starting point for every message they sent; the Japanese Navy sometimes assigned individual clerks different starting points, evenly spread throughout the additive book, and told them to work straight through, that way ensuring that the traffic was spread evenly throughout the book. Thus a well-used enciphered code offered few opportunities for building depth to start with.

Even if messages did overlap, as they eventually would when an additive book had been in use for a while, there was no obvious way to discover the fact. To tell the intended recipient of a message the page and line number of the additive book where the encipherment began, the code clerk buried within the message an additional code group that served as an "indicator." The indicator was usually enciphered with a special system used only for that purpose. If that indicator system could be cracked, then the code breaker was in business; each message could be definitively placed in the additive book. But the indicator was invariably disguised to look just like every other cipher group in the transmitted message; it was placed in a prearranged position in the message and nothing called attention to it or in any way announced that it was other than just part of the message text.

The only other recourse a cryptanalyst had was the approach that came to be called "brute force." This was a search not for a needle in a haystack but for a straw in a haystack. Or, rather, a search for the same straw in two haystacks. The idea at first sounds like a triumph of optimism over common sense. But over the course of thousands upon thousands of messages, eventually it will happen not only that two messages overlap—meaning that both messages contain code groups that have been enciphered with the same additive—but that they will contain the *same* code group enciphered with the same additive. When that happens, the resulting message groups are identical. Of course it is entirely possible that two messages containing identical message groups do *not* overlap a common stretch of additive key at all; pure random coincidence will from time to time cause the sum of one code group and one additive group to equal the sum of an entirely different code group and an entirely different additive group. What a cryptanalyst employing the brute force method is really looking for, then, is an even longer shot—the same *pair* of straws in each of two haystacks. Such a "double hit" is far less likely to arise from random coincidence and is much more likely to indicate a true overlap; for instance:

| *message 1:* | 3419 | 2100 | **7364** | 5642 | 9468 | **2316** |
| *message 2:* | **7364** | 7130 | 0072 | **2316** | 0924 | 7464 |

is a near sure sign that these two messages can be properly placed in depth thus:

| *message 1:* | 3419 | 2100 | **7364** | 5642 | 9468 | **2316** | | |
| *message 2:* | | | **7364** | 7130 | 0072 | **2316** | 0924 | 7464 |

Tens of thousands of messages might have to be searched to find a handful of such double hits, however. And many fairly simple measures could be taken by code makers to thwart such a search, by making the straws smaller and the haystacks bigger. Japanese Navy code clerks were instructed always to start encoding messages in the middle of the text, thereby shifting any possible stereotyped addresses or signatures out of a position where they would be obvious. Commonly used words, phrases, or punctuation marks likely to generate double hits were assigned multiple synonyms in the code book so no single code group appeared too often. A place name like "Manila" could be encoded at least five different ways: as a single code group standing for the entire word, as three code groups standing for the three kana syllables that make up the word, as several code groups standing for the Japanese words that form a homonym for the word, as six code groups each representing the letters of the Roman alphabet that form its English spelling, or as a special coded area designator.

————

On a steamy Washington morning in June of 1930, William Friedman appeared in the vault of the Munitions Building where his three "junior cryptanalysts" had spent the last two months struggling their way through badly smudged carbon copies of a typewritten draft of Friedman's latest course in cryptanalysis.

Assuming a more than slightly theatrical air of mystery, Friedman summoned Kullback, Sinkov, and Rowlett to follow him, and silently led the way down the stairs, along a corridor, and into what appeared to be a deserted wing of the building. There he opened two huge steel doors secured with combination and key locks. The windowless vault within was pitch black. As if continuing to play out a scene from a bad gothic novel, Friedman struck a match in the darkness, revealing in the dim glow a room thick with dust and crammed with double rows of filing cabinets. "Welcome, gentlemen," he intoned—by this point in the performance they probably wouldn't have been surprised had he said "to Castle Dracula"—"to the secret archives of the American Black Chamber!"

During their first few weeks on the job the three had been told they would be shot if they even went near the filing cabinets in their own office. Now they were being let in on the most closely held secrets of the American government, the records of Yardley's Black Chamber, which had been packed up and taken into possession by the Army's MI-8 when the bureau was shut down the year before.

Friedman explained that they were to start working through the file cabinets and see what they could find about Japan's diplomatic codes. There was a small chance that some of the Japanese codes cracked by Yardley's team were still in use, but Friedman's larger hope was that even if the codes had changed there might be enough continuity—in their underlying code books, in their encipherment method, or in the way the indicators were used to specify a key—that the Black Chamber's work on earlier systems would at least provide useful clues that might shave the impossibly long odds against breaking the current codes. Maintaining continuity was a cardinal principle of cryptanalysis. Rarely do codes change completely.

But in Japan's case, they had. Some low-level Japanese diplomatic codes, to be sure, showed the same general characteristics of codes that had been solved by the Black Chamber, and some slow progress against them was made by Friedman's team of neophyte cryptanalysts. But the intercepts now starting to arrive via Mauborgne's basement radio station in San Francisco were mostly of a very different ilk. Japanese diplomatic traffic exchanged between Tokyo and the major capitals of the world—Washington, London, Rome, Paris, Berlin, Moscow, Warsaw, Ankara—now consisted of a series of five-letter groups preceded by a five-digit indicator, and there was nothing in Yardley's files that remotely resembled this.

It was an almost exact replay of the setback the Biuro Szyfrow had experienced a continent away and just a few years earlier in attempting to read the codes of the German armed forces. And the reason was the same, too. The Japanese had put into practical effect the lessons of Yardley's revelations and had adopted a machine cipher, at least for their most sensitive diplomatic communications.

One recourse when faced with such a cryptanalytic challenge, of course, was simple thievery. The U.S. Office of Naval Intelligence had made breaking into Japanese consulates, businesses, and cultural organizations practically a full-time occupation, and not without results. Faced with what would have otherwise been an almost insurmountable challenge of breaking into the Japanese Navy's first enciphered code from scratch, Safford and Rochefort had received an enormous assist from copies of the Red code book photographed by an ONI black-bag team during a whole series of break-ins at the Japanese consulate in New York in 1920 and again in 1926 and 1927. That left only the task of breaking the encipherment system, a considerable shortcut indeed. When the Red naval code was superseded at the end of 1930 by a new, two-part enciphered code (dubbed "Blue"—both colors referred to the binders that the Navy code breakers used to record recovered code groups in), there was enough continuity that a team led by Miss Aggie was able to crack the encipherment system, although only after nearly two years of unremitting labors.

Breaking into foreign embassies was so outrageous a breach of diplomatic obligations that it made reading other gentlemen's mail look like kid's stuff by comparison. It was completely lacking legal cover. And this was at a time when both the Army and the Navy were still desperate to keep the State Department from knowing even that they were tuning in to foreign diplomatic radio traffic. In the early 1930s the Army still pretended that its intercept activities were for research and training only. Shortly after Mauborgne began his intercept operations, a near fiasco had occurred when the head of the Signal Corps' War Plans and Training Division learned of SIS's plans to expand its monitoring of foreign diplomatic traffic. He innocently referred the matter to the State Department, which immediately objected to any Army involvement in such activities. This brought a swift under-the-table kick in the shins from the Navy, which had been holding unofficial meetings with the Army to discuss coordination of the two services' radio intelligence efforts. "If the Army can not be trusted to use a semblance of discretion in disclosing these matters outside the military services then we will lose everything by our efforts to cooperate," a subordinate fumed in a memorandum to the Chief of Naval Communications on April 10, 1933.

But the virtues of theft were about to run their course in any event. By the time the Japanese introduced a new (and radically altered) fleet naval code to replace the Blue Book a few years later, Japanese naval attachés had been issued their own separate cryptographic system, which employed a cipher machine; consulates were unlikely to keep copies of the new codes book used by the Navy proper lying around. And break-ins, law and diplomatic immunity aside, were always a two-edged sword. There was no guarantee of success, and a bungled job was far worse than having done nothing. An attempt in 1935 to steal a look at the Japanese naval attaché machine was, in fact, a complete failure. Surveillance of the naval attaché's luxurious Washington apartment in the Alban Towers building at Wisconsin and Massachusetts avenues had detected a mysterious clicking noise. While the attaché and his wife were at dinner one July evening—by prearranged plan they had been invited by a U.S. naval officer who had gotten to know the couple—Lieutenant Commander Jack Holtwick and a Navy chief radioman entered the building posing as electricians and slipped into the attaché's residence and office suite. Despite the claims of many writers that Holtwick actually made off with one of the machines, in fact he found nothing at all that looked like a cipher device, and that was probably just as well. Nothing would have more surely prompted the Japanese to switch with alacrity to a new and more secure code. A particularly heavy-handed black-bag job subsequently showed the wisdom of leaving well enough alone. Around 1938 a somewhat shady private detective named Seeman Gaddis—"Soapy" to his friends—was hired by the Twelfth Naval District in San Francisco to pick locks and break into various Japanese offices on the West Coast. In May 1941, posing as a customs agent, he boarded a Japanese merchant ship in San Pedro harbor, "discovered" a stash of drugs he planted in the captain's safe, and consequently seized the safe's contents—including a copy of the Japanese merchant ship code book. The Navy photographed the book before returning it, but the Japanese could hardly have failed to realize what was happening. They promptly changed the code. By that time the Army had become so alarmed that further break-in attempts by the Navy might compromise years of work on the much more important diplomatic cipher machines that they pleaded with the Navy to try to resist its cloak and dagger urges, and the Navy seems to have complied.

The machines the Japanese had introduced for their naval attachés and high-level diplomatic traffic were considerably simpler than the Enigma. But the cryptanalysts of OP-20-G and Friedman's Signal Intelligence Service had no way of knowing that at first because, unlike the Enigma, the Japanese "Red" (diplomatic) and "M1" (naval attaché) machines were of an entirely unknown design. Even their principle of operation was a mystery. But several oddities were immediately apparent from an examination of the enciphered traffic itself. Most notably, the six vowels A, E, I, O, U, Y accounted for almost half the letters that appeared. The SIS team suspected that this was because the machine enciphered vowels as vowels and consonants as consonants. Such a scheme would make the cipher text form into pronounceable words, which was also a feature of many codes employing alphabetic code groups; that reduced the chances of transcription errors, and cable companies also charged less for handling coded telegrams made up of words like LOVVE ZOXIL YUMUP as opposed to XHDGH WJDQW KJGDG.

For months Rowlett and Kullback had tried to tackle the Red machine traffic with the statistical techniques they had been so thoroughly tutored in. That was the systematic, scientific approach. But late one night as Rowlett lay sleepless in bed, it occurred to him that among the messages they had amassed were three very long ones, all of which had been transmitted the same day over the same network. If they had all been enciphered using the same setting of the machine, then it might be possible by hunch, intuition, and guesswork to spot some patterns of vowels and consonants that looked like specific Japanese words. It was clear that the Japanese were probably using a Romanized spelling of Japanese words in the system, and while Kullback and Rowlett were hardly proficient in the language, they had learned some basics and knew how the syllables characteristically combined.

Rowlett and Kullback were an odd team. "Kully," as he was always known, was a voluble, verbal bulldozer of a New Yorker. A slightly dumpy appearance was exaggerated by the fact that he always looked more than slightly disheveled, and he was about the last man on earth to stand on ceremony. A trademark feature colleagues would remember was the roll of toilet paper, lifted from the men's room, that he plunked on his desk to use for blowing his nose when he had a cold. He was always giving offense unintentionally, but balanced that gruff exterior with a fundamental decency and an inability to take offense himself that earned him the enduring loyalty of those who later served under him. Rowlett was far more reserved and fastidious in his personal bearing, as well as his work. He was outwardly soft-spoken, courteous, a southern gentleman, with the precise habits of a craftsman—as indeed he was, for Rowlett was also the mechanical adept of the group who could assemble precise machinery whenever it was needed. He was also prickly about his sense of his due, which, especially in later years, grew to obsessive proportions as he fought a decades-long battle to attain credit, and financial compensation from the government, for the cipher machines he invented while working under Friedman. The one gentile of the four SIS professionals, his resentments of slights, real and imagined, at times took on an anti-Semitic hue. In their different ways neither Kullback nor Rowlett were easy men to get along with.

Yet both rose to an intellectual challenge the same way, and the day after Rowlett's midnight epiphany they plunged into the three long messages looking for likely words. By noon they had spotted the repeated places in the messages where the word *oyobi*—"and"—appeared, its distinctive vowel pattern shining through the sea of letters. From there the crucial discovery quickly followed: The substitution pattern by which the vowels were scrambled went through a repetitive, cyclical rotation; in fact it was basically a 6 × 6 Vigenère tableau. The table shifted down one key position as every successive letter of text was enciphered. If *a* were enciphered as E in the tenth letter of the message, than an *a* appearing as the sixteenth letter of the message would also be enciphered as E. Other "isomorphs" then began to be evident in the text—that is, two strings of cipher text that appeared to represent the same plain-text word. For example, the strings:

QIVVDA

TUZZHY

would seem both by their vowel pattern and by the doubled letters to be isomorphs. They exhibit another striking property: In an alphabetical sequence of the twenty consonants that remain after the six vowels have been removed, T is three letters beyond Q, Z is three letters beyond V, and H is three letters beyond D. In other words, the cipher alphabet had simply been slid over three positions. The consonants were thus being enciphered by a Vigenère tableau, too, this one 20 × 20; it too shifted down one key position as each successive letter was typed in. The tell-tale doubled letters in the cipher text were thus the result of a plain-text pattern such as *ts* or *sr;* when subjected to the shifting key these alphabetically adjacent letters produce the same cipher character:

plain text

b c d f g h j k l m n p q r s t v w x y z

key position	1	B C D F G H J K L M N P Q R S T V W X Y Z
	2	C D F G H J K L M N P Q R S (T) V W X Y Z B
	3	D F G H J K L M N P Q R S (T) V W X Y Z B C
	4	F G H J K L M N P Q R S T V W X Y Z B C D

etc.

Apparently the machine was a quite simple rotor device. The twenty-six letters of a keyboard fed into two separate sets of rotors, one for the six vowels and one for the twenty consonants; as each letter was typed in the rotor turned one position, effecting a shift of one line down the substitution table. A few extra wrinkles in the operation of the machine were quickly nailed down by Rowlett and Kullback. A plug board apparently could be used to swap the identities of the vowels and consonants among themselves before they entered the rotor. That changed the identities of the letters being shuffled but it could not alter the underlying patterns that the rotors used to ring those changes. Once the pattern by which the cipher letters were shuffled about by the Vigenère tableau was established, identifying the identities of those letters for each day's plugging became a fairly simple matter of frequency counts.

They also discovered that the stepping was not always constant; sometimes the rotor would apparently turn two positions instead of one in going from one letter to the next, skipping down an extra row in the Vigenère tableau. In the actual Red machine the stepping pattern was controlled by a "breakwheel" with forty-seven pins, some of which were removable to cause such a double step. But the Japanese procedures for setting up the daily key specified a fairly narrow range of options, and Rowlett and Kullback discovered that the pattern of stepping through the rows of the Vigenère tableau always followed a cycle of forty-one, forty-two, or forty-three.

In a single day, they had cracked a cipher machine, sight unseen. Friedman had been at Fort Monmouth, New Jersey, at the Army's Signal Corps Laboratory for a few days, and was suitably astonished at the news that greeted him on his return the next morning.

The success temporarily quelled some ill winds that had been stirring between the Navy and the Army, though not for long. OP-20-G had just a few months before solved the naval attaché machine, which appeared to share many of the basic characteristics of the Red machine. But Lieutenant Joseph Wenger, OP-20-G's head, had refused to divulge the Navy's results. The Navy claimed that the Army leaked like a sieve because it let too many civilians in on too many secrets, but that was probably more justification than a real explanation. The real explanation was that the two services were locked in a bitter rivalry. The most Friedman could coax out of Wenger was the suggestion that they hunt through the cipher texts for a cycle of more than forty and less than fifty in the stepping of the alphabetic substitutions. When the Army's intelligence staff subsequently began to routinely circulate decrypts of Red traffic to the Office of Naval Intelligence, the Navy on more than one occasion tried to beat the Army to the punch in delivering some particularly hot item to the (now post-Stimson) State Department, which left the SIS cryptanalysts fuming.

The breaking of the Red machine and of the Red Book and Blue Book codes yielded several pieces of intelligence of the first importance. One prewar coup was a series of Japanese naval messages decrypted during the summer of 1936 that reported the results of sea trials of the battleship *Nagato*. This was a refurbished First World War dreadnought, which the U.S.

Navy believed was incapable of making better than 23½ knots. But equipped with new turbines, *Nagato* reported that it had run at 26 knots. New American battleships then in the blueprint stage were being designed with a maximum speed of 24 knots. Upon receiving OP-20-G's report, the Navy immediately ordered designs for the *North Carolina* class battleships to be redrawn so they could make 27 knots, and the later classes, 28 knots. It was, as Laurance Safford later noted with some justification, a piece of intelligence that had paid for all the expense of radio intelligence "a thousand times over."

But as the clouds of war gathered, the outlook began to degenerate quickly. In November 1938, a new code replaced the Blue Book as the main Japanese naval code. Although this new "AD" code was similar to the Blue Book, all of the underlying code groups had been changed. It was back to square one. On February 18, 1939, a Red message was decrypted announcing that in two days a new "B" machine would take its place. Two days later the system that SIS would dub Purple appeared; it was a vastly more complex cipher than the Red machine had generated, and it was clear at once that no quick solution was possible. And on June 1, 1939, a second completely new naval code appeared, unlike any of the other Japanese codes that OP-20-G had ever encountered. It was clear from traffic analysis that this "AN" code, later to be known as JN-25, was the chief high-level operational code of the entire Japanese Navy. It was completely impenetrable.

By the summer of 1939, America was blind.

CHAPTER 3

"Il y a du Nouveau"

BRITAIN was blind, too, at least when it came to reading the radio traffic of Germany's armed forces and diplomats. Through the 1920s and early 1930s the staff of GC&CS grew at an almost imperceptible crawl. Promotion from the ranks of "junior assistant" to "senior assistant" was a rare event; it was literally a matter of waiting for a senior to retire or die and free up an allocated slot. This was the source of much discontent and not a little grim humor. Another drag on morale, Nobby Clarke recalled, was the irritating smugness of the code compilers in the British military services, who maintained that GC&CS was all a waste of time and money since new codes were becoming unbreakable. In 1937, however, Sinclair took the growing possibility of war seriously enough to seek, and obtain approval for, a substantial increase in both pay and number of positions, and the total establishment (counting thirty clerical workers and typists) grew to an authorized level of more than a hundred. Top pay for a senior assistant was increased to £1,000. By the summer of 1939, GC&CS had about thirty people working on diplomatic traffic, and they had solved the diplomatic codes of just about every country of the world worth paying attention to—except for Germany, Russia, and (in the case of the unsolved Purple machine) Japan.

German naval and military signals were proving equally impervious to attack. Only in 1936, when German ships began sailing in the Mediterranean, did GC&CS even begin to receive German naval traffic in any quantity. Dilly Knox led a team to examine the messages; he was able to determine that it was the product of a steckered Enigma, but could get no further than that. An interest was expressed in trying to gather lower-grade German naval traffic transmitted by ships and coastal stations; the Admiralty pleaded it lacked enough men and radio receivers at its listening posts to handle the job, and the matter was dropped. The following year signals from Germany's burgeoning Army and Luftwaffe began filling the air-waves, and the British Army's intercept station at Chatham, hidden behind the ramparts of an old fort on the Thames estuary, began supplying GC&CS with intercepts; again Knox determined the signals were the product of a steckered Enigma; again he was soon at an impasse.

If GC&CS was blind to Hitler's plans, at least the scales had fallen from Prime Minister Neville Chamberlain's eyes, and the eyes of many others who out of horror of war and a lack of imagination would not, or could not, imagine that Hitler failed to share their horror. At Munich Hitler had con-temptuously answered, "We want no Czechs!" when the British Prime Min-ister had pressed the Führer if he would be satisfied once he had the German-speaking areas of western Czechoslovakia under his control. The Sudetenland, Hitler said, was "the last territorial claim I have to make in Europe." On that promise Chamberlain had sold out the Czechs, lock, stock, and barrel; on September 29, 1938, while the Czech representatives were banished to an adjacent room to await their fate, Hitler, Chamberlain, Mus-solini, and French Premier Édouard Daladier presided over the dismem-berment of their nation. In exchange for abandoning the strongest fortress line in Europe, held by thirty to forty divisions of well-equipped troops, the Czechs would receive a promise of protection from the same powers that had just deserted them in the moment of crisis. From the back benches of the House of Commons Winston Churchill's growling voice rose in protest at the betrayal: "The Czechs, left to themselves and told they were going to get no help from the Western Powers, would have been able to make better terms . . . they could hardly have had worse." The First Lord of the Admi-ralty, Duff Cooper, resigned in protest. But theirs were voices in the wilder-ness, drowned by the cheers of joyous relief that "peace for our time" had been secured. As Chamberlain alighted from the airplane that bore him home from Munich, a courier in royal livery emerged from the crowd to hand him a message from the King, summoning him to Buckingham Palace "so that I can express to you personally my most heartfelt congratulations." Crowds literally thronged the streets as he drove to the palace, shouting themselves hoarse. "Do not suppose that this is the end," Churchill warned.

It took no more than six months to prove him right, and this time even Chamberlain was shocked into agreement. At six o'clock on the morning of March 15, 1939, German troops poured into what was left of Czechoslova-kia. Hitler flew to Prague and declared a "protectorate" over the state. Churchill, who had seen firsthand the wizardry of Room 40 in the First World War, could not believe that British intelligence had failed to know in advance of Hitler's move on Czechoslovakia; he blamed the politicians:

After twenty-five years' experience in peace and war, I believe the British Intelligence Service to be the finest of its kind in the world. . . . It seems to me that Ministers run the most tremendous risks if they allow the information collected by the Intelligence Department and sent to them, I am sure, in good time, to be sifted and coloured and reduced in consequence and importance, and if they ever get themselves into a mood of attaching weight only to those pieces of information which accord with their earnest and honourable desire that the peace of the world should remain unbroken.

There was truth to the accusation of wishful thinking in Chamberlain's circles, but Churchill had no idea how badly British intelligence had deteriorated since the war, nor how blind the code breakers in particular were to German military and diplomatic signals.

Chamberlain was genuinely stunned by the news of Hitler's betrayal of the Munich pact, and reacted at first with a pusillanimous statement to the House of Commons: Since Czechoslovakia had indeed—in Hitler's words—"ceased to exist," the Prime Minister weakly explained, His Majesty's Government was no longer obligated to come to her defense. After meeting the Führer at Berchtesgaden the previous September, Chamberlain had remarked privately that "in spite of the hardness and ruthlessness I thought I saw in his face, I got the impression that here was a man who could be relied upon when he had given his word." Now at last Chamberlain saw he had been lied to, and his resolve was awakened. Two days later, riding the train to Birmingham where he was to deliver a long-planned speech that would be broadcast around the world, Chamberlain threw out his prepared text and began work on a statement that would, as he said in his opening words, correct "the very restrained and cautious" remarks he had addressed to the House of Commons. Chamberlain recited Hitler's broken promises that had left hopes of peace "so wantonly shattered." How, he acknowledged, "can the events this week be reconciled with those assurances which I have read out to you?" Two weeks later the Prime Minister announced to the House of Commons that His Majesty's Government was committed to lend all support in its power to Poland should she become the next target of Nazi aggression.

Here at last was action, as Churchill said, but action under the worst possible circumstances, action only after all advantages had been cast aside, action on behalf of the very state that had colluded with Hitler in the events that now threatened her own existence. The Polish guarantee placed Europe's fate in the hands of a junta of politically inept colonels in Warsaw; it pledged Britain and France to come to the defense of a country they had no way of reaching. The loss of Czechoslovakia had meant the loss of thirty-five fully mobilized divisions that had tied down the full strength of Germany's mobile land forces. On April 27 Britain approved conscription of two hundred thousand men, but it was years behind Germany by this point. Germany had over two hundred divisions by the summer of 1939, including five heavy panzer divisions and four light armored divisions. Britain had five regular divisions and one mechanized division. British military expenditure edged up to an annual total of £304 million; Germany's was at least five times as great.

The only possible counterweight was Russia, but the Poles still would hear nothing of it. On April 16, Maxim Litvinov, the Soviet commissar for foreign affairs, had made a startling proposal to the British and French, delivering to their ambassadors the draft of an iron-clad treaty pledging that all three nations would stand firm, with all of the military might at their disposal, against any new Nazi aggression in Europe. The French gave their cautious assent. But Chamberlain's aversion to treating with the Bolsheviks was almost as great as the Poles'; moreover, Chamberlain doubted the military might of the Red Army, its three hundred divisions notwithstanding. Litvinov's proposal was met with stony silence from London. Two weeks later Litvinov was out, fired by Stalin. A week later Chamberlain publicly responded to the Soviet offer with an unenthusiastic suggestion of further discussions. Almost simultaneously, Litvinov's successor, Vyacheslav Molotov, began secret negotiations with his German counterpart, the Nazi Foreign Minister Joachim von Ribbentrop, on a nonaggression pact—as well as on a formula for dividing Poland between them, for the fourth time in that country's sorry history.

Amid belated preparations for a war no longer unthinkable and impossible, but now become unthinkable and inevitable, the summer of 1939 for many in England took on a strange air of unreality, a slow and helpless descent into blackness. Churchill brooded at Chartwell, his country estate, correcting galleys of his 460,000-word *History of the English Speaking Peoples,* meeting with exiled Czechs and those few British officials who were willing to risk their careers by associating themselves with Chamberlain's chief antagonist, writing blistering newspaper columns urging an alliance with Russia, trying to fight off the defeatism that lay heavy in the air. Churchill railed against the British mien of unhurried imperturbability when a sense of urgency was what so obviously was required—urgency to press the negotiations with Russia to conclusion, urgency to get on with rearmament. Britain's ruling class continued "to take its weekends in the country," Churchill acidly observed, while "Hitler took his countries in the weekends." But Chamberlain did not like to be bothered with telephone calls after dinner or on weekends, and he saw no reason to alter his habits just because a war was about to start. (Chequers, the Prime Minister's country residence, at least *had* a telephone; other officials remained literally unreachable on weekends.) And so the world slid toward slaughter and calamity in an almost eerie calm. Leonard Woolf, the writer, publisher, and husband of Virginia Woolf, remembered the summer of 1939 as "the most terrible months of my life." One late summer day he was out working in the garden of his cottage when Virginia called out from the window that Hitler was making a speech on the radio. Leonard shouted back: "I shan't come. I'm planting iris and they will be flowering long after he is dead." But that act of defiant optimism was balanced by an act of equally defiant pessimism. Understanding perfectly well what his fate, as a Jew and an intellectual, would be at Nazi hands, Woolf equipped himself with poison so that he could kill himself if German troops landed in England. Woolf, and the iris, survived. But in the summer of 1939, as darkness fell over Europe, neither seemed like a sure bet.

———

Throughout the late spring and early summer of 1939 Hitler's propaganda chief Paul Joseph Goebbels busily generated rumors of an impending coup in Danzig: The German populace, and the Nazi-dominated Senate of the Free City, were supposedly preparing to seize power from the city's Polish administrators. It was largely a ploy to test British and French resolve—and to distract attention from the fact that Hitler did not want Danzig, he wanted Poland, and was determined to get it by war; this time there would be no repeat of the Czech capitulation that had robbed him of the military triumph he so dearly coveted. In late June the rumors intensified. On June 30, amid this tense atmosphere of mounting crisis, Lieutenant Colonel Gwido Langer, head of the Biuro Szyfrow, wired his French and British counterparts. Using an agreed code phrase, *Il y a du nouveau*—"there is something new"—Langer requested a meeting in Warsaw of British, French, and Polish cipher specialists.

What was "new" was not any cryptologic development; what was new was Poland's desperation. With war just weeks away, the Chief of the Polish General Staff had given the Biuro Szyfrow permission to reveal to Poland's allies its most closely guarded secret. Simply, since 1933 it had been reading German Enigma traffic.

The two sides had met earlier, in Paris on January 9 and 10, 1939, but the Poles had been guarded. Langer was under orders to feel out the French and British and only reveal the full truth of the Polish success if it was clear that the French and British had made similar progress themselves, and only if they were willing to swap information. Major Ciezki, the only cryptanalyst from the Polish side to attend the meeting, discussed with Dilly Knox in general terms how, in principle, the daily setting of the Enigma could be solved and what it would take to deduce the new internal wirings of rotors that the Germans had adopted for the military version of the machine. Both sides had clearly been thinking along the same lines, but Knox could see no way around one fundamental problem: There were simply too many unknown factors that had to be solved simultaneously. Although Knox had worked out a mathematical procedure for recovering the daily settings, it depended on first knowing the internal rotor wirings, and there just seemed to be no way to isolate that part of the equation. The Poles politely agreed, and the meeting had ended with an official report concluding that "reconstruction of the machine solely through the study of [cipher] texts must be regarded as practically impossible." The Poles, however, knew that conclusion was untrue for one simple reason: That was precisely what they had already done.

On July 24, Denniston and Knox flew to Warsaw from London, accompanied by Commander Humphrey Sandwith, the Admiralty official in charge of radio interception and direction finding. The head of the French radio intelligence service, Captain Gustave Bertrand, and Captain Henri Braquenié, a top French cryptanalyst, arrived by train, crossing through Germany. At a welcome lunch hosted by the Poles, the conversation was limited to noncommittal chit-chat in German, the only language the French, the British, and the three young Polish cryptanalysts, Rejewski, Zygalski, and Rozycki, had in common. But it was clear the Poles had something up their sleeves, and the British and French retired to their hotel rooms that evening "all anxious about the morrow," Bertrand recalled.

The next morning the visitors were driven out to Pyry, a small village six miles south of Warsaw where the German Section of the Biuro Szyfrow had been moved, in part to conceal its activities from the German agents who were swarming through Poland. After a short tour of the facility, the French and British were led into a room where several objects lay on a table cloaked with a sheet. Langer removed the cover to reveal what the visitors immediately recognized as German Enigma machines.

"Where did you get these?" Bertrand asked.

"We made them ourselves," Langer said.

Denniston and Knox, Bertrand recalled, were struck speechless. But Knox then recovered sufficiently to ask the question that had thwarted his efforts. In the commercial Enigma, the keys of the keyboard had been wired to the entry ring in the same order that they were found on the keyboard itself. The German typewriter keyboard has a slightly different arrangement from the English keyboard; the top row of keys is QWERTZUIO (as opposed to the English QWERTYUIOP). And so in the commercial Enigma, Q was wired to the first contact, W to the second, E the third, and so on around the circle of twenty-six contacts on the entry ring. Knox had dubbed the wiring sequence of the entry ring "the QWERTZU." The "QWERTZU" for the commercial Enigma was, thus, QWERTZU. Knox's analyses of intercepted traffic had gotten far enough to establish that the entry ring of the military Enigma had been changed, however, and given that there were $26 \times 25 \times 24 \times 23 \times 22 \times \ldots \times 4 \times 3 \times 2 \times 1$, or about 400 million million million million possible ways the Germans could have rewired it, the odds of making much progress seemed dim. Without knowing which one of those myriad possibilities was correct, the wiring of the rotors themselves remained completely disguised.

So the very first words out of Knox's mouth were: "What is the QWERTZU?"

Rejewski's answer infuriated him. The Germans had wired the keys to the entry ring in alphabetical order. A was wired to the first contact, B to the second. . . . It was not so much that Knox was mad at himself for not having figured it out; it was that he was indignant that a challenging mathematical puzzle had a trick answer. It was a swindle. It was too simple, and that was an affront to his sense of the universe.

Denniston wanted to send word at once to the British Embassy in Warsaw for a draftsman to be dispatched to make a copy of the Poles' Enigma replica, but Langer explained that would not be necessary; they had built two spares, one for the French, one for the British, to take home with them. The replicas, or "doubles" as the Poles called them, would be turned over at once to Bertrand to send via diplomatic pouch to Paris.

Knox was still smarting over the QWERTZU, as well as the obvious fact that the Poles had been holding out on them at the January meeting in Paris, and riding back in a taxi with Denniston and one of the Poles, Knox—not realizing that the Pole spoke English perfectly well—began making "very derogatory remarks" about the Poles, Denniston recalled, to his considerable embarrassment. But at some point Knox seems to have recovered his humor and taken heart from the breakthrough he now had to work with. A story made it back to GC&CS that Knox had later that day begun chanting to one of the Frenchmen, *Nous avons le QWERTZU, nous marchons ensemble!*—"We have the QWERTZU, we march together!" And a few days later, upon his return to London, Knox sent a gracious thank you note to the three Polish mathematicians. In Polish, Knox had written *Seredecznie dziekuje za wspolprace i cierpliwosc*—"My sincere thanks for your cooperation and patience"—and below that inscription he added in French, *Ci inclus (a) des petits bâtons (b) un souvenir d'Angleterre.* The *petits bâtons,* one set for each of the three Poles, were paper strips inscribed with letters of the alphabet, the abortive method that Knox had been working on to solve the Enigma. It was, the Poles gathered, a chivalrous gesture of surrender, akin to handing one's saber to a rival who has vanquished him. The "souvenir d'Angleterre" for each of the mathematicians was, somewhat incongruously, a scarf depicting a horseracing scene.

The Polish Enigma replicas were duly shipped to Paris; from there on August 16 Bertrand and the Paris representative of the British Secret Intelligence Service, Wilfrid "Biffy" Dunderdale, accompanied by a bodyguard of three of Dunderdale's assistants, hand-carried the machine destined for GC&CS aboard the Dover ferry. The entourage was met on the platform of Victoria Station by Colonel Stewart Menzies, the deputy head of SIS who was soon to succeed Admiral Sinclair as "C." Menzies was in black tie, which Bertrand thought a fitting gesture. *Accueil triomphal!* he said, a triumphal reception. Menzies tactfully refrained from explaining that he was simply on his way to an evening party.

Sixteen days later, five German armies, more than one and a half million men supported by a thousand bombers, attacked Poland from the north, the south, and the west.

————

The Polish mathematicians' feat of cryptanalysis was so astonishing that, for decades afterward, even many of GC&CS's experts on the Enigma were convinced that the Poles had gotten hold of a German Enigma machine through espionage or luck and examined the wiring of the new rotors; they could not believe the Poles had deduced the wirings through pure mathematical insight. Espionage did play a part, but not in the way the British thought; in fact, it *was* an act of almost pure mathematical insight. The crucial break was made by Marian Rejewski, working alone, over the course of just a few weeks in late 1932.

Rejewski had precious little to go on. There was the commercial Enigma machine the Poles had purchased; there was the mounting stack of intercepted Enigma traffic that arrived each day; and there were two German manuals that Rejewski was handed that contained photographs of the military Enigma and that described the German operating procedures. Practically the only solid information Rejewski could find in all of this was the fact that the first six letters of each message served a special function—they were an indicator of how the Enigma machine had been set for enciphering that particular message. The procedure, as described in the purloined German manuals, went like this: The German operator would first set his machine to a daily setting taken from a printed list. This specified the wheel order, steckers, and ring settings, plus an initial starting position, a "ground setting," for the three rotors. Then, for each message sent that day, the operator would pick at random a different three-letter starting position to use for enciphering the actual text. With the machine at the ground setting, he would type in his chosen message setting twice and write down the enciphered text that resulted. If he had chosen WJC, for example, he would type in WJCWJC, which the Enigma might encipher as DMQVBN. Those six letters formed the indicator that was transmitted at the start of the message. Then the operator would twirl the rotors so that WJC showed in the windows, and begin enciphering the body of the message.

The "double encipherment" of the message-setting indicator was apparently done as a garble check, but it bore some ever so slight mathematical implications that Rejewski immediately realized might be exploitable. Rejewski began with two fundamental observations. First, all of the indicators for a single day had been enciphered with precisely the same ground setting. In other words, every indicator was produced by the same six cipher alphabets. They were in depth, that is. The first letter of every six-letter indicator for a single day was the product of one monalphabetic substitution; likewise the second letter was the product of a second monalphabetic substitution; and so on.

The second observation was pure genius. Rejewski realized that for any given ground setting, certain characteristic patterns would begin to appear in the enciphered letters that fell in the first and fourth positions of the transmitted indicators, and likewise in the second and fifth positions, and the third and sixth positions. These patterns were there in the indicators for anyone to see—anyone who knew to look for them, that is. But Rejewski's startling discovery was that these patterns gave away nothing less than the complete monalphabetic substitutions that the Enigma had performed at the first six positions of the day's ground setting.

The patterns arose like this. The first and fourth letters in any message were the encipherment of the same letter at two different positions of the Enigma machine. At any given position, the total effect of the Enigma's rotors, steckers, and reflector was to swap letters in pairs in a unique monalphabetic substitution; thus at positions 1 and 4, for example, the substitutions might follow this schema (to use a simplified alphabet of only six letters to illustrate the idea):

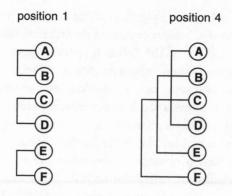

position 1   position 4

When the operator typed in a message setting—say, CDFCDF—the enciphered indicator that would result from these substitutions would have to be of the form D– –F– –, since C is swapped with D in position 1 while C is swapped with F in position 4. Working through all possible first letters, the universe of possible enciphered indicators for this particular ground setting would be:

message setting	enciphered indicator
A – –	B – – D – –
B – –	A – – E – –
C – –	D – – F – –
D – –	C – – A – –
E – –	F – – B – –
F – –	E – – C – –

The pattern that Rejewski then observed was that the letters of the enciphered indicators formed sets of chains. In this example, if there is a B in the first position there is a D in the fourth; a D in the first occurs with an F in the fourth; and finally an F in the first occurs with a B in the fourth, closing the loop. A, E, and C form a second closed loop of their own. In Rejewski's notation, this particular pattern of chaining would be written as:

(bdf)(aec)

Rejewski's crucial discovery was that the pattern of chaining that resulted in the first and fourth letters of the enciphered indicators in each day's traffic was a direct function—a "characteristic," he termed it—of the two monalphabetic substitutions the Enigma machine performed in position 1 and position 4. A different pair of monalphabetic substitutions in positions 1 and 4 would yield an entirely different pattern of chaining, perhaps (de)(cf)(a)(b) or (adfe) (bc). In fact, the connection between the cipher sub-stitutions in positions 1 and 4 and the resulting pattern of chaining in the en-ciphered indicators could be described precisely by a mathematical relationship that was well known from the field of mathematics known as permutation theory. It meant that by collecting the message indicators from a single day's traffic, Rejewski could in theory recover the complete monal-phabetic substitutions that the Enigma machine performed in position 1 and in position 4. Precisely the same procedure would work for recovering the cipher substitutions for positions 2 and 4 and positions 3 and 6.

In other words, knowing nothing of the machine's internal wiring, he had devised a scheme to break the first six positions in each day's setting. Re-jewski calculated that with about eighty messages, all twenty-six letters of the alphabet would occur in all six possible positions of the enciphered in-dicator, permitting him to construct the complete set of characteristics, which would in turn give him the six cipher alphabets.

Those six cipher alphabets described the total effect of all of the ma-chine's components: how the steckers, three rotors, and reflector *combined* to scramble the alphabet at each position. Though an almost unbelievable coup, it still did not allow messages to be read. For that, Rejewski had to fig-ure out a way to split apart the machine's components. The problem, in other words, was now to deduce the internal wirings of the three rotors.

In an unsteckered machine that would have been immediately possible. Knox had early on realized the same thing, for one flaw of the Enigma was that it placed the "fast" rotor in the rightmost position. That this was a flaw was far from obvious, but under such an attack as Rejewski and Knox con-ceived, it was a flaw of absolutely fatal consequences, for it meant that all of the other adjustable components of the Enigma could be lumped together into a single unknown factor, greatly simplifying the mathematical problem.

The point was this: Input signals from the keyboard fed into the right-most rotor, which advanced one step with every letter typed in. Once per revolution, it triggered a turnover of the middle rotor. But between those turnovers, for twenty-six letters at a stretch, all of the components of the ma-chine to the left of this first rotor—the two other rotors plus the reflector—remained fixed. The *only* thing that moved was the fast rotor; the rest of the machine could be viewed as a single unit that performed some unknown, but constant, monalphabetic substitution on the output of the fast rotor. To the right of the fast rotor, at least in the unsteckered Enigma, the machine was a known quantity; to the left it was an unknown, but fixed quantity, which Rejewski called Q.

If one knew the complete substitutions the machine performed in six consecutive positions for a given day's traffic, one had a set of six equations and only two unknowns: the monalphabetic substitution Q, and the monal-phabetic substitution effected by the internal wiring of the fast rotor. Knox had brilliantly exploited this very same fact in devising a scheme for break-ing messages enciphered with the unsteckered commercial Enigma ma-chine, which was used by the Italians and by the Fascist forces in the Spanish Civil War. Indeed, that was precisely what his "petits bâtons" were for. Knox was not the only one to have figured this out, though: The cryptanalysts of the German Foreign Office's code breaking section also successfully broke commercial Enigma traffic during the Spanish Civil War, which affirmed the wisdom of the decision made by the German military services a few years earlier to add the stecker board to their models.

It was a prudent step, for in the steckered Enigma there were too many unknowns to solve for. Besides the fast rotor, and Q to its left, there were two unknowns to the right: the entry ring sequence (the "QWERTZU") and the steckers. Rejewski had got off to a flying start only to hit a tree.

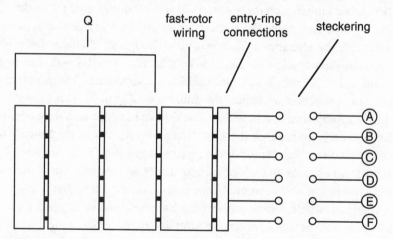

The four unknowns in Rejewski's equation for the Enigma machine

Rejewski nonetheless presented his results to Major Ciezki, who promptly ordered him to drop all of his other work and concentrate solely on the Enigma. Instead of coming in for two hours in the afternoon after everyone else had gone home, Rejewski would be given a private office so he could work on the project all day long while still keeping what he was up to tightly under wraps, even from his colleagues. But it was immediately apparent to Rejewski that the snag he had hit was a formidable one. It was theoretically possible that the six equations with four unknowns could be solved (in fact it remains an unanswered question in permutation theory to this day, ac-cording to comments Rejewski made five decades later), but as a practical matter he could see no way through.

Then sometime around mid-November Ciezki handed Rejewski another two German documents. They were the daily settings, including the steckers, for the months of September and October 1932. It is unlikely that a deus ex machina was ever better timed. Rejewski did not ask where the documents came from. Had he been told, he probably would not have believed it: Basically they had come in over the transom. In June 1931 a man claiming to be a well-placed employee of the Reichswehr Chiffrierstelle, the German cryptographic agency, contacted the French Embassy in Berlin offering to sell some documents. The French quite naturally assumed it was a setup by German intelligence, or, even more likely, a scam by a freelance con artist. But soon Bertrand was brought in on the case and a bit of rummaging in the French intelligence files confirmed that the German, one Hans-Thilo Schmidt, was at least what he claimed to be. A meeting was finally arranged in November 1931, Schmidt delivered the goods, and Bertrand immediately recognized the documents as genuine—or at least extremely elaborate forgeries. Among them were the two operating manuals for the Enigma that Rejewski had been supplied when he began work on the Enigma in October 1932.

Schmidt, who was given the code name HE (in French the letters are pronounced ahsh-ay, and so "Asché" became another version of his code name), was a simple mercenary, unburdened by political or ideological motives. This was months before Hitler came to power; Schmidt, said Bertrand, "was fond of money; he also needed it because he was fonder still of women." Schmidt was certainly well rewarded for his work by the French. But neither the French nor the British could make much of the Enigma materials he handed over. Copies of the documents were sent to GC&CS, which dutifully studied them and dutifully filed them away on the shelf, concluding that they were of no help in overcoming the Enigma's defenses. Bertrand's code breakers came to the same conclusion. It was then that Bertrand contacted Colonel Langer in Warsaw.

Langer, unlike his French and British counterparts, was ecstatic over the windfall and agreed with Bertrand that the Poles would share any results they managed to obtain; the French in return agreed to share with the Poles the further fruits of Asché's treasonous endeavors. The harvest proved abundant: Asché, ever hard up for cash, continued to meet secretly with his French handlers at Verviers and Liège, towns in Belgium just across the German border, and turn over additional information. He was paid with suitcases full of currency left at the baggage room of a Berlin train station, the claim ticket mailed to him, poste restante, under a false name. At a meeting some time in late summer or early fall of 1932 Asché turned over the setting lists that, in November, finally landed on Rejewski's desk.

By revealing the steckers for each daily setting, the stolen lists eliminated one of the four variables in Rejewski's set of equations. Now all that was left was Q, the entry ring sequence, and—the jackpot—the fast rotor's wiring. Rejewski tried QWERTZU for the entry ring but got nowhere. But, perhaps more imaginative than Knox, or perhaps better grasping the Germans' lack of imagination, Rejewski decided to take a wild guess that the order was alphabetical. As if "by magic," the solution to the rotor's wiring began emerging on the paper before him.

The Germans' procedure at this time was to change the steckers every day but to change the left-to-right wheel order only every three months. September and October fell into different quarters, which meant a different rotor was in the rightmost position during each of the two months for which he had the key lists. That bit of luck allowed Rejewski to recover the wiring of not one but two of the rotors. And with two solved, the only unknown left for any given day's ground setting was the leftmost rotor and the reflector; Rejewski quickly solved the wirings for those two last components. It was a true irony that a procedure intended to enhance the Enigma's security had undermined it; were it not for the change in rotor order, Rejewski would have hit another impasse and would never have recovered more than the wiring of a single rotor, which by itself would have ranked as a mathematically impressive but practically useless accomplishment. Similarly, it was only by virtue of the Germans' decision to encipher the message starting position at all that Rejewski had anything to work with. Had they transmitted the starting position in plain text, Rejewski observed, they actually would have been better off. By enciphering the indicator, they exposed the characteristics of the encipherment process itself.

The brilliance of Rejewski's method was that it defied all conventional thinking about how to break a code. It was fighting fire with fire. The Enigma was designed to defy conventional cryptanalytic weapons such as frequency counts, and so it largely did. Rejewski's attack probed flaws that had nothing to do with the Enigma's active cryptographic defenses; it struck at coincidences and quirks of the machine itself. Of course some of those flaws were the result of almost arbitrary decisions in the Enigma's design, especially the placement of the fast rotor in the rightmost position. From a mechanical point of view it was no harder to make one of the other rotors the fast one, and indeed that was done in the American SIGABA machine— which was never successfully broken by cryptanalysts from another country. Likewise, were it not for Asché's weakness, or the Germans' methodical entry ring connections, or their thoroughness in doubly enciphering the indicators, Rejewski would have gotten nowhere at all. Luck played a huge part in creating the conditions that permitted Rejewski to use his formidable mathematical skills.

Langer did not exactly keep his end of the bargain with Bertrand. Although the Poles did provide the French with some intelligence derived from enigma decrypts, these were disguised in the form of general intelligence summaries that did not reveal their source; only at Pyry did Langer at last redeem his promise, and Bertrand was as surprised as Knox and Denniston by the full account of the Poles' work they received there. Schmidt, for his part, suffered a rather harsher and more permanent betrayal. One of his handlers in the French intelligence services, captured and interrogated when the Germans seized Vichy France, fingered Schmidt as a spy, and in July 1943 Schmidt was arrested, and shot.

———

Recovering the rotor wirings sight unseen was a phenomenal accomplishment. But it was only half the job; Rejewski and his coworkers (Zygalski and Rozycki were at last let in on the secret) now had to devise a scheme for recovering the daily settings of the machine. Several extremely laborious techniques were used at first, but eventually an efficient solution was developed—efficient once a huge amount of ground work had been laid. The basic stratagem followed directly from Rejewski's initial insight about the chaining patterns, or cycles, in the enciphered indicators. Each initial rotor setting and wheel order gave rise to a unique pattern of cycles. If they could catalogue the cycle patterns for all 105,456 possible combinations of six wheel orders and 17,576 initial rotor settings, they could simply look up each day's pattern. The cycle patterns were entered on index cards with their corresponding rotor setting and wheel order, and then placed in numerical order according to their lengths. For example, the cycle pattern:

> *positions 1,4:* (dvpfkxgzyo) (eijmunqlht) (bc) (rw) (a) (s)
> *positions 2,5:* (blfqveoum) (hjpswizrn) (axt) (cgy) (d) (k)
> *positions 3,6:* (abviktjgfcqny) (duzrehlxwpsmo)

had cycles of length 10-10-2-2-1-1, 9-9-3-3-1-1, 13-13, and the card would be catalogued under that numerical heading.

Steckering would have the effect of substituting the identities of some of the letters in each cycle, but could not alter the number of cycles or their length—that pattern was purely a function of the wheel order and initial setting. So once the code breakers had accumulated about eighty messages of a day's traffic, they could construct the cycle patterns, look for a matching

pattern in the catalogue (which would give wheel order and initial rotor setting), and see which specific letters within the matching pattern had been
swapped for others (which would give the steckering).

The mathematicians devised an electromechanical device they called a
cyclometer to automate the work of compiling the catalogue. It was essentially two Enigma machines wired together, with the rotors set three positions apart. Pressing a letter key would send current into the first Enigma;
the output of the first would then flow into the second, and continue in a
loop through all letters of one permutation chain for that setting. By counting the number of lights that illuminated, the length of the chain could be
determined. Pressing the key for a letter that had not lit up in the first chain
would activate another chain for that setting. It was mind-numbing work,
and for security no one but the three mathematicians was allowed to under-
take it. They often scraped their fingers raw and bloody by the end of the
day from turning the rotors of the cyclometer by hand.

In January 1933, as soon as Rejewski had determined the three rotor
wirings, the Biuro Szyfrow commissioned a Warsaw electronics firm, AVA
Radio Manufacturing, to begin building the Enigma "doubles" that would
be needed to actually decipher traffic. As a model, the firm was given the
commercial Enigma machine that the Biuro Szyfrow had purchased years
before, photographs of the military version from Asché's stolen documents,
and the rotor wirings written by Rejewski himself "on a scrap of paper."

Rejewski and his coworkers never did see any of the hundreds of other
documents Schmidt turned over, even though they included the setting lists
for dozens of additional months. Langer apparently decided the mathematicians should be forced to develop a method for recovering the daily settings
that did not depend upon such handouts. If so, his strategy paid off hand-
somely. By early in 1933 the Biuro Szyfrow was in business, reading German
military traffic the same day it was transmitted. The three mathematicians
had the task of recovering the daily setting and passing that on to the staff
that operated the decoding machines. The main determinant of success or
failure each day was whether they could accumulate the required eighty or
so messages needed to build up the indicator characteristics. When they
could, recovering the actual setting was a matter of only ten or twenty minutes' work. In January 1938 the General Staff ordered a test of how well the
Biuro Szyfrow could handle the mounting flow of Enigma traffic, and over a
two-week period the team of ten cryptanalysts and clerks successfully read
75 percent of all intercepted messages from beginning to end.

Most of the time Rejewski and the others did not actually read the contents of the messages whose settings they broke, though years later Rejew-

ski would still vividly recall one intercepted signal from June 30, 1934, the "night of the long knives" when leaders of the SA, the Brown Shirts, were rounded up and shot. Addressed to "the commanders of all airfields," the Enigma message ordered SA Chief of Staff Karl Ernst apprehended and transported to Berlin "dead or alive." Ernst, on his honeymoon in Bremen, was caught by SS troops who opened fire on his car. Knocked unconscious in the attack, Ernst was flown to Berlin, stood up against a wall at the Cadet School, and shot along with 150 other high officials of the storm troopers, whose radicalism and street brawling tactics were no longer of use to Hitler now that he had seized power.

For five years the Poles broke a steady stream of Enigma messages. The Germans tightened up their operating procedures during this time, introducing a series of changes that made recovering the daily settings more difficult, though Rejewski's basic techniques were still robust enough to handle them. Starting on January 1, 1936, the wheel order was changed monthly rather than quarterly. On October 1, 1936, wheel order changes began appearing daily; at the same time, the number of stecker plugs used for each day's setting changed from a constant six to a variable five to eight. On November 2, 1937, a new reflector was introduced whose wiring had to be recovered. But all of this was manageable.

The change that occurred on September 15, 1938, was not. Without warning, the indicator method was scrapped. Instead of enciphering the message setting indicator using an initial ground setting that was in force for an entire day, operators now began picking an initial rotor setting for each message separately. Daily setting lists still specified the stecker, wheel order, and ring setting for each day. But now the messages bore a nine-line indicator. The first three letters were the initial rotor setting, followed by the twice-enciphered message setting indicator. This meant that Rejewski's entire system was useless. With every message setting now enciphered at a different initial setting, there was no way to build up the eighty messages needed to establish the characteristic. The fact that the initial setting was transmitted in the clear at the start of each message was no direct help, either, since the variable ring setting on each wheel disguised the actual position of the wired rotors; the "tires" might be at setting XRA, but the wheels could be turned anywhere, depending on the ring setting chosen for that day. The problem was thus to figure out which of the six different wheel orders and which of the 17,576 different ring settings was in use each day.

Rejewski immediately saw that the initial *relative* rotor positions of all the messages sent in one day was one thing he did know. If one message began with the three-letter indicator XRA and another XRG, the absolute po-

sition that the rotors had been set in each case to encipher the message indicator was unknown, but it was certain that XRG, whatever it really was,
was six positions past XRA.

The message setting indicator was still enciphered two times in a row as
before. Thus letters 1 and 4, 2 and 5, 3 and 6 still represented the same plain
text letter enciphered three rotor positions apart. A certain fraction of the
time, Rejewski realized, two different positions of the Enigma will encipher
a given plain text letter with the identical cipher letter. That is, once in a
while an enciphered indicator might appear in the form:

PSD**P**WR   *or*

A**G**WJ**G**E   *or*

KW**Y**OP**Y**

A simple calculation showed that this was an unusual event, but not that unusual; two monalphabetic substitutions pulled out of a hat at random will,
close to half the time, encipher at least one plain text letter with the same cipher letter. It was time to make another catalogue: This time, however, the
catalogue would indicate which of the 105,456 combinations of initial rotor
settings (17,576) and wheel orders (6) could give rise to such repeated cipher letters three positions apart. Henryk Zygalski quickly worked out a
system for encoding the information onto sheets. A sheet would be made for
each initial position of the left rotor, A through Z; on each sheet the hori-
zontal axis would indicate the initial middle rotor position and the vertical
axis the initial right rotor position. A hole would be punched at any coordinate corresponding to a rotor position that could give rise to a repeat. A typical sheet would have about half the possible holes punched out. Separate
sets of sheets were required for each of the six wheel orders and for each of
the twenty-six possible positions of the left rotor. All told, that meant 6×26
$= 156$ sheets would be required, each containing $26 \times 26 = 676$ separate
data points. This task made compiling the original catalogue a snap by comparison. The possibility of making a mistake in transcribing some 100,000
data points was enormous. Zygalski and his coworkers had to slice out each
hole by hand using a razor blade. Once the sheets were completed, however,
they could be used as a sort of semiautomatic calculator. For each repeat
discovered in a day's batch of Enigma signals, the sheet corresponding to
the initial left rotor setting of that message would be placed on a light table.
Each sheet would be aligned so that the square corresponding to the initial
position of the middle and right rotor was at the upper left-hand corner of
the light table. That way all of the sheets would be offset relative to one another by the relative settings of their middle and right rotors. Scanning left

to right or up or down over the stacked sheets would then be the equivalent of shifting the middle and right ring settings of all messages simultaneously. Anywhere light shone through corresponded to a ring setting that would permit the repeats to have occurred at all of those relative initial rotor settings. The sheets could then be restacked twenty-five more times per wheel order to try all possible ring settings of the left rotor. (For example, if messages containing repeats had initial rotor settings of AXB, DGH, EHQ, RVN, and YZA, then first the A, D, E, R, and Y sheets would be stacked; then the B, E, F, S, and Z sheets; then the C, F, G, T, and A sheets; and so on.) In about a hundred messages, the laws of chance showed, there would be about twelve with a repeated letter in the enciphered indicator; with twelve repeats to work from, the sheets would rule out all but about four possible ring settings for each of the six possible wheel orders, leaving about twenty-four possibilities to be tested. That could be done quite simply by trying the Enigma replicas at each possible setting, typing in a bit of the actual message text, and seeing whether German or gibberish emerged.

Simultaneously, the Polish mathematicians began work on a second and more automatic way of testing for which ring settings could give rise to the observed repeats in each day's enciphered indicators. This was basically an adaptation of the cyclometer. For reasons that will probably forever remain obscure they dubbed their device the *bomba*—"the bomb." For the *bomba* to do its job, the code breakers needed three indicators in which the repeat had occurred with exactly the same letter, such as:

RTJ	**W**AH **W**IK
DQW	D**W**J M**W**R
HPN	RA**W** KT**W**

Thus, at the correct ring settings and wheel order, an Enigma set to RTJ would encrypt R as W (and vice versa); so would a second Enigma with the same ring setting and wheel order set to RTM (three positions farther forward). Likewise, Enigmas set at DQX and DRA (the second and fifth positions following the initial rotor setting DQW) would connect Q and W; and Enigmas set at HPP and HPS (the third and sixth positions following HPN) would connect N and W.

The stecker, however, threw a monkey wrench into the problem. A cyclometer-type device could reproduce the transformation that the rotors and reflectors generated, but the stecker was an unknown for each day's setting. The Poles got around this problem with two stratagems, one mundane, one profound. The mundane stratagem was to assume a bit of luck and hope that W was one of the ten to sixteen letters that were left unsteckered that day. Thus W really was W. The profound stratagem was to recognize that while R might be steckered to some other letter whose identity was a mystery, that mystery letter was a constant regardless of rotor position. At the correct ring setting, the rotors and reflector of the RTJ Enigma would transform W into the mystery letter α; but so would the RTM Enigma. Likewise,

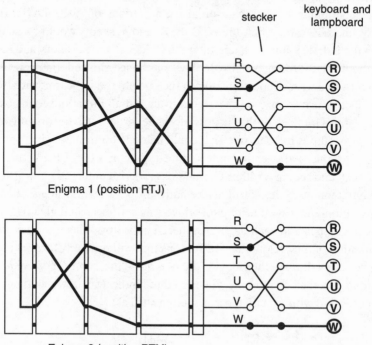

The case of the repeated indicator: At the correct ring setting the rotors and reflector of the Enigma will connect the repeated indicator letter (W in this case) to the same letter in both positions. The actual identity of this "mystery letter" depends on the steckering of R, which is unknown to the code breaker.

at the (same) correct ring setting and wheel order, the DQX and DRA Enigmas would both transform W into a second mystery letter β; and the HPP and HPS Enigmas would both transform W into a third mystery letter γ.

So this was the procedure: Six Enigmas would be set to their relative initial rotor positions. A current would be fed into the W wire in all six Enigmas. Then an electric motor would rotate all six in synchrony through all 17,576 possible positions, thus keeping their relative positions fixed while all possible ring settings were tried. When the three pairs of machines simultaneously lit up their output bulbs in three matching pairs, they had hit the jackpot.

The procedure not only gave them the ring setting, but an added bonus as well. Whichever bulbs the first two machines lit was the mystery letter α, thus establishing what letter R was steckered to. Likewise β was steckered to Q and γ was steckered to N. It is likely that rather than have a person sit and monitor the light bulbs while the *bomba* clicked away, the test was performed automatically by wiring up the six machines in series. At the "jackpot" position a current would flow from one end of this series to the other, tripping a relay and halting the motor. Twenty-five-wire cables running between each pair of Enigmas allowed for all possible identities of the mystery letters α, β, and γ:

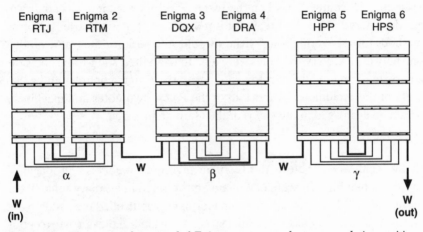

The bomba: *The rotors of six coupled Enigmas are set to the correct relative positions, then turned through all 17,576 possible positions by an electric motor. At the position which corresponds to the correct ring setting, each pair of Enigmas will connect the test letter (W in this case) to a common "mystery letter" (α, β, and γ), allowing current to flow from one end to the other.*

The AVA factory was able to produce six *bomby,* one for each of the six wheel orders, by November 1938. The machines could crank through all 17,576 ring settings in about two hours; that was thus the maximum time it would take to break the day's setting once a series of indicators with suitable repeats was found. Zygalski's team had meanwhile punched fifty-two of the sheets, covering two of the six wheel orders. Then the final blow fell: On December 15, 1938, the Germans introduced two new rotors. The number of possible wheel orders jumped by a factor of ten, from six to sixty. That would mean building fifty-four new *bomby* and punching an additional 1,500 sheets. The Poles were overwhelmed; they could now read traffic only for those one in ten days when the settings happened to use one of the wheel orders they already were equipped to deal with. On January 1, 1939, the number of stecker pairs used in each daily setting increased to ten, meaning that only six of the twenty-six letters of the alphabet were unsteckered, and that essentially defeated the *bomby* altogether.

But the Polish mathematicians had managed one final triumph before being stymied. The SD, Reinhard Heydrich's "Security Service," which had begun as an intelligence branch of the SS but which expanded to become, like the Gestapo, an organ of internal repression with a network of a hundred thousand or more informers, used the Enigma for its own coded messages. In a remarkable lapse of security, the SD continued to use the old indicator procedure even after September 15, 1938, and, more important, even after December 15, when the two new rotors were introduced. It was another incredibly lucky break, for it meant that Rejewski's tried and true methods for recovering rotor wirings would still work for the traffic on that one network. At Pyry, the Poles were able to hand their soon-to-be allies the complete wiring of all five rotors along with the Enigma doubles and a diagram of the cyclometer. Almost all of the Poles' own notes and machinery would be destroyed in the chaos of the German invasion.

———

In the final weeks before the German invasion Goebbels's propaganda campaign had hit a crescendo. William Shirer noted in his diary some of the hysterical headlines that appeared in Germany's controlled press: COMPLETE CHAOS IN POLAND—GERMAN FAMILIES FLEE—POLISH SOLDIERS PUSH TO EDGE OF GERMAN BORDER—IN CORRIDOR MANY GERMAN FARMHOUSES IN FLAMES!— WHOLE OF POLAND IN WAR FEVER!

Just before midnight on Monday, August 21, Berlin radio interrupted its musical program with an announcement that was even wilder than any of Goebbels's inventions. "The Reich government and the Soviet government have agreed to conclude a pact of nonaggression with each other. The Reich Minister of Foreign Affairs will arrive in Moscow on Wednesday, August 23, for the conclusion of the negotiations." This, however, was nothing but the truth. A secret annex to the pact stipulated how Hitler and Stalin would divide Poland between them. To ensure that there would be no last-minute concessions to spoil his victory this time, Hitler telegraphed a final ultimatum to the Poles, then ordered the cables shut down before the Poles could reply. At dawn on September 1 he struck.

The Polish generals' plan for defending their country was their final delusional act. They deployed six armies right at the frontier with only a single reserve group. The plan was to counterattack directly against any German breakthroughs. In a few instances this was carried out with horse cavalry, pennants flying and lances leveled at panzers.

Chamberlain, vacillating again at the moment of crisis, waited until September 3 to act; then, in a reedy, tired voice, took to the air waves that sweltering Sunday morning to announce that a state of war existed between Britain and Germany. By that time the Corridor was already cut off, the Polish air force destroyed, the roads clogged with fleeing refugees.

The coming of the war that Britain had so long avoided set in motion plans that Sinclair and Denniston had prepared for mobilizing GC&CS. In 1937, "C" had given instructions to Denniston "for the earmarking of the right type of recruit to reinforce GC and CS *immediately* on the outbreak of war." The Treasury had approved allocating positions for "56 senior, men or women, with the right background and training (salary £600 a year) and 30 girls with a graduate's knowledge of at least two of the languages required (£3 a week)." Denniston later explained the procedure he had followed; it was the old boy network par excellence in action:

> To obtain such men and women I got in touch with all the universities. It was naturally at that time impossible to give details of the work, nor was it always advisable to insist too much in these circles on the imminence of war.
>
> At certain universities, however, there were men now in senior positions who had worked in our ranks during 1914–1918. These men knew the type required. Thus it fell out that our most successful recruiting occurred from these universities. During 1937 and 1938 we were able to arrange a series of courses to which we invited our recruits to give them a dim idea of what would be required of them.
>
> This enabled our recruits to know the type of man and mind best fitted and they in turn could and did earmark, if only mentally, further suitable candidates.

These "certain universities" were—no surprise—mainly Oxford and Cambridge, and mainly the latter, though the Universities of London and Edinburgh and a few other smaller institutions were also approached. Most of the earmarked recruits were classicists, historians, and scholars of modern languages. Discreet inquiries were made, and those thought to be of "the type required," professors and bright undergraduates alike, were tactfully approached. The connection that Frank Birch and Dilly Knox had with King's College was milked with particular zeal. Frank Adcock, another Room 40 alumnus and Fellow of King's, as well as a lieutenant commander in the RNVR, became in effect the chief recruiter. Adcock was professor of ancient history at Cambridge; he officially joined GC&CS on September 4, the day after war was declared. But for the previous two years he had been inviting members of the scholarly community to discreet dinners. Professor E. R. P. Vincent, an Italian and German scholar, was summoned to one of these tête-a-têtes in spring 1937:

> We dined very well, for he was something of an epicure, and the meal was very suitably concluded with a bottle of 1920 port. It was then that he did something which seemed to me most extraordinary; he went quickly to the door, looked outside and came back to his seat. As a reader of spy fiction I recognised the procedure, but I never expected to witness it. He then told me that he was authorised to offer me a post in an organisation working under the Foreign Office, but which was so secret he couldn't tell me anything about it. I thought if that was the case he need not have been so cautious about eavesdropping, but I didn't say so. He told me war with Germany was inevitable and that it would be an advantage for one of my qualifications to prepare to have something useful to do.

The course for the chosen took place at Broadway Buildings in London, the GC&CS headquarters. The building, opposite St. James's tube station, was shared by the real spies of GC&CS's sister organization, the SIS, and this occasioned more antics taken straight from the pages of spy novels. One recruit to GC&CS in the spring of 1939 recalled the lecture he was given by the bowler-hatted War Office official who showed him to his office at Broadway Buildings his first day on the job: "All you ever say here is 'Third' as you get into the lift. Nobody must know where you work. If you have lunch with someone who insists on walking back to your office with you, you must proceed to the War Office, round the corner to the staff entrance, say good-bye, count 120, and then walk back here."

On September 3, 1939, Denniston forwarded to the Foreign Office a list of eight "men of the professor type" who had already been called up during the last weeks of August. A week later Denniston reported an additional ten hires, and a week and a half after that another three. Only three of the twenty-one "professor types" were mathematicians; all the rest were historians of art, or professors of medieval German, or lecturers in ancient Greek, or other distinctly nonscientific professions.

Partly this simply reflected Denniston's and Adcock's networks of contacts. But partly it reflected the persistent British public-school attitude that properly educated boys studied Latin and Greek; science, even mathematics, were tainted by their association with useful things and thus of "trade." Perhaps by way of overcompensation the mathematics that was suffered to rub shoulders with history and art at Cambridge was of an extremely refined pedigree, so abstract and theoretical that no one could mistake it for anything that might be associated with steam engines or bridge building. Cambridge did, to be sure, have a tradition of scholarship in what was termed "applied mathematics," but the term was apt to be misleading to those not familiar with it. Applied mathematics was really mathematical physics, theoretical excursions into the unseeable worlds of fluid dynamics and quantum mechanics and cosmology, almost as abstract in their own way as the studies of proofs and the properties of numbers that formed the core of the mathematics degree program, known as the Mathematical Tripos. So thoroughly did the mathematicians of Cambridge and Oxford succeed in eschewing the practical that the first to be recruited by GC&CS, Peter Twinn, who joined in 1938, was told that there had been doubts about hiring him because mathematicians "were regarded as strange fellows notoriously unpractical"; if having someone with scientific training "were regretfully to be accepted as an unavoidable necessity," as Twinn put it, some at GC&CS wondered whether "it might not be better to look for a physicist on the grounds that they might be expected to have at least some appreciation of the real world." Twinn may have been acceptable because he at least had done graduate work in physics.

The three mathematical "professor types" who came aboard in September were, however, pure mathematicians of the purest order. Gordon Welchman, a lecturer in mathematics and a fellow of Sidney Sussex College at Cambridge, specialized in the algebraic geometry of multidimensional space. Unlike Vincent, he had not been wined and dined by Frank Adcock; Welchman had merely received what he called "a polite note" from Denniston in 1938 asking if would "be willing to serve our country in time of war." Welchman said he would and was summoned to a "very informal" couple of weeks' indoctrination in London where he was instructed in the rudiments of code making and code breaking. The second of the mathematical trio was John Jeffreys, a lecturer at Cambridge who had worked closely with Welchman.

The third, who had also attended the course at GC&CS headquarters in 1938, and who thereafter began showing up regularly at the Broadway Buildings to consult on the work, was a man who on the surface seemed to confirm all the worst fears Denniston might have had about mathematicians. Alan Turing was on the one hand almost a caricature of the absent-minded professor. Born in 1912, the son of an Indian civil servant, he had been packed off to boarding school in England at age nine and suffered the usual miseries of a bright and sensitive boy who does not fit the rigid pattern prescribed by a rigid educational system. "Not very good," wrote his mathematics teacher. "He spends a good deal of time apparently in investigations in advanced mathematics to the neglect of elementary work. A sound groundwork is essential in any subject. His work is dirty." His work, even in mathematics and science, was indeed disorganized and full of errors. Schoolmates remembered Turing, and his papers, perpetually covered in ink blotches. Commenting on the results of one mathematics examination, a master wrote, "A. M. Turing showed an unusual aptitude for noticing the less obvious points ... and for discovering methods which would at once shorten or illumine the solutions. But he appeared to lack the patience necessary for careful computation of algebraic verification, and his handwriting was so bad that he lost marks frequently—sometimes because his work was definitely illegible, and sometimes because his misreading of his own writing led him into mistakes." Turing managed to go on to King's nonetheless, where, besides joining a "politically rather communist" organization called the Anti-War Council, he focused his mathematical interests in an area as far removed from practical application as mathematics could be: It was a mathematical examination of the nature of mathematics itself, an exploration of whether it was possible to formulate a method of deciding whether mathematical assertions were true or false. He published some precociously brilliant papers on the subject, while remaining as personally unkempt and disorganized as ever. William Filby, the science librarian at Cambridge before the war (he would later join GC&CS himself), recalled repeatedly crossing paths with Turing: University members were permitted to use the library after hours provided they locked up after themselves. "But whenever I found the doors wide open, the lights on, and books strewn about, I guessed it was Turing," Filby wrote. "I usually found him working, unwashed and unfed, in his room, quite unaware of the trouble he was causing."

But Turing had another, almost contradictory, side. An otherworldly scholar like Dilly Knox was almost comically inept at mechanical things. Knox had twice severely injured himself with motor vehicles, in 1904 by blowing up an acetylene lamp, which left him seriously burned, and in 1931 by breaking his leg while riding a motorcycle. He was notorious for being a terrible driver, too distracted to pay any attention to his surroundings. He had a theory that the way to avoid accidents was to traverse every intersection at maximum possible speed; during the war he commuted home on weekends in his Baby Austin and once commented to a colleague, "It's amazing how people smile, and apologize to you, when you knock them over."

But Turing, for all his clumsiness and ineptness in some things, had a mechanical mind that fully complemented his abstract mind. He was good with machines, and what was more he thought in terms of mechanical analogies. His most enduring contribution to mathematics was his brilliant conception, in 1936, of what would become known as the "Turing Machine." This was not a literal machine, but rather a way to think about the question of whether mathematical assertions were computable by breaking them down into a series of mechanical steps. His conceptual machine would scan a strip of paper tape that was divided into a series of squares and detect whether the square it was looking at had a mark on it or not. The machine, as Turing conceived it, could then be programmed to perform a sequence of just four different operations: It could move the tape one square to the left or one square to the right; it could place a mark on the square or erase the mark. Turing's remarkable mathematical discovery was that *any* calculation that was calculable at all could be solved by such a machine. So powerful was Turing's insight that this concept would later become the foundation of the logical operations used in all digital electronic computers.

Turing actually did build, or at least start to build, two other machines. Even before he had any contact with GC&CS, he had been thinking about cryptology. While at Princeton University as a visiting fellow in 1937, he conceived of a cipher machine that would work by multiplying two long binary numbers together. One number would represent the code text, the other would be the key; their product would be the transmitted cipher text. To decode the text one would need to divide by the key, which, Turing calculated, could be made sufficiently long so as to foil all normal attempts at decryption. Eager to see his if his idea would work, he got a colleague in the physics department at Princeton to lend him a key to the department's machine shop. Turing fabricated and wound his own relays and built a small prototype of an electrical binary multiplier. "To our surprise and delight," recalled the physicist Malcolm MacPhail, "the calculator worked."

Back at Cambridge in 1939, Turing teamed up with MacPhail's brother Donald, a mechanical engineering student at King's, on another computing device. This was to be an elaborate—and beautiful—assemblage of gears, pulleys, and weights that determined at what values the sum of a complex set of functions equaled zero. Each one of the thirty functions was represented by a weight that was cranked around by gears and strings; their sum was zero when they were in such a position that they precisely balanced a counterweight. A fellow of King's who visited Turing in his rooms in the summer of 1939 found the floor littered with gear wheels; as before, Turing personally took a hand at grinding and finishing the parts. The work had only just begun, however, when, on September 4, Turing was summoned to report to GC&CS.

––––––

During the 1930s the Royal Air Force had become the world's leading proponent of the doctrine of strategic bombing. The terror that bombers could

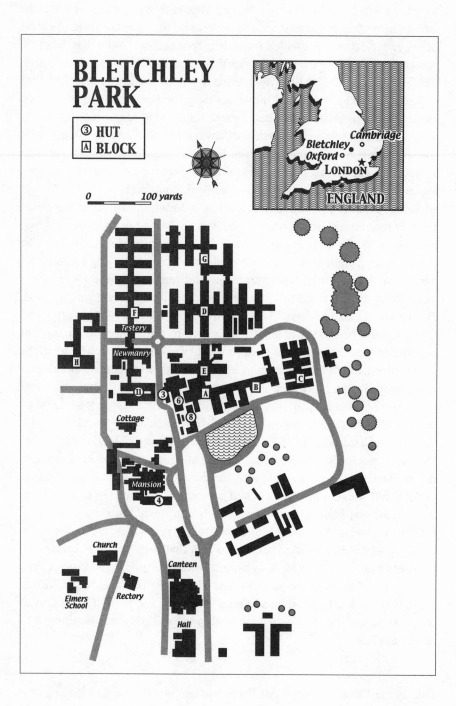

rain upon enemy population centers, the RAF's strategists said, would act as a deterrent to war; if war nonetheless broke out, air power could deliver a knockout blow that would force a quick capitulation. It was a notion that appealed equally to British politicians eager to avoid a replay of the slaughter of the trenches and to air commanders who sought to justify the expansion of the RAF as an independent military service. Convinced that the Luftwaffe had the same ideas, British air power proponents widely popularized the idea that the next war would be a war of the bombers. It was generally believed that London would be targeted in the first hours of a conflict, possibly even with poison gas bombs. The populace was issued gas masks and plans were made to evacuate children from the capital. Plans were also made to disperse whatever vital military operations could be moved from London to secret "war stations" in the countryside.

On May 28, 1938, there appeared a curious story in the *Bletchley District Gazette,* a weekly newspaper that served that small and dreary brick town in Buckinghamshire about fifty miles northwest of London. The story was about Bletchley Park, a mansion and grounds of fifty-five acres; someone, presumably acting for the government, had bought the property under great secrecy:

MYSTERY OF ITS FUTURE USE:
NO STATEMENT OBTAINABLE

Special Telephone Cable Being Laid

Following the signing of the contract during the week, immediate possession was taken and on Thursday, Post Office officials were busy laying a telephone cable. Further rumors say that this cable is being laid to provide a direct line to Whitehall.

A stop press item in the same edition reported a new development: A statement had been obtained that the site had been purchased "in connection with the air defence of Great Britain." Within a few days, however, that story quickly began to unravel. First the Air Ministry said it had nothing to add to what had been reported. The local MP made inquiries, whereupon the Air Ministry officially and somewhat huffily denied it had anything to do with the site; perhaps the War Office could explain what was going on. The War Office said it knew nothing about Bletchley Park; perhaps the Home Office could shed some light. At that point someone seems to have given the newspaper's publisher some friendly advice to drop the story.

The mysterious buyer was in fact Admiral Quex Sinclair, "C." Frustrated by the Treasury's refusal to come up with £7,500 to purchase what he be

lieved were suitable wartime accommodations for GC&CS and SIS, Sinclair had, in an act of noblesse oblige, simply bought it himself. And it was Sinclair, visiting the site, who was met by the local newspaper's reporter and had come up with the not terribly well thought out cover story about air defenses.

Bletchley Park, or "B.P." as it would quickly come to be known, was a bad parody of a stately home. It had begun life as a red brick farmhouse. But then in 1882 the property had been purchased by a nouveau riche London stockbroker named Herbert Samuel Leon, a friend of Lloyd George's who sought to announce his arrival in the time-honored English fashion by acquiring a country estate. Leon expanded the house in a cacophony of architectural styles, mostly a sort of Victorian mock-Tudor. Visitors pulling into the circular driveway were assaulted by a garish brick-and-stone façade in which arches, pillars, gables, domes, and parapets competed with one another for attention. The interior was equally overdecorated in an unsettling combination of carved oak and red plush. The overall effect was, at a minimum, tacky; "ugly" and "hideous" were the words that more than one member of GC&CS who served there during the war used.

Bletchley Park was not only hideous; it was also, as Nobby Clarke observed, "inconvenient" and "quite unsuitable for the purpose for which it was intended." From a security standpoint it was a disaster—the house and grounds stood just off a main junction of the busy North Road. The only obvious advantage it possessed was that it was easily accessible, not only from London but also from Oxford and Cambridge; Bletchley stood just about halfway between the two universities along the branch railway line that connected them at the time. The Park itself was a short walk from the train station. And, perhaps most important, it was close to the main Post Office telephone cables that ran along the North Road, permitting the direct connection to Whitehall offices that the *Bletchley District Gazette* had correctly reported.

During the Munich crisis in September 1938 Sinclair had ordered a test evacuation to Bletchley Park, which exposed more farcical lapses in security. Some 150 civilians and Army and Navy officers descended on Bletchley and neighboring towns where they were billeted at the best hotels and shuttled about by a fleet of hardly inconspicuous high-powered cars. Clarke ended up in a group of six GC&CS employees who were staying at the Bridge Hotel in Bedford—three men in their late fifties and three younger women, all of whom would vanish together in a car each morning and return late at night, scandalizing the hotel staff. A chambermaid who was complaining of overwork was admonished by one of the visitors that times were

serious and she should pitch in and do her part. "It's all right for you," she retorted, "but some of us have to work."

Sinclair, never one to sacrifice creature comforts, brought in his favorite chef from the Savoy Hotel in London to prepare the midday meal, which was served by waitresses to the staff at long tables in the mansion's dining room. This indulgence almost caused another security breach when a few days later the chef had a mental breakdown and attempted to kill himself. Clarke had to send for the Chief Constable of the district and explain that the matter had to be kept secret to avoid drawing attention to the house and its odd inhabitants. After a month GC&CS packed up and returned to the capital, leaving the hotel bills to be paid by a single officer at Bletchley Park in whose name all the rooms had been reserved, another absurd security lapse.

On August 1, 1939, "C" ordered the military services sections of GC&CS to take up their war stations at B.P.; two weeks later the diplomatic section was dispatched as well. The staff crowded into the mansion and surrounding outbuildings in a state of barely controlled chaos. GC&CS occupied the ground floors; a telephone exchange was run into the billiard room; the library was divided into offices for the various subsections; furniture was so minimal that books and papers were piled in huge stacks on the floor and the staff spent considerable parts of the day on all fours sorting through documents. A boys' school ten minutes' walk away was snapped up to help accommodate the diplomatic section.

The Enigma section, consisting of Knox plus GC&CS's four mathematical recruits, Twinn, Jeffreys, Welchman, and Turing, were ensconced in "The Cottage," a low building in the stable yard. But nothing was close to anything; access to the vital teleprinters that connected Bletchley Park to London was inconvenient; and office space was "grossly overcrowded." The espionage branch, SIS, had been moved to Bletchley Park as well, and there was much grumbling about the disparity of treatment between GC&CS and "the other side," as the SIS was called by the code breakers. The staffs were billeted at hotels and inns throughout the surrounding area, some miles away; official cars were provided for the SIS officials but the GC&CS staff had to organize car pools with their own vehicles.

One of the few redeeming features of Leon's grounds, a rose garden and maze that stood adjacent to the library, was shortly leveled to make way for a wooden hut, one of many that were springing up on the grounds to accommodate the rapid expansion. Amid the chaos Sinclair continued to see to it that a lavish midday meal was served (prepared now by a new, and presumably less suicidal chef)—beefsteak puddings, chicken, ham, sherry tri-

fles, bowls of fresh fruit. It was all somehow surreal; the ornate show house
of another era, itself a distorted fantasy of a venerable English manor house
of yet another era, now overrun with a fantastic assortment of spies and
Cambridge dons and eighteen-year-old girls away from home for the first
time. For surreal contrasts, however, nothing could match the setting that
B.P. offered for the section led by Lieutenant Colonel John Tiltman, assigned to investigate the hand ciphers used by the German police and SS. In
what had been the nursery room, on the second floor, down the corridor to
the right, still decorated with Peter Rabbit wallpaper, Tiltman and his staff
began the work that over the next several years would decipher the matter-
of-fact tallies of daily atrocities committed on the eastern front and the precise reports of "discharges" from the concentration camps.

———

September came and went and the Luftwaffe did not bomb London, with
gas bombs or with anything else. The RAF struck Berlin with bales of
leaflets. The French Army, still the strongest in the world, sent nine divisions
creeping forward five miles into Germany along a fifteen-mile front in the
Saar, occupied twenty deserted villages, declared success, and withdrew vic-
toriously back behind the Maginot Line.

The Poles had drawn up an orderly plan for relocating their three radio
intercept stations if the German attack required their temporary evacuation; only one would move at a time so that two would always be in operation picking up German signals and forwarding them to the code breakers
at Pyry. The plan lasted less than two days. By then all three stations were
quickly fleeing the advancing German armor. By September 5, Warsaw
itself was threatened, and the staff at Pyry was ordered to pack up any indispensable equipment and documents, destroy the rest, and prepare to
evacuate. So began a nightmarish journey for Rejewski, Zygalski, and Rozycki. A special evacuation train carried them from Warsaw to Brzesc on the
Bug River, where the Polish Supreme Command was establishing its headquarters. The railway was repeatedly bombed along the way; when they finally arrived in Brzesc three days later, the Supreme Command had already
abandoned its headquarters and the new orders were to flee to the southeast. The crates carrying the Enigma replicas, *bomby,* and documents were
loaded onto trucks; as the trucks ran out of fuel or broke down the crates
were abandoned one by one and destroyed. On September 17 they reached
the Rumanian border with but a single truck still running, which the Rumanian border police confiscated. Polish military officials crossing the border were being ordered by the Rumanians to report to an internment camp,
but Rejewski, Zygalski, and Rozycki mingled with a group of civilian

refugees who arrived a few minutes later and slipped through amid the chaos and confusion. Making their way on foot to a train station, they managed to exchange currency and buy tickets and arrived at Bucharest, at the other end of the country, twelve hours later.

There, the Polish military attaché heard their tale and frankly advised them go directly to the British or French Embassy where they might have better luck securing the necessary documents to spirit themselves out of Rumania. The three arrived at the British Embassy just as a bus carrying the entire evacuated staff from Britain's Warsaw embassy pulled up; a harried official told the Poles to "come back in a few days"—*mañana,* as Rejewski said to his colleagues. Off to the French Embassy they went. There the Poles took a gamble; telling the official who asked their business that they wished to speak to a representative of the French Army, they added, "Please just tell your superior we are friends of Bolek." Bolek was Bertrand's code name, and that did the trick. The official returned a moment later in a flurry. "*Mais oui, messieurs!* The colonel will see you just as soon as he has finished speaking with Paris." The colonel appeared, said he had just received instructions to help them leave at once for Paris, and, without any further questions, instructed his aide to arrange for passports for the three.

A few days later, at the end of September, the mathematicians were on their way once again, another fourteen hours by train through Yugoslavia and Italy to France. Bertrand had meanwhile gone to Rumania to search the internment camps for the three code breakers and other members of the Biuro Szyfrow's German Section. He arrived back in Paris on October 2. Four days later all organized resistance in Poland ceased.

CHAPTER 4

Fighting Back

LEONARD Woolf described wartime England as grey boredom; "endlessly waiting in a dirty, grey railway station waiting-room." The boredom was personal and cosmic, petty and universal. Adding to the inevitable boredom of wartime, of rationing and lines and shortages of everything and endless waits, was the incomprehensible boredom of no news. American war correspondents had flocked to the Continent to cover the war in the West and promptly dubbed it the "phony war." The French hunkered behind the Maginot Line and waited. Blacked-out London waited, too, for the Luftwaffe bombers that failed to appear night after night. Whatever slight novelty the blackout had at first provided was soon replaced by only more boredom and irritation; one in five Britons, a Gallup poll reported a few months into the phony war, had been injured in the blackout, tripping and falling on the sidewalk, being knocked down by a car, colliding with other pedestrians. The stifling heat wave that had gripped London in September melted away to a cold grey fall and then to the bitterest winter in forty-five years. Coal was in short supply, water pipes burst, trains were buried in snow drifts, telephone lines were knocked down, the Thames froze to solid ice from Teddington to Sudbury.

At Bletchley Park, everyone sat at their desks bundled in coats and mittens. The house was bad but the wooden huts were worse, with their bare concrete floors and erratically functioning electric heaters. Though the erratic electric heaters were at least better than the coal stoves, whose undersized metal flues, poking straight up through the asbestos roofs of the huts, refused to draw in a high wind. The backdrafts would send flames shooting from the stoves, or simply extinguish the fire and fill the rooms with smoke.

Back in August, after receiving the Enigma replica from the Poles, Knox had set to work at once to have additional replicas made; the quickest route was to adapt British Typex cipher machines. The Typex worked much the same way as the Enigma, with the added advantage that it printed out the results of its encipherment or decipherment on strips of paper. New Typex rotors were accordingly wired to duplicate the Enigma rotors.

The other task was equally straightforward: begin cranking a cyclometer through all 17,576 settings in all fifty-eight remaining wheel orders, note down the ones that permit repeated letters in the doubly enciphered indicators, and punch two new complete sets of 1,560 Zygalski sheets—one for GC&CS, one for the French and Poles. The repeats were known as "females"; Knox and his team devised a cyclometer that would automatically print the results onto grid sheets, which could then simply have a hole punched at each mark and be ready for use. That eliminated the great danger of transcribing errors. It also promised to speed up an otherwise overwhelmingly slow and tedious chore. On November 1, 1939, the mathematicians reported that a small manual cyclometer was continually breaking down, but that the new automatic device was "promised in a fortnight." Because its task was to sort settings into "males" and "females," they dubbed the new machine a "sex-cyclometer."

Twinn, Turing, Welchman, and Jeffreys—GC&CS's four Oxbridge mathematicians—were now firmly established in the Cottage as members of Knox's Enigma team. They had not had an easy start of it. Knox, said Welchman, was "neither an organization man nor a technical man; he was, essentially an idea-struck man." Knox's habit of solitary contemplation in the bath, for which he had been famous in Room 40, continued at Bletchley Park; at his billet, a colleague later recalled, Knox "once stayed so long in the bathroom that his fellow-lodgers at last forced the door. They found him standing by the bath, a faint smile on his face, his gaze fixed on abstractions, both taps full on and the plug out."

Knox was also an extremely difficult man who "disliked most of the men with whom he came in contact," Welchman found, and he was also possessive in the extreme about what he viewed as his ideas. Knox told Welchman next to nothing about the Enigma, gave him some menial tasks for a couple of weeks, then subjected him to "some sort of test and appeared to be, if anything, annoyed that I passed," Welchman said. Whereupon Knox banished him from the Cottage and Welchman found himself deposited in a barren room in Elmers School with instructions to make a study of the call signs that appeared at the beginning of each intercepted Enigma message.

It was not a useless task by any means, though it was hardly higher mathematics. Studying the call signs and other "externals" of radio traffic, such as message length, radio frequency, and time of transmission, could reveal how many different networks were in operation. Each network used its own daily settings of the Enigma machine, so being able to find a quick and sure way to sort the batches of incoming Enigma traffic into these separate networks was vital to avoid a lot of wasted effort and confusion when it came to breaking it. The process was known as "traffic analysis," and it could even be used to extract intelligence about the enemy order of battle by studying which stations communicated with which others. Welchman found that colored pencils were the easiest way to chart the various separate networks, and the practice of calling Enigma nets by color names stuck. The system quickly became so dependent on Welchman's color scheme that when the stocks of red, green, yellow, orange, brown, and several other colored pencils ran out shortly thereafter, Bletchley Park sent an urgent request to America for new supplies.

Working alone in the School, Welchman began to wonder if he could do something a bit more challenging than marking charts in colored pencil. And over the course of several weeks in October, to his growing excitement, and utterly unaware that he was reinventing exactly what Rejewski and Knox had already done, Welchman cracked the Enigma indicator system, even conceiving the idea of using punched sheets to isolate the day's ring setting and wheel order. He hurried to Knox with his "discovery." Knox furiously told him to get back to his call signs—without telling him that punched sheets were already in the works.

———

But by the following month Welchman was in the picture. During those two short months Turing had performed a feat that if anything surpassed Rejewski's. He had laid out the entire theoretical framework for tackling the Enigma in all its variations. Turing eventually wrote up his analyses in a sprawling, jumbled, confusing, and at places illegible 150-page "Treatise" on the Enigma, known at B.P. as the "prof's book." Its typed pages were covered with X-ed out words and lines and entire sections; emendations and table upon table of cipher alphabets were scrawled in by hand; and the whole was liberally spotted with the ink smears and blots that Turing's schoolmasters had, apparently quite in vain, tried to cure him of. At first it looks like the work of a hopelessly disorganized mind. But on closer examination a very different picture emerges; it is the work of someone who grasps things in an instant and who is merely impatient with having to set forth the detailed explanations by which he arrived at his results.

Knox and Turing recognized from the start that the Zygalski sheets had serious limitations. With the increase from three to five rotors, there were now ten times as many wheel orders to examine; using the rule of thumb that a hundred messages gave about four possible ring settings to try for each wheel order, that meant that about 240 settings would have to be tested by hand even after the Zygalski sheets had done their work. While not impossible, that represented a huge amount of labor. It might be possible to devise ways to eliminate certain wheel orders, but even under the best of circumstances the success of the punched-sheet method was wholly dependent on the Germans' continuing to double encipher the message setting indicators—a procedure that, as Knox observed during the meeting in Warsaw, "may at any moment be canceled."

For the next two months Turing accordingly concentrated on developing a theory for the recovery of the daily settings that would be proof against any such setback. A general approach to narrowing the possibilities in any cipher is to use a crib—that is, to match a string of cipher text letters from the body of the message itself with the probable plain text letters that they represent. For an unsteckered Enigma, Knox had already developed a scheme that employed this approach; this was what his paper strips—the "petits bâtons"— were for. Given the number of ways the Enigma could scramble cipher alphabets this might not seem like a very promising avenue. But once again the placement of the fast rotor in the right-hand position had left a back door gaping wide open. The crucial point was that a short crib would be unlikely to trigger a turnover of the middle rotor. Thus the effect of the reflector and the left and middle rotors could be taken as constant. The fast rotor would advance one position as each successive letter of the crib was enciphered; in one position the fast rotor might turn A into B, in another position E might become B. But what the rest of the machine did to that "B" would not change, and that was the crucial fact. For example, in position 1:

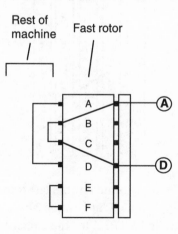
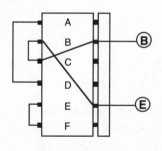

Position 1   Position 2

the Enigma links A and D together: Each is the encipherment of the other. The fast rotor converts A to B and D to C; the rest of the machine acts to couple B and C to complete the circuit. At position 2, B and E are linked; the fast rotor converts them to C and B, and again the fixed effect of the rest of the machine is to couple B and C on the left side of the fast rotor. What that means is that for the length of a short crib, whenever matching pairs of plain and cipher letters are fed into the right side of the fast rotor, then for *any* position of that rotor the set of letters that emerges at the left side must form a single, consistent, monalphabetic substitution that couples pairs of those left-side letters. If in one position of the fast rotor a plain text letter emerges from the left side of the rotor as B and its matching cipher text letter as C, then it is impossible that in another position a given plain text letter will emerge as B while its matching cipher text letter emerges as, say, D. That fact made it possible to figure out the identity and starting position of the fast rotor through a process of elimination.

The method worked like this: Knox's batons (they were also called "rods" and so the process was called "rodding") were strips of paper, each representing a single plain text letter. Down their length was listed, in sequence, the transformation of that letter effected by the fast rotor at each of its twenty-six possible positions. The technique was to line up, side by side, the batons that made up the plain text letters of the crib, and then to do the same for the matching cipher text. The diagonals of the resulting matrices then represented the output of the fast rotor at every one of its possible initial settings. Thus the plain text crib KRAUT and its matching cipher text AZBRY:

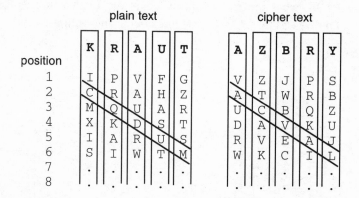

would, enciphered starting at position 1, be transformed by the fast rotor to IRUSS for the plain text (K at position 1 becomes I; R at position 2 becomes R; A at position 3 becomes U; and so on) and VTBKJ for the cipher text.

Starting at position 2, the fast rotor would transform the pair to CQDUM and ACVAL.

At the correct starting position, the fast-rotor transformations of the cipher text and its plain text crib would have to be isomorphs: That is, they could not contain any contradictions to a simple monalphabetic substitution. Both positions 1 and 2 could be eliminated on these grounds. The fast-rotor transformations starting at position 1:

$$
\begin{array}{c c c c c}
I & R & U & S & S \\
V & T & B & K & J
\end{array}
$$

require S to be coupled to K in one position and to J in another. Likewise

$$
\begin{array}{c c c c c}
C & Q & D & U & M \\
A & C & V & A & L
\end{array}
$$

contain a contradiction of C and A versus C and Q (and likewise, A and C versus A and U). If none of the twenty-six positions yielded a valid pair of isomorphs, then that particular rotor could be eliminated from contention as being in the rightmost position and another rotor tried.

Turing and Knox developed an extremely complex pencil-and-paper method, similar in theory though far more involved in practice, that would accomplish this same process of elimination by assuming certain letters in a crib were unsteckered. Then, to aid the subsequent process of establishing the wheel order and initial setting of the remaining two rotors, Turing began to develop a catalogue of all the couplings generated by the reflector, left, and middle rotors at all possible positions. Jeffreys and Turing devised a system for punching this information into sheets, dubbed "Jeffreys sheets." Like the Zygalski sheets, these could be superimposed to act as a form of semiautomatic calculator; in this case, however, the sheets served to eliminate settings of the left and middle rotors that did not produce the required monalphabetic couplings established by the rodding process.

Turing's next insight came with remarkable speed. The Polish *bomba* had automated the process of finding the correct initial rotor position by turning a series of coupled Enigma machines through all 17,576 possibilities. The *bomba* however had very limited applicability. It depended on there being three doubly enciphered indicators in which the same letter was repeated in all three; it also would work only if that repeated letter happened to be unsteckered, so its true identity was known.

But with a text crib Turing at once saw that a series of Enigmas could be linked together in a different architecture to perform an automated search.

And it would be an incredibly powerful method. A given string of plain text and its matching cipher text formed a characteristic pattern that was preserved regardless of the steckering. By plugging a series of Enigmas together and rotating them through all 17,576 positions, it would be possible to search for the position that produced the characteristic pattern required by the crib. There was no need to make any assumptions about the steckering at all. It was a way to defeat not only a change in the indicator system, but also the hundred million million permutations that the stecker introduced.

The characteristic patterns were a sort of geometric property that arose from repeated letters in the plain text–cipher text strings. For example, if the following crib and cipher text had been matched up:

relative position	1	2	3	4	5	6	7	8	9
plain text	M	I	T	S	C	H	L	A	G
cipher text	H	M	I	X	S	T	T	M	I

then when the Enigma was at position 6 it transformed H to T and vice versa. At position 3, it transformed T to I; at position 2, I to M; and at position 1, M to H. That formed a closed loop of letters. A geometric diagram of the chains and loops formed by this sequence would look like this:

Thus, if A were fed into an Enigma set to relative position 8, M would emerge; if M were then fed into an Enigma at position 1, H would emerge; and so on. The snag in trying to find the correct initial setting of the rotors, however, was that the true identities of A, M, and H—their identities, that is, when they actually reached the entry ring and were fed into the scramblers of the Enigma—were unknown and unknowable because of the mixing-up of their identities performed by the stecker board. Yet Turing saw that it didn't really matter: This *same* geometric relationship implied by the crib–cipher text match-up would have to be preserved regardless of whatever "mystery letters" A, M, H, and the others were replaced with by the steckers. The trick was thus to find an initial setting in which the set of scramblers produced that same geometric pattern, with *any* set of letters; for instance one setting might link these letters together in the prescribed pattern:

$$A \overset{8}{-} S \overset{1}{-} J \qquad C \overset{5}{-} M \overset{4}{-} G$$
$$\underset{2}{|} \qquad \underset{6}{|}$$
$$I \underset{3}{-} T \underset{7}{-} Z$$
$$\underset{9}{|}$$
$$X$$

There might be several different rotor settings that would connect letters together in the required geometry, but if the number was not too large, each could be tested by trying it out on the actual message and seeing if readable text emerged. And some candidate settings could be immediately ruled out if the letters in their pattern implied an inconsistency in the steckering. In the example above, the "mystery letter" that replaces the M in the original pattern is S, implying that M has been steckered to S. In a real Enigma, that immediately implies that S has been steckered to M as well, and so the letter S in the original must be replaced by M—as is indeed the case here. But if S had been replaced by any other letter, the setting could be rejected at once as invalid.

The Polish *bomba* had one crucial innovation that Turing incorporated into his "bombe" (the French version of the word was the one that stuck)—the use of multiwire cables to connect a series of Enigmas set to the proper relative positions. As with the *bomba,* the multiwire cables allowed for all possible identities of each "mystery letter." The geometric relationships arising from the matching of text cribs with cipher text were called "menus." To search for this portion of the geometric pattern:

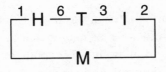

for example, four Enigmas could be plugged together with cables thus (again, for simplicity, only six of the twenty-six wires in each interconnecting cable are shown):

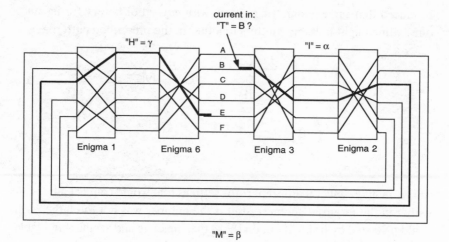

The four Enigmas would be set in the proper relative position and cranked together through all 17,576 possible start positions. To test the hypothesis that T is steckered to B, a current would be applied to the B wire on the "T" cable, which feeds into the input of the Enigma set at relative position 3. At any given rotor setting, that Enigma would transform B to some letter α; the multiwire "I" cable would then feed all possible identities of α into the next Enigma, set to relative position 2, from which would emerge an unknown letter β; that letter in turn would be transformed to γ by Enigma 1; and finally γ would be fed into Enigma 6. The question then was what would come out of Enigma 6. If the steckering hypothesis and rotor settings were correct, it would *have* to be B—since in the actual intercepted message both the input of Enigma 3 and the output of Enigma 6 are the same letter. In such a case, the current would start at B on the "T" cable and end at B on the "T" cable.

But the genius of Turing's design was that when the bombe reached the correct rotor setting, it did not even matter which letter one had chosen to apply the initial current to; all twenty-six hypotheses for the steckering of T were in fact tested for simultaneously thanks to the way the machine was wired. If B happened not to be the right hypothesis—and the odds were 26 to 1 against any given steckering hypothesis being correct—then the output of Enigma 6 would be a letter other than B. For the rotor position being tested in the example above, such is the case: The current emerges from Enigma 6 on wire E. That contradictory result is then automatically fed back into Enigma 3 through the "T" cable; the current would travel around the loop again, emerge from Enigma 6 as yet another letter, and so on around and around the loop.

At an incorrect rotor setting, this feedback process would usually continue until all twenty-six wires on the "T" cable ended up carrying current. Since electricity travels through a wire at a speed approaching that of light, for all practical purposes this would occur instantaneously. When the bombe detected such a state, it rejected that rotor setting as impossible—since it gave rise to a contradiction no matter what steckering was assumed—and went on to try the next rotor setting.

But if only one circuit was energized, that indicated a possibly valid steckering and rotor setting. Likewise, if all but one wire were energized, that indicated that there existed one isolated, closed loop at that rotor setting that would yield a consistent steckering result. (Such is the case in the example above: Tracing the A wire around the loop reveals that it connects back only to itself. So if the A wire happened to have been chosen as the entry point for the current, then at this rotor setting only the A wire would end up carrying a current; if *any* other wire was chosen as the entry point, then all wires *but* A would wind up energized.) Turing noted that the bombe could readily be equipped with relays that would trip in either condition—one wire energized or twenty-five wires energized; when the relays tripped the bombe would stop turning and the setting of the rotors could be noted. Any possibly valid setting identified by the bombe was thus called a "stop."

The concept of the bombe owed several ideas to the Poles: the idea of rotating interconnected Enigmas together in synchrony through all possible rotor settings, and notably the idea of using multiwire cables to allow for all possible identities of letters whose steckers are unknown. But the fundamental mathematical insight behind the British bombe was wholly Turing's. The discovery that matching strings of plain and cipher text defined a characteristic geometrical relationship that depended only upon the wheel order and starting position, regardless of steckering, was Turing's; so was the idea of feeding a contradiction back into an interconnected loop of Enigma machines. That idea was really the crux of Turing's invention, and it was an idea that went beyond ordinary brilliance; it was one of those revelations that simply cuts across conventional thinking. The Germans had believed that the stecker would defeat any attempts to penetrate the Enigma even if an enemy had matching plain and cipher text to work with—and they were right, so far as conventional cryptanalytical approaches went.

By early October 1939 Turing had already drawn up specifications for the bombe and GC&CS had contracted with the British Tabulating Machine Company at Letchworth to manufacture the device. Although the breaking of the daily settings would be an unrelenting struggle throughout the war, the winning strategy, at least in its essential elements, was conceived and set into motion in just over a month after the first shot was fired.

The remains of the Polish Army tried to piece itself back together in France that fall, adding to the ranks of refugees with tens of thousands of newly recruited men from the Polish émigré community in France. In late October the members of the Biuro Szyfrow were detailed by the Polish Army's command to work directly under Bertrand. The French radio intelligence headquarters had been moved out of Paris with the start of the war, to the town of Gretz-Armainvillers, about twenty-five miles northeast of the capital. The French Section d'Examen had done itself rather better than GC&CS. Its new home was a luxurious three-story chateau, the Château de Vignolles; officially the site was known as Poste de Commandement Bruno. By the end of December GC&CS had finished punching at least most of two sets of the Zygalski sheets and had dispatched part of the second set, covering twenty-four of the sixty wheel orders, to P.C. Bruno along with a set of the Jeffreys sheets. But GC&CS then began to drag its feet. Knox, who by this point nursed a furious loyalty to the Poles, sent a blistering note to Denniston on January 7, 1940: "As you remember on our journey to Warsaw I promised to assist the Poles and the French in producing statistics," he began. Knox noted that the sheets had been produced in a third the time expected and he could not understand why a complete set had not yet been given to their allies. "My personal feelings on the matter are so strong that unless they leave by Wednesday night I shall tender my resignation," he concluded.

Part of the reason for the delay was that a snag had developed. The sheets weren't working, and the British were not sure why. The British had lobbied for the three Poles to come to B.P. for consultations. The French demurred, apparently arguing that since the French government was paying the costs of the Polish army in exile, the Poles had to stay under their wing.

Knox did not resign; instead, Turing was dispatched to France at once to deliver the sheets—and to see if the Poles cold help explain the source of difficulty in making them work. The problem was this: Knox and Turing had been developing a method to speed up the use of the Zygalski sheets by eliminating certain wheel orders from consideration. The method depended upon identifying the right-hand rotor by its turnover pattern—the point during its rotation that it caused the middle rotor to advance one position. This method was made possible by the convenient fact that each of the five rotors had its turnover notch at a different position on the ring. The Germans had no doubt intended that as a baffling security measure, but like the encipherment of the indicators, like the changeable wheel order, it had backfired, and actually gave away more information than it concealed.

Upon arriving in France, Turing soon learned why their system was running into trouble: the information on the turnover notches of the two new rotors, IV and V, which the Poles had recovered in December 1939, had been interchanged.

With that kink ironed out and the new sheets in hand, the Poles on January 17 succeeded in breaking the first wartime Enigma setting—the "Green" key, an Army administrative system, for October 28. Within a week Knox had broken three more days, Green for October 25, and Red (the Luftwaffe's operational key) for January 6 and 17.

Meanwhile a fledgling staff of four intelligence officers at B.P. had been given a bleak room in "Hut 3," equipped with a table and three chairs, and told to stand by and be ready to translate the Enigma decrypts as they came in. Amid an air of great anticipation, the first texts arrived one snowy morning. F. L. Lucas, a fellow of King's College who had been recruited for the job, recalled the deflation they felt: "Here lay a pile of dull, disjointed, and enigmatic scraps, all about the weather, or the petty affairs of a Luftwaffe headquarters no one had heard of . . . the whole sprinkled with terms no dictionary knew."

It was a start, at least. But the recovery of the daily settings continued to be slow and frustrating work. The main problem was simply a lack of raw material. The phony war dragged on; the German Army was communicating mainly by landlines; and the Allies' interception system was not up to coping even with the traffic that was passing by radio. The Germans were taking pains to make interception as difficult as possible: changing frequencies, using low power, changing the call signs of every station daily, sending dummy traffic, having different networks use the same frequency; it required constant compilation of the habits of the German operators and searching up and down the radio band to keep up with all their tricks.

Adding to the trouble was that the War Office was beginning to grumble loudly, wondering why its intercept station at Chatham, which was bearing the brunt of the Enigma work, should be handling Air Force traffic. GC&CS pleaded with the powers-that-be not to be shortsighted and parochial: Yes, most of the Enigma traffic was on the Red key, but there was no way to easily draw a neat line when both Army and Air Force networks were using the same frequencies, and in any event Chatham had the experience to handle the job. Trying to make a highly complex and technical task conform to bureaucratic lines of authority would be "disastrous," GC&CS warned: "Enigma traffic must be treated as a whole. . . . Also, it is certain that sooner or later the Germans will make some alteration in their machine, and if the enigma section failed to detect this alteration the chances of decoding any of the traffic will vanish." Allowing Army radio operators to continue the job of intercepting German Air Force traffic was a small contribution to make to the greater cryptanalytic effort, on which all depended.

The French and the British were interchanging intercepts but, a memorandum noted, the material sent to GC&CS from France "arrive in a muddle about a week late." Communication was improved when a teletype link was established between P.C. Bruno and B.P. to exchange recovered keys and decrypts, but with the lines running over hundreds of miles of France and under the Channel there was the danger of tapping by enemy agents, so some form of encryption was necessary. Sharing Typex with the French and Poles raised policy issues that no one was prepared to deal with. But the French still had the Enigma double supplied at Pyry, and Langer had managed to escape Poland with two more; the obvious solution was to use the one cipher system all parties had easy access to—the German Enigma. The French officer who handled communications amused himself by rigorously following German traffic procedures in preparing the cross-Channel messages, closing each with the words, "Heil Hitler."

———

Welchman, an organization man, had by late fall 1939 seen what his boss Knox, the "idea-struck man," had not: that no one had yet given a moment's thought to the practical problem of how Bletchley Park was going to sort and decode hundreds or even thousands of Enigma messages a day once the theoretical problem of recovering the daily settings was solved. Knox was still operating in the mode of the Cambridge don, the lone intellect delving into an arcane scholarly problem. Yet if he succeeded, B.P. would at once have to transform itself into something more akin to a General Motors assembly line. Large staffs would be required to log and sort the intercepts as they came in, to shuffle and stack the Zygalski sheets (the "Netz" as they came to be called), to operate the bombes, and, finally, to punch the coded messages into the converted Typex machines to produce the actual decrypts. In addition there would have to be constant coordination of all of the work, especially in providing guidance to Chatham and the other intercept stations on what traffic was the most valuable. Welchman sent a proposal to Edward Travis, GC&CS's deputy head, for building up such a complete organization; Travis immediately agreed, and on November 18 forwarded to Denniston an outline for the organization that would shortly be known as Hut 6—the name was used both for the building they would occupy, one of the new wooden structures about a hundred yards from the mansion, and for the work that went on within its drafty walls. Welchman was to direct the "registration section," which would sort the incoming traffic, search for "females" in the indicators, and coordinate the whole operation; Jeffreys would head the "Netz" party that would find the daily settings by stacking up the punched sheets. A deciphering section would operate the converted Typex machines to test the recovered settings and then produce the actual decrypts. Knox and Turing would be eased over to a new research section, leaving the way clear for the people who actually knew how to get the work done each day. Welchman was told to go out and start hiring the staff that would be needed.

There were few precedents for the kind of people who would be good at these jobs. Many of the lower-level tasks, to be sure, were purely clerical; they demanded absolute accuracy but little or no imagination. But many of the jobs defied categorization. An almost infinite tolerance for drudgery and repetitive detail was a requirement for almost every position, but in some cases it seemed that what was required was to combine those qualities with their exact opposites, with a capacity for imaginative, even mad leaps of insight. The ideal cryptanalyst was Beethoven with the soul of an accountant; or vice versa. Cryptanalysis required an ability to work systematically through endless permutations; it required a certain kind of compulsive drive, a determination that one was *going* to solve the problem no matter what; but it also required a certain mode of thought that many highly intelligent and driven people do not have, and that was an ability to sense—at times almost intuitively or even irrationally—underlying analogies or patterns. Time and again, a code that had defied a systematic, brute force attack would yield to a foray from a totally unconventional angle. Knox became famous at Bletchley for quoting the *Alice in Wonderland* sort of riddle, "Which way does a clock go round?", and anyone foolish enough to answer "clockwise" would be sternly told off, "Not if you're the clock, it doesn't!" (Being able to figure out in which direction the relative position of the Enigma rotors changed for a given shift in the ring setting required just this sort of thinking.)

Most of the actual work of code breaking was not really a matter of mathematics or science, but the mental habits the job required were ones that mathematicians and scientists often tend to possess. A psychological study of successful scientists found that beyond the traits commonly attributed to the profession—shyness, detachment, intensity—scientists, or at least the best scientists, tend to think in ways that separate them from the mass of men. Their thought processes, the study found, were strikingly similar to those of a creative artist, marked by an "overalertness to relatively unimportant or tangential aspects of problems," which leads them to "look for and postulate significance in things which customarily would not be singled out." It is a mental attitude that borders on the "autistic" and even "paranoid": "The difference between the thinking of the paranoid patient

and the scientist comes from the latter's ability and willingness to test out
his fantasies or grandiose conceptualizations . . . and to give up those schemes
that are shown not to be valid." Paradoxically, the very fact that the system
in which they work is governed by hard rules of inflexible logic gives them
the freedom to indulge in such "unrealistic, illogical, and even bizarre thinking" that, in a less bounded world, would become a literal threat to one's
sanity. Most ordinary people do not think in this fashion precisely because it
is, in a very real sense, mad.

But this described the mental process of cryptanalysis to a tee. The "paranoid" attitude was really the essence of what it took to be a cryptanalyst; not
"paranoid" in the sense of having a persecution complex, but paranoid in
the sense of believing that hidden in some almost irrelevant detail lay a
grandiose truth; paranoid in the sense of believing that one could see what
everyone else had failed to see.

These mental traits are also characteristic of musicians and chess players,
which may explain the oft-noted connection among these talents. Music,
chess, and mathematics are all worlds of absolute bounds, of set rules and
order, of logic and structure. Yet the minds that excel within these spheres
are precisely the ones capable of looking at a problem in a way that is not at
all customary or bounded or logical; a brilliant chess move is, almost by definition, one that seems on its face to be illogical and bizarre, one that could
not have been arrived at by systematic analysis or logical elimination of possibilities; a brilliant musical composition is one that, even as it obeys the
rules of melody and harmony, combines them in a way that the rules themselves would never on their face suggest. In all of these fields, the self-disci-
pline necessary to master complex rules—to put up with the drudgery, in
other words—is the price of admission one pays for the privilege of letting
one's mind run wild and free.

One of Welchman's first recruits, early in the new year, was Stuart Milner-
Barry, a London stockbroker and a classmate from his days at Trinity College. Welchman explained that he "shamelessly recruited friends," but there
was more to it than that. Milner-Barry said later, "I was not a mathematician, and why Gordon thought I would be useful to him I don't really know
except he thought chess might go well with cryptanalysis." Milner-Barry was
no ordinary chess player, though: He had been in Buenos Aires representing Britain in the International Team Tournament when war broke out. The
British team had just qualified for the finals, but "with visions of London in
flames" Milner-Barry and his teammates decided they could not continue
and took passage on a boat for London the very night war was declared.
Back home, Milner-Barry was rejected for military service and was "looking
for something to do" when Welchman invited him to join Hut 6. It was an act

of faith on both sides, for Welchman could not tell potential recruits anything about what the work might involve. But Milner-Barry concluded that "I knew Gordon well enough to know if he said it was an important job, it was an important job."

Milner-Barry in turn recruited Hugh Alexander, another member of the chess team—and a mathematician as well, though not a first-rate one. Alexander had missed being elected a fellow at Cambridge and had accepted a job as a mathematics teacher at Winchester, the boys' school. In 1938, the owner of a London department store, himself a chess fanatic, convinced Alexander to join the firm, though as Milner-Barry observed, Alexander "was not cut out to be a businessman" and "looked singularly incongruous in a black jacket and striped pants." Within a short time Milner-Barry would be head of Hut 6 and Alexander head of the corresponding naval cryptanalysis section, Hut 8.

As the war progressed, Bletchley Park's search for cryptanalytic talent never exactly became systematic, but it did broaden somewhat. Lieutenant Colonel Tiltman initiated a three-month cryptanalysis course, which took place in a showroom above the Gas Company's offices in the city of Bedford, fifteen miles from Bletchley; in an adjacent room an intensive six-month course in Japanese was being administered to a succession of mostly young and mostly classics majors from the universities. (In his usual determined fashion, Tiltman had simply ignored the experts he had consulted at the London School of Oriental and African Studies, who adamantly insisted that it was impossible to teach even the rudiments of Japanese in less than two years. As usual, Tiltman proved the experts wrong, and the graduates of his six-month crash course were in enormous demand both by Bletchley Park and for naval intelligence postings overseas.) The route by which students were selected for the courses remained decidedly informal; candidates were summoned for an interview and Tiltman and a few other officers asked a few innocuous questions—"What are your hobbies?" "How is your health?" "Do you have any religious scruples about reading other people's correspondence?" But the process at least spread the net far wider than the King's College old boys' network, and the course served as a quick initial test of aptitude that made it feasible to take a chance on a broader range of candidates. The "A" and "A−" students from one cryptanalysis class, deemed to possess "decided ability, capable of independent research," included a thirty-nine-year-old schoolteacher, a twenty-seven-year-old physicist, a twenty-eight-year-old accountant, and two twenty-year-old Oxford undergraduates, one in classics and the other in modern languages.

Perhaps the oddest, yet perhaps also the most imaginative, recruiting device that Bletchley Park tried during the war was to tap the champion

solvers of the *Daily Telegraph* crossword puzzles. Crossword puzzles had
come into vogue in the 1920s, and toward the end of 1941 the *Telegraph,*
with much fanfare, published its five thousandth puzzle. That set off one of
those typically British series of letters to the editor in which various readers
claimed never to have failed to solve each day's puzzle. That in turn led to
the *Telegraph's* holding a competition in its offices one Saturday afternoon.
Twenty-five contestants were invited and given a puzzle to solve, a winner
was declared (winning time, 7 minutes, 57.5 seconds), and everyone was
then given tea in the chairman's dining room and sent on their way. A few
weeks later the contestants received a letter inviting them to come and see
a Colonel Nichols "on a matter of national importance."

In the United States, Friedman and Safford also used contests to spot
talent; starting in the 1920s Safford circulated cryptograms and puzzles
throughout the Navy and earmarked winners for future duty in OP-20-G as
positions became available. The American Cryptogram Association, a group
of puzzle hobbyists, regularly held competitions in its magazine *The Cryptogram,* and winners on more than one occasion received letters signed by
William F. Friedman of the U.S. Army Signal Corps inviting them to enroll
in an Army correspondence course on cryptography. As a result of its unprecedented mobilization and drafting of civilians in the First World War,
the U.S. Army had come to place great faith in the use of aptitude tests to assign recruits to the proper task, and it had "courses" for everything. It was
easy to parody this, and indeed the utter mismatches between civilian experience and military assignments that sometimes resulted was a standing
Army joke, but the system nonetheless did have the effect of casting the net
extremely widely; everyone who came into the Army was in principle eligible for every position. William P. Bundy, an American officer who would
later serve at Bletchley Park, recalled that the Signal Corps took just such a
completely democratic approach when it came to recruiting potential cryptographers: "Everybody who had no visible talent or aptitude whatsoever
for electrical work or communications in the technological sense," and who
had achieved a certain minimum score in the appropriate categories of the
Army General Classification Test, was dumped in the "Crypt School" at the
Signal Corps' training camp at Fort Monmouth, New Jersey. Friedman's
course was used at Fort Monmouth as well as in a special ROTC cryptography program at the University of Illinois that served to spot talent.

The strategy of seeking out undiscovered talent from a huge pool of candidates, rather than recruiting from within an elite academic circle, made a
certain amount of sense. It really was more a matter of finding someone
with the right sort of mind as opposed to the right sort of education, and
even the connection between cryptanalytic talent and such vocations or av-

ocations as science, chess, and music only went so far. The American crypt-analytic agencies rarely sought out top mathematicians or scholars the way the British did—or the way the Manhattan Project recruited the best minds in physics and engineering, men and women with international reputations in their fields. There was no such *thing* as a civilian cryptanalyst, the way there were civilian physicists or radio experts or for that matter pilots and automobile mechanics. Civilian occupations or training told little about who would do well at the job, and several recruits who stumbled into the Signal Intelligence Service almost by chance before and during the war ended up devoting their entire careers to code breaking. A 1944 U.S. Army study found that "no particular background of training" was "concretely indica-tive" of cryptanalytic ability; "there were cases of high school graduates who showed a surprising aptitude for difficult cryptanalytic assignments; like-wise there were cases of individuals with five and six years of specialized university training who were strangely limited in their aptitude for this par-ticular type of work."

———

The growing pains at Bletchley were painfully obvious during that first win-ter of the war. By the end of March 1940 about fifty days' worth of Enigma keys had been broken, about twenty of those the work of the French and Poles at P.C. Bruno. The traffic was mostly the Red key used by the Luft-waffe, little of it bearing any immediate military significance; worse, the breaks were coming for the most part a week to several weeks late. The sys-tem for turning decrypts into usable intelligence was primitive in the ex-treme. Hut 6 and Hut 3 were adjacent to each other but not interconnected. Bletchley's carpentry crew, under the command of a fox-hunting local builder named Hubert Faulkner, was handed an office tray and told to build a small tunnel that would link the two buildings through which the tray could slide. At first a string was fastened to each end and the tray was pulled through; later this system was replaced by a broom handle used to shove the tray along.

Even when the fledgling organization did come up with something of military importance it was exceedingly hard to get anyone in the Admiralty or War Office or Air Ministry to pay attention. Quex Sinclair had died, of cancer, on November 4, 1939, and an acrimonious succession struggle en-sued for the post of "C," which did nothing to help GC&CS secure an in-stitutional foot in the door at Whitehall. Menzies, who had been Sinclair's deputy, was the likely heir, but Churchill, at this time back in government as First Lord of the Admiralty, was pushing his own candidate—apparently Admiral Godfrey, the Director of Naval Intelligence. Two other high officials of the SIS also coveted the job, Valentine Vivian, the head of counter-intelligence, and Claude Dansey, the espionage chief.

Menzies was charming and moved in the right circles, as Sinclair had; he belonged to a string of London clubs, hunted with the Duke of Beaufort's hounds, had been married three times—to an earl's daughter, a baron's granddaughter, and a baronet's daughter—and did nothing to discourage the false but widespread rumor that he was the illegitimate son of Edward VII. He had been a star athlete at Eton. He had also been a mediocre scholar, and that did not inspire confidence in many quarters. But the day after Sinclair's death he produced a sealed letter from his former chief recommending him for the job. In the weeks of wrangling that ensued, that factor eventually tipped the scales, and on November 29 Menzies was officially named to be the new "C." But in a compromise that aimed to satisfy everyone while doing precisely the opposite, Vivian and Dansey were given the titles of vice chief and assistant chief, and proceeded to use their new positions of power to snipe at each other with renewed vigor, while both working to undermine Menzies's authority at almost every turn. Dansey, for his part, could never quite decide which he loathed more, counterintelligence officers or intellectuals—"long haired 'planners' injecting themselves into everything." It was not a happy ship.

While Welchman was recruiting mathematicians and chess champions, Hut 3 and its naval counterpart (which would become known as Hut 4) were building up their staffs with German scholars who could read the decrypted messages and pass along anything of value to the appropriate authorities. But it was proving very much a rerun of Room 40's experiences from the First World War. The Admiralty still saw no reason to pay attention to a bunch of professors who presumed to tell them how navies operated. Christopher Morris, another of the fellows of King's College recruited to Bletchley Park, had been assigned to work on one of the lower-level hand ciphers used by the German navy. In the first week of April 1940 he read a message that ordered ships heading for Bergen to report their positions to the German War Office in Berlin. He passed this finding on, only to be told impatiently by the Admiralty that ships report to naval headquarters and that he obviously didn't know what he was talking about. "The ships were of course troopships," Morris later wrote, "and the signal would have given advance warning of the invasion of Norway."

In the face of a vastly superior British fleet, the Germans landed troops up and down the coast of Norway with "surprise, ruthlessness, and precision"—Churchill's description—in the predawn hours of April 9. Certain that no one would be so foolish as to attempt a landing against such odds, the Admiralty was convinced that signs of German naval activity presaged an attempted breakout by the battle cruisers *Gneisenau* and *Scharnhorst* into the Atlantic; the commander of the British Home Fleet, Admiral Sir Charles Forbes, ordered his forces to the northeast where they would be in a position to block such a move. That left the approaches to Norway uncovered. Though the fiasco earned the Commander in Chief the nickname "Wrong Way Forbes," it still did little if anything to alter attitudes about the value of signals intelligence among British commanders.

Nor, even, did a veritable flood of intelligence that now sprang forth from Hut 6. As Nazi troops descended on Norway, a new Enigma key crackled to life. Dubbed Yellow by Hut 6, it was used by the German forces for army–air coordination, and suddenly for the first time in the war GC&CS was reading Enigma traffic of immediate operational importance, and reading it almost as soon as it was intercepted. The first break came April 10, the day after the invasion, and each day's haul from the intercept stations brought several times the hundred messages required to recover a daily key using the Zygalski sheets. Yellow continued in use through June, for the duration of the Norway campaign, and continued to provide a wealth of details about German operations. "This was something unprecedented in the history of warfare," observed Ralph Bennett, a Cambridge historian who the following year would join Hut 3 as an intelligence analyst. "No spy ever imagined in his wildest dreams that he could discover so many of the enemy's secrets and deliver them to his own side within a few hours, yet here loomed a future in which this might be done, not once or twice, but regularly and as a matter of routine. . . . How would generals, admirals and air marshals, accustomed to distrusting intelligence and paying little heed to it, react to a flood of novel and undoubtedly reliable information?" The answer was, exactly as they had always reacted to intelligence: hardly at all. The chaos and catastrophes of the Norway campaign for the British made it a far from ideal theater to inaugurate a new source of intelligence, to be sure; commanders were too distracted by a series of crises to focus on something as new and unprecedented as Enigma intelligence. There was a practical bureaucratic problem as well. The Admiralty, which actually commanded ships at sea directly from Whitehall, had established secure channels of communication between England and its forces engaged in combat. GC&CS had teleprinters that it could in theory use to send verbatim texts of broken messages directly to the Admiralty. But Yellow dealt only tangentially with naval operations. The War Office and the Air Ministry on the other hand were not in a parallel position in relation to the commanders of armies and air forces; those ministries did not give orders or control troops or squadrons, and neither they nor Bletchley Park had any mechanism to rapidly communicate with commanders on the ground, especially when the information to be passed required protection worthy of the most closely guarded state secrets. But behind all of these practical obstacles lay the yawning indifference of commanders who simply didn't know what to do with something so new and different. And so Bletchley Park's diamonds spilled out onto the ground, trampled unseen into the dust of the battlefield.

———

The British collapse in Norway marked the end of a long road for Prime Minister Neville Chamberlain. On April 4 he had incautiously asked why Hitler had failed to strike in the west when he had the chance; the Allies were now in an incomparably stronger position than they had been in September and one thing was certain: Hitler had "missed the bus." A month later, with the Germans securely lodged in Norway, a hostile and mocking Parliament threw those unlucky words back at Chamberlain. Lloyd George took to the floor and observed that the Prime Minister had demanded sacrifice. "I say solemnly that the Prime Minister should give an example of sacrifice, because there is nothing which can contribute more to victory in this war than that he should sacrifice the seals of office." On the day that German troops set upon Holland and Belgium, May 10, Chamberlain resigned and Churchill, the voice in the wilderness, took office at the head of a government of national unity. His claim to authority was utterly unique: "My warnings over the last six years had been so numerous, so detailed, and were now so terribly vindicated, that no one could gainsay me," he wrote.

The long-feared Nazi attack was heralded by another long-feared German move. On May 1, the twice-enciphered Enigma indicators disappeared from all keys except for Yellow. The "Netz" method was useless.

This was the development the bombe was supposed to neutralize. The only trouble was that the bombe, so far, was a near-total flop. The first bombe had been delivered by the British Tabulating Machine Company to Bletchley on March 14. It was installed in Hut 1 in a twelve-by-fifteen-foot room that, even within B.P., was known by several cover names designed to conceal its true purpose: "The Office," "The Technical Room Maintenance Shop," "The I/C Ops Room," "The Power Room." The other half of Hut 1, separated by a solid wall, was the sick bay. The bombe had been at once pressed into service for the naval Enigma problem, which from the start had been impervious to the Netz method because of its far more complex indicating system. But it was already apparent that some radical improvement in the design would be needed if it was going to work at all, on the naval problem or anything else. Attacking a single message was taking a week of twenty-four-hour-a-day operations of the machine. The bombe in its current form would be of essentially no help in getting back into the Red traffic.

What Welchman called two "astonishingly bad habits" on the part of German operators, however, miraculously stepped in to save the day. With the dropping of the twice-enciphered message setting, the preamble of Enigma messages now consisted of just two three-letter groups. The first group, as before, specified the initial setting of the three rotors. The second group specified the message indicator, disguised (also as before) by being enciphered by the Enigma at that initial rotor setting. But as it was enciphered only a single time, there was no longer anything in the indicators that would mathematically reveal the underlying machine setting—the wheel order and ring settings.

The solutions that Hut 6 found owed far more to psychology than mathematics, however. John Herivel, one of the young mathematicians recruited by Welchman in January, discovered the first of the astonishingly bad German habits of the Luftwaffe operators. Herivel was musing over what it would be like to be a German operator preparing to send his first message of the day. The procedure required taking out the selected rotors, releasing a clip on each, and rotating the outer ring—the "tire"—so that the designated letter for the day's ring setting was at the zero position on the inner wheel, and then inserting each rotor into the machine in the specified left-to-right order for the day.

Herivel suddenly realized that the natural tendency would be to place the rotor into the machine with the day's ring setting letter facing up. Then, if the operator were appropriately lazy, he might not bother to change the initial rotor setting for his first message. Herivel proposed gathering the first messages sent by each station after midnight, when a new day's setting would take effect, and looking for any clusters. If there was more than one lazy operator, a cluster such as SWE, TVG, SXF, and RXG might appear, suggesting that the ring setting of the first rotor was R, S, or T; the second V, W, or X; and the third E, F, or G. Those ring settings could be tested at each possible wheel order. The idea did not at first bear out in the period of relatively low traffic during the winter. But with the German strike in the west, the amount of traffic on the Red key exploded, and the pressures of communications in combat apparently did the trick. It was an astonishing insight, and Welchman dubbed it "the Herivel tip" in honor of its discoverer.

The other stopgap that Hut 6 relied upon before the first bombes became available was known as "cillis." The term was said to derive from one operator who repeatedly used what were apparently the letters of his girlfriend's name as his message-setting indicator (though in memorandums in early 1940 Welchman also referred to them as "sillies," which was an apt description as well). Another form of cillis that Hut 6 began to discover were indicators derived directly from adjacent letters of the keyboard. Enigma operators had been forbidden to use such obvious combinations as AAA or BBB, but some took refuge in the next best thing.

The Enigma keyboard was laid out like this:

```
Q   W   E   R   T   Z   U   I   O
  A   S   D   F   G   H   J   K
  P   Y   X   C   V   B   N   M   L
```

As Welchman explained it, sometimes a three-part message would come in with preambles such as these:

Part 1	QAY	MPR
Part 2	EDC	LIY
Part 3	TGB	VEA

What had happened was that for his initial rotor settings, the operator had chosen alternate rows of diagonal letters starting at the left side of the keyboard: QAY, EDC, TGB. Given such laziness, it seemed highly probable that the operator had likewise chosen for his corresponding message settings the adjacent diagonals: WSX, RFV, ZHN.

That discovery provided a crib: At relative position QAY, W was enciphered as M; at the next position S was enciphered as P; at the next, X was enciphered as R. With enough such cillis to work with, Turing's and Knox's paper-and-pencil method for using a crib to recover wheel orders and steckerings could be employed. (Later, "cilli menus" were sometimes run on the bombes when text cribs were unavailable.)

It took until May 20—twenty days—to break back into the Red key using these methods. But when the break at last came, it produced a flood of hundreds of signals a day. And since the Luftwaffe's major role was to support the swiftly advancing panzers, the Red key traffic provided extremely current intelligence of the first operational importance. But once again the battle itself had stolen a march. At seven-thirty on the morning of May 15, Churchill was awakened by an aide who informed him that Paul Reynaud, the French Premier, wished to speak to him urgently. Churchill picked up the telephone by his bedside and heard a distraught voice saying in English, "We are beaten; we have lost the battle." Churchill tried to reassure Reynaud that it could not be that bad. "We are defeated; we have lost the battle," Reynaud repeated. It was the simple truth. In two days the spear point of the German army had cut through the supposedly impassable Ardennes Forest, where it had been opposed only by the Chasseurs Ardennes, a unit of Belgian forestry workers hastily issued uniforms and rifles. The French Ninth Army, at the west side of the forest, was splintered. By the morning of May 14 there was a fifty-mile-wide gap in the French defensive line; behind the panzers, motorized divisions came pouring through. Churchill flew to Paris to consult with French generals. *"Où est la masse de manoeuvre?"* he asked in his execrable French accent: Where are the reserves, to counter-attack against the German breakthrough? General Maurice Gamelin shrugged. *"Aucune,"* he replied. There are none. It was, Churchill said, the greatest shock of his life.

On the twenty-first, the first day the Red key was read currently by Bletchley Park, the German armies reached the sea at Abbeville. The Allied armies in Belgium and northern France were cut off. It was no longer a bat-tle to save France, but now only a battle to save as many British and French troops as possible before they were destroyed or captured. Five days later the order to begin the evacuation from Dunkirk was issued.

Once again, the Polish cryptanalysts found themselves trying to stay ahead of advancing German troops. P.C. Bruno was evacuated in a series of stages across the country. On June 17 the aging Marshal Henri Philippe Pé-tain, the hero of the First World War, assumed the premiership. "I give the gift of myself to the people of France," he grandly announced. "The fighting must stop." Two days later, German Army engineers began demolishing the walls of the small museum in the Forest of Compiègne that housed Marshal Foch's railway carriage, in which the German–French armistice had been signed on a nearby siding in 1918. On June 21 Hitler arrived to cap France's humiliation. The engineers had moved the railroad car to the exact spot in a clearing where it had stood on that day of German humiliation twenty-two years before. The French delegation arrived and was handed Hitler's terms for a cessation of hostilities. French POWs would not be released even after the fighting stopped. Any French national bearing arms against Germany in the forces of another country would be considered a "franc-tireur," and shot if captured. The French fleet would be demobilized in ports under German supervision. The government of "unoccupied France," which was shortly to take up residence in the health-resort town of Vichy, would, in other words, be a hostage of Germany. The following day the French signed. Foch's rail-way carriage was hauled off to Berlin. A granite monument at the site had marked the 1918 armistice with an engraved inscription:

HERE ON THE ELEVENTH OF NOVEMBER 1918 SUCCUMBED THE
CRIMINAL PRIDE OF THE GERMAN EMPIRE — VANQUISHED
BY THE FREE PEOPLE WHICH IT TRIED TO ENSLAVE.

Hitler ordered it blown up.

Alarmed over what might happen if Bertrand and his staff fell into German hands, Denniston cabled to Bertrand seeking to have the Poles brought to England. Bertrand replied on June 26 that they had been safely evacuated before the British cable was received, and that the Army command had for now forbidden all further movement by officers. On June 24, Bertrand had managed to secure three airplanes to fly the Poles and other members of the cryptanalysis staff to Algiers. Bertrand also, surprisingly and defiantly, announced that the "current plan is to continue work under similar conditions." Denniston replied: *"Mon chef compte sur vous pour la securité de notre travail"*—"My chief counts on you for the security of our work"—and inquired *where* Bertrand was planning to continue the job. The answer was audacity itself: in Vichy France. The armistice forbade the French military command to engage in any intelligence operations against Germany, but Bertrand managed to secure the approval of his superiors to reestablish his code breaking station in the utmost secrecy. Operating under the alias of Monsieur Barsac, Bertrand purchased the Château des Fouzes in the small town of Uzès. It was near the Mediterranean coast, a location chosen to increase the odds of being able to escape successfully to North Africa if the Germans moved to seize Vichy. By October the Poles had been brought back from Algiers, and "P.C. Cadix" was in operation. Bertrand finessed the extremely delicate matter of securing intercepts. The French military intelligence service was disbanded under the armistice, but the Vichy government was permitted to create a small radio monitoring service called the Groupement des Controles Radioélectriques de l'Interieur. Its purpose was supposed to be tracking down clandestine radios operated by the anti-Nazi resistance. But its director was a close friend of Bertrand's, who managed to carefully select a few trusted officers to monitor German Army and SS signals and forward the intercepts to P.C. Cadix. Four short-wave receivers were also installed at Cadix itself for picking up long-distance communications.

In March 1941 Bertrand arranged a rendezvous in Lisbon with his old SIS contact "Biffy" Dunderdale. Bertrand wanted to "resume cooperation" on Enigma. GC&CS, reassured by the caution Bertrand had shown so far, agreed to take what was nonetheless a considerable risk and worked out a scheme for sending daily Enigma settings to Cadix.

———

The German strike in the west had forced the hasty end of the ill-fated venture for the British forces in Norway. A planned counterattack against Narvik, designed to secure a foothold for the government and the king in the north of the country, and to cut off German supplies of iron ore, had begun the night of May 12, 1940; now the sixteen thousand French, British, and Polish troops, the destroyers, and the hundred antiaircraft guns were urgently needed in France. The Allies seized the port of Narvik on May 27 only long enough to cover their withdrawal and blow up the ore-loading piers. The royal family and the country's gold stocks were loaded on ships and carried off to Britain.

A week later, in what to the public seemed to be just another tragic footnote to the hapless Norway campaign, there occurred an event that would finally shock the military services into paying attention to what the professors at GC&CS were trying to tell them. Neither the Yellow nor the Red Enigma traffic had yet done so. Reading the Enigma messages was a feat of true wizardry, but circumstances had conspired against the generals' and admirals' being impressed—the generals because they were too busy and the admirals because the contents did not affect them. What finally did get their attention was a technically far less impressive feat, but one to which the fortunes of war gave extraordinary prominence within the top circles.

GC&CS had gotten almost nowhere in breaking the naval Enigma. The last resort in such cases was traffic analysis; even if one could not read the texts of messages, it might be possible to deduce some things about enemy forces, command structure, and movements by studying the pattern of signals—where they came from, who they were transmitted to, how much traffic was carried at different times of day. This was not a glorious occupation in the realm of code breaking, and GC&CS had given the task to a twenty-year-old Cambridge history student, Harry Hinsley, who had been recruited to B.P. in October 1939. By the end of the year he was, in his own words, "the leading expert outside Germany on the wireless organization of the German Navy"—a claim that, he hastened to add, did "not amount to much" given how little radio activity there was to monitor. Hinsley's only communication link with the Admiralty was a direct telephone line that he had to activate by energetically turning a handle. This would cause a bored naval officer at the other end to answer and listen to Hinsley's deductions with something ranging from indifference to contempt.

On June 3, Hinsley had noticed a sudden upsurge in radio communications that appeared, by the stations involved, to indicate a substantial fleet movement from Kiel in the Baltic to the Skagerrak. Hinsley cranked his telephone and reported his findings, with the usual result. It would have been well if the Admiralty had paid attention for once. Unaware of the British evacuation of Narvik, the German naval commander, Grand Admiral Erich Raeder, had ordered the battle cruisers *Scharnhorst* and *Gneisenau* and the heavy cruiser *Hipper* to attack Allied ships he believed still to be anchored near there. As the Germans approached the Norwegian coast, however, they received Luftwaffe reports that the British had abandoned Narvik and that convoys were at sea, and the German squadron commander decided to move against the convoys at once. At 4:00 P.M. on June 8 the two battle cruisers found their quarry: a midshipman on watch atop *Gneisenau*'s foremast sighted smoke on the distant horizon. Within minutes the Germans had identified it as an enemy aircraft carrier.

The chain of miscalculation, incompetence, and misconduct that led to the disaster that quickly befell the British may have begun with the Admiralty's intelligence failures, but others were far more to blame. The British carrier *Glorious*—for that was the ship the Germans had sighted—had sailed in the company of only two destroyers. The official reason *Glorious* had refused to wait for a stronger escort was that it was short of fuel. In fact, its commander, Captain Guy D'Oyly-Hughes, was racing back to England in order to press court-martial charges against the ship's air commander, with whom he had had a violent dispute and whom he had put ashore. D'Oyly-Hughes was an authoritarian at best and mentally unbalanced at worst: A former submarine captain with no experience in air operations, he was sailing on that afternoon of the eighth with no air patrols and the crew on a "make and mend" day—a half-holiday. Even though *Glorious* lacked radar, it had no lookout posted. The ship spotted its pursuers a full twenty minutes after being spotted. It never had a chance. *Glorious* was carrying not only its usual complement of aircraft but also ten Hurricane fighters that, against all conventional wisdom about the impossibility of landing high-performance aircraft on carriers, it had managed to rescue from Norway. None were ready to take off, however, and the ship's armorers were still struggling to change bomb racks on her Swordfish biplanes to torpedo racks when an eleven-inch shell fired by *Scharnhorst* crashed into the hangar, setting the Hurricanes ablaze. D'Oyly-Hughes was killed by another shell that scored a direct hit on the bridge. Within an hour and a half *Glorious* went down. Of the 1,511 men aboard there were 37 survivors.

Where the Admiralty had ignored GC&CS's successes, it responded to its own failure with a stunning swiftness. It immediately invited Hinsley to spend a month at the Admiralty. Hinsley was sent to Scapa Flow where he was the guest of the Commander in Chief of the Home Fleet aboard his flagship. A scrambler phone replaced the antique crank model. When he called now, someone on the other end listened. Several months later the Home Fleet sent a message to the Admiralty querying an intelligence report it had received: "What is your source?"

The Admiralty's reply was a single word, and it spoke volumes about the change that had occurred:

"Hinsley."

There were still jealousies and friction, but Bletchley Park would not be ignored again. It certainly would not be ignored at the top, for as the GC&CS staff was soon to discover, the new Prime Minister had an almost schoolboy enthusiasm for secret weapons, and he was about to become the code breakers' chiefest advocate. At the end of the summer, "C" received the following letter from Churchill's top aide and confidant at 10 Downing Street:

MOST SECRET.

September 27, 1940.

Dear C.

In confirmation of my telephone message, I have been personally directed by the Prime Minister to inform you that he wishes you to send him daily all the ENIGMA messages.

These are to be sent in a locked box with a clear notice stuck to it "THIS BOX IS ONLY TO BE OPENED BY THE PRIME MINISTER IN PERSON".

After seeing the messages he will return them to you.

Yours ever,

Desmond Morton

Churchill did not realize how much things had changed since the First World War: There was no way one person could read and comprehend hundreds of messages a day. Before long the routine evolved into a system whereby Churchill received a selection of the more important messages. But he continued to insist on seeing the actual texts, not summaries, and frequently complained when he suspected "C" was holding back. Even when traveling he expected to be kept informed. When Churchill flew to the Casablanca conference in January 1943 he sent a rocket to "C": "Why have you not kept me properly supplied with news? Volume should be increased at least five-fold." In his trademark fashion Churchill began to bombard the military service departments, the chiefs of staff, the theater commanders with questions and proposals based on this new source of information. He was sometimes off-base and overenthusiastic, but the impact was every bit as great as that of *Scharnhorst's* eleven-inch shells. To defend themselves against Churchill's proddings, the top British officers began making sure that they were fully briefed themselves about the Enigma traffic. Signals intelligence, at long last, was beginning to make it onto the map.

CHAPTER 5

Impossible Problems

IN the early morning hours of February 12, 1940, a British minesweeper, HMS *Gleaner,* was on a routine patrol off the west coast of Scotland when its hydrophone picked up the sound of a diesel engine churning through the water nearby. For the next half-hour *Gleaner* tracked its unseen quarry with sonar. Suddenly the ship's searchlight caught a white flash to starboard, possibly the spray from a periscope, that disappeared at once. Lieutenant Commander Hugh F. Price ordered *Gleaner* hard to starboard, full ahead.

The procedure for attacking a submerged target was to fix its location with sonar, which cast a search pattern ahead of the bow, then steer for the target, pass over it, and drop depth charges from off the stern. On *Gleaner's* first pass it was already over the target before the crew could collect sufficient sonar data; the ship turned, regained sonar contact, and at 3:53 A.M. dropped a pattern of four charges.

Lurking in the waters below was *U-33,* out of Wilhelmshaven, its mission to lay mines in the Firth of Clyde. When the four depth charges exploded a few meters above its hull the U-boat was plunged into darkness, its electrical system crippled by the blast. Gauges cracked and water began pouring in from multiple small leaks. His torpedo tubes filled with mines, the U-boat's captain knew he would hardly have a fighting chance on the surface. His only hope was to lie on the bottom and maintain absolute silence. But over the next hour *Gleaner* dropped five more depth charges, and *U-33* had had enough; the order was given to surface and run for it. In preparation, the U-boat's lieutenant took the eight rotors from the ship's Enigma machine and distributed them among the crew: If forced to abandon ship, they were to drop the rotors overboard the first chance they had.

At 5:22 A.M. *U-33* blew her ballast tanks and erupted to the surface. Almost at once the submarine was caught in *Gleaner's* searchlight; the British

ship fired five rounds with its four-inch gun and turned to ram. At that moment the U-boat's crew appeared on the deck holding their hands in the air. *Gleaner* turned hard to starboard, coming parallel to the U-boat a cable length away. But before *Gleaner* could lower her boats into the choppy seas, a shower of sparks erupted from the submarine's conning tower, its stern heaved into the air at a forty-degree angle, and *U-33* disappeared below the surface. The U-boat's senior engineering officer, ordered to set off scuttling charges to destroy the vessel, did not have time to escape and was blown into the air by the explosion. Other members of the crew, including the captain, drowned in the bitterly cold water. Only three officers and fourteen men out of the forty-two aboard were rescued.

Aboard *Gleaner,* the exhausted survivors stripped off their soaking clothes and hung them to dry. Lieutenant Heinz Rottman, the senior officer to survive the ordeal, asked if he could visit his men. When he entered the compartment where the crew was being held, Machinist First Class Fritz Kumpf came up to him with a stricken look. "Herr Oberleutnant!" he said. "I forgot to throw the wheels away!" Rottman went to the bulkhead where Kumpf's pants were hanging: the pockets were empty.

———

The unexpected trophies from the sinking of *U-33* proved to be two Enigma rotors that GC&CS did not yet have in its possession. The wiring of the five rotors used since December 1938 by the Army and Luftwaffe had been recovered cryptanalytically by the Poles. These new rotors bore the numbers VI and VII, and, Hut 8 realized, were apparently used exclusively by the German Navy.

But extra rotors were far from the only problem to be conquered in breaking the naval Enigma, and as great as the *Gleaner*'s coup undoubtedly was, it did nothing immediately to change the conviction at Bletchley Park that the naval system was insoluble. Turing began to work on it in late 1939, but only, he explained, "because no one else was doing anything about it and I could have it to myself." The difficulty was that the German Navy, far more security conscious than the other services, had early on adopted a far more complex indicating system for its Enigma messages.

The weakness of the Army and Luftwaffe system was that the message indicators had themselves been enciphered by an Enigma machine set with the day's wheel order, stecker, and ring setting. That fact was what permitted Rejewski's method of "females" to work: The characteristic patterns that showed up in the twice-enciphered indicators betrayed the underlying machine settings for the day. But as early as May 1, 1937, the German Navy had used an entirely different method. The indicator that told a recipient where to set his three rotors to start deciphering the message text was not enciphered using the Enigma at all. For that, the Navy used a separate code, a set of bigram substitution tables that changed every month. (The naval Enigma indicating system is described more fully in Appendix B.) The operator would pick a three-letter group from a list, encipher it with the bigram tables, and transmit that resulting indicator at the start and end of the message. The actual message setting—the position in which the rotors were set to begin enciphering the text for each message—was determined by enciphering those same three letters on the Enigma with its rotors set to an initial position specified by a daily key list. (As with the Army and Luftwaffe Enigmas, the key list also specified the stecker, wheel order, and ring setting to be used each day.)

This was a devilish complication. With the "Standard Indication System" of the Army and Luftwaffe networks, breaking one message would automatically break all other messages for the same day on the same network. But breaking one naval message got you nowhere. Even with the bigram tables, all that could be extracted from each subsequent message was the original unenciphered three-letter group; it was hard to see a way around having to try enciphering that group at all 17,576 possible initial rotor settings and seeing which one produced a message setting that worked. Without knowing both the bigram tables and the daily initial rotor setting—the "ground setting," or *Grundstellung* in German—it was impossible to convert the enciphered indicator into the actual message setting for each message.

Even breaking a single message was an enormous challenge in the case of the naval Enigma. What was obviously needed was a text crib, but the German Navy ran a tight ship when it came to basic cipher security. The procedure manual issued to Enigma operators in the German Navy strictly warned against repeatedly using the same abbreviations or other stereotyped phrasing. Frequently used addresses were to be varied as much as possible; *Befehlshaber der Unterseeboote*—Commander of U-boats—was to be written as BDUUU or BEF. UNTERSEEBOOTE or BEFHBR. UUUBTE, or so on.

The Poles were able to break a few naval messages before the war using what Bletchley Park later referred to admiringly as "the earliest known form" of cribbing: Based on external evidence it appeared that messages were frequently continuations of one another, and the Poles had learned enough about general German radio procedures to know that subsequent messages usually began with the word FORT (literally, "forward from," hence continuation) followed by the time of origin of the preceding message. Numbers were written out by using the keys in the top row of the keyboard—Q stood for 1, W for 2, E for 3, and so on—bracketed by Y . . . Y.

One message attacked this way was the continuation of a message with a time of origin of 2330, which yielded the pronounceable crib FORTY WEEPY, and so the method was named.

The first step in trying a crib was to get it positioned properly opposite the cipher text. The Enigma's cryptographic property of never enciphering a letter with itself proved of enormous help here. The procedure was to slide the crib along a likely stretch of cipher text until a position was found where there were no "crashes"—that is, places where the same letter appeared in both the crib and the cipher text at the same spot.

Applying much the same pencil-and-paper cribbing method that Knox and Turing later devised—which hinged on assuming that certain letters were not steckered—the Poles were able to break about fifteen messages from each of the first eight days of May 1937, enough to convince them that the indicating system was based on bigram tables, and was probably impenetrable.

Taking up the problem at the end of 1939, Turing again with remarkable speed worked out the entire theoretical foundation for a successful attack. There were three keys to breaking the naval Enigma. Foremost was the bombe, which would automate the process of breaking the first message of the day, using a crib. Second was a scheme Turing devised for speeding up the breaking of subsequent messages from the same day; this was a way, once the wheel order, ring setting, and stecker for the day had been found from the first bombe run, to recover additional message settings without having to do a full bombe run with a new text crib. And third—the ultimate goal—was to break the bigram tables themselves. Turing estimated that with two hundred messages from one day, this might be possible. If the bigram tables could be cracked, it still would not be possible to read the message-setting indicator for each message directly—that would first require finding the day's *Grundstellung*. But with the bigram tables it would be possible to recover each day's *Grundstellung* by running a special short bombe menu after perhaps only a dozen messages had been broken. Equally important, the bigram tables would make it possible to employ yet another intricate scheme of Turing's, one that could in theory vastly speed the breaking of each day's first message, the bottleneck in the whole system.

This last idea, which came to Turing in a single night, was a way to eliminate from consideration many possible wheel orders even before a single message was read. With the eight rotors that were eventually in use in the naval Enigma, there were $8 \times 7 \times 6 = 336$ possible wheel orders. Each would have to be run separately on the bombe. A run took at least half an hour, when the time required to mount the rotors and plug all the jacks was counted. That meant that running all possible wheel orders could easily take over 150 hours, a week of full-time operation. Turing's method, which became known as "Banburismus," used punched sheets to place messages in depth and brilliantly exploited a subtle quirk in the naval indicator system. In theory it could reduce the number of wheel orders that required testing from 336 to 112, or even 42, or with a lot of luck, 6. (Banburismus is described in detail in Appendix B.)

The complexity of the entire program was daunting, even overwhelming. Turing acknowledged he was not even sure it was practical. But at least there was now a map to follow, even if it pointed to a way that led through some murky swamps.

While waiting for the first bombe to arrive, Turing, aided by Twinn and two girls, started in early 1940 trying to break some traffic using the hand crib method. A German prisoner interrogated in November 1939 had revealed that numbers were now spelled out in full in Enigma messages, and Turing and his staff spent two weeks trying to apply the FORTY WEEPY method with these new cribs to traffic from November 1938—a period when there were still only six steckers used at a time. At last they managed to break five days. To break subsequent messages for these days—the second key to Turing's grand program—he invented what came to be known as the "EINS" catalogue. Breaking the first message gave the wheel order, stecker, and ring setting for the day. Subsequent messages for the same day shared those same basic settings but had of course a different starting position of the rotors. There were 17,576 positions possible, a large but hardly astronomical number, and Turing saw that a brute force method was not beyond the realm of feasibility. In reading the 1938 traffic, Turing discovered that the word *eins* occurred in about 90 percent of all messages. So one of the modified Typex machines was set to the day's settings and the letter E enciphered at all 17,576 possible positions; the process was then repeated for I, N, and S. A catalogue of all possible encipherments of EINS was then assembled from the strips of paper that the Typex had printed out. The next step was to search, by eye, through the cipher text of each message for an appearance of any of the four-letter groups listed in the catalogue. A statistical calculation had determined that on average a message would contain four such hits, three of which were due to chance, but one of which corresponded to an actual *eins* enciphered at that actual rotor setting.

When the first bombe arrived in March 1940, it was at once put to work running jobs for Turing's group in Hut 8, but progress proved almost as tedious by machine as it had been by hand. Bombe No. 1, unimaginatively christened "Victory" (later arrivals received more creative and facetious names, including "Otto," "Eureka," and "Agnus Dei" or "Aggie" for short), was an ungainly and temperamental beast. Six and a half feet high, seven feet wide, weighing a ton, and perpetually leaking machine oil, it contained thirty Enigma machines driven from a common shaft. The rotors were not positioned side by side as in an actual Enigma, but rather were individually mounted flat against the face of the cabinet:

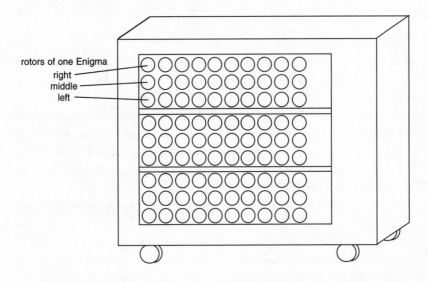

rotors of one Enigma
right
middle
left

This configuration made it easier to take the rotors on and off when setting up a new wheel order and also simplified the mechanics of driving the rotors in tandem. As the rotors spun, four concentric circles of twenty-six wire brushes on each rotor's face contacted four corresponding circles of fixed brass contacts on the cabinet face. Four circles were needed because the logic of the bombe circuitry required that the input and output of each Enigma be kept separate. As the current entered each rotor for the first time—traveling what would be the right-to-left path in a real Enigma—it entered via one circle of contacts and exited via a second; on the return path, after passing through the reflector and now traveling "left" to "right," the current was directed into each rotor's third circle of contacts and exited via a fourth. The internal wiring in each rotor connected the first and second circles just as the left and right faces of an actual Enigma rotor were connected, then duplicated those connections between the third and fourth circles.

To form the electrical contacts between the rotors and to duplicate the effect of the reflector, about ten miles of wire snaked through the back of the cabinet, joining the fixed contacts together with tens of thousands of soldered connections. Another tangle of movable, plugged cables joined the Enigmas together in the manner prescribed by the menu for a given run.

The bombe operators remembered the wire brushes on the rotors with a particular horror: Before a run each brush had to be meticulously preened with tweezers to ensure it did not spread out and short the contacts. Another enduring memory for operators was the sound of a spindle "being whacked with a strictly forbidden weapon" to remove a rotor that had become stuck in position. It was a rough and not very promising beginning.

———

Bombe No. 1 was barely running when the German attack on Norway came; Turing and his Hut 8 group had to drop work on the naval Enigma temporarily while they were pressed into service to help with the rush of Army and Luftwaffe traffic. Their initial results had been extremely discouraging, however. Running a single crib in all wheel orders took a week. Hut 8 was working 10:00 A.M. to 6:00 P.M.; the bombe would be run all night and if it detected a stop, it would be ready for testing by the Hut 8 staff when they arrived. Most of the time the bombe did not stop.

Taking a week to discover that the crib was wrong or mispositioned was an agonizingly inefficient process. It was nothing but a series of vicious circles. To get good cribs, one had to read enough traffic to know the habits of Enigma operators and the contents of typical messages, but the only way to read the traffic was to have good cribs. To speed up the bombe operations, one had to have the bigram tables so that Banburismus could be employed, but the only way to get the bigram tables was by first breaking several hundred messages a day, which required efficient bombe procedures.

One ray of hope glimmered briefly on April 26, 1940, when another "pinch" had taken place at sea. The British destroyer *Griffin* was ordered to search for survivors from a German trawler that had been sunk off Narvik that morning in an encounter with another British warship. The trawler had been disguised as a Dutch vessel; it was flying the Dutch flag and had the word HOLLAND painted on its side; when ordered to stop it had suddenly run up the German ensign and rammed the British ship. At 10:15 A.M. *Griffin* spotted another trawler similarly costumed. This trawler stopped when ordered and a boarding party put out in the ship's boats in rough and choppy seas. The trawler's captain and fifteen men were all dressed as fisherman, but it was obviously another German patrol boat. On the side it bore the name *Polares;* in reality it was the armed German trawler VP 26. A prize crew was sent over in another boat to take charge, and the German crew was warned that if they tried to scuttle the ship with explosives the survivors would not be picked up.

While the prize crew was rowing over, they noticed a canvas bag floating nearby. The bag was on the verge of sinking, so F. H. W. T. Foord, a ship's gunner, jumped in holding a line and grabbed hold of the bag. As the boat's crew tried to haul him back in, the line broke and Foord went under. The crew threw him another line. Still clutching the bag, Foord grabbed the line with one hand, lost his grip, and went under again. On the third try he managed to get the line over his shoulders and was hauled to safety. In the canvas bag and among other papers scattered on deck were the naval Enigma settings for April 23 and 24, plus the operator's log, which gave the complete, matching plain text and cipher text for messages sent April 25 and 26.

The papers arrived at Bletchley a few days later. Hut 8 immediately set to work and was able to break a message for April 23 on May 11, and for April 24 within another day or two. Thanks to the Germans' practice of using the same wheel order for two days running, solving "paired" days required only recovering a new stecker and message setting, a somewhat simpler problem, and a message for April 25, which was paired with April 24, was also soon read. After several weeks of futile attempts to break traffic for April 26 on the bombe using the cribs obtained from the operator's log, the bombe finally scored its first solid triumph in late June when Turing's crew tried rejiggering the menu. April 27, which was paired with the twenty-sixth, was also broken.

But the ray of hope quickly faded. More messages were "EINSed" out for the six days that had been broken, and all hands in Hut 8 went to work building up the bigram tables. But different tables were used on different days, nine different ones in all. The twenty-second and the twenty-seventh shared the same table, and enough progress was made at bigram recovery to attempt Banburismus on other days that used this same table, but months of frustration ensued as the method produced "depressingly few results."

It had meanwhile come out that the capture of VP 26 had possibly been a huge opportunity lost. The captain of the *Griffin* admitted that he had permitted members of his crew to "take away souvenirs," and when the German ship reached Scapa Flow under its prize crew it was a shambles. Papers, including some cipher forms, were strewn about the deck. One was later found under a mine. The *Griffin*'s captain was sent a reprimand and ordered to return two pairs of binoculars that he had claimed for himself.

Again, the lack of cribs was the fundamental obstacle. The Naval Section at B.P. peppered Turing with a steady stream of suggestions of words to try, but the Hut 8 cryptanalysts had their own ideas, and as the months passed with no progress an atmosphere of ill feeling between the two groups festered until it erupted into something approaching a state of war. Hut 8 began making sardonic jokes about "Hinsley's *certain* cribs." Frank Birch, the head of the Naval Section, shot back with a memorandum to Travis on August 21:

Turing and Twinn are like people waiting for a miracle, without believing in miracles. . . . We supply [cribs], we know the degree of reliability, the alternative letterings, etc. and I am confident if they
were tried out systematically, they would work. Turing and Twinn
are brilliant, but like many brilliant people, they are not practical.
They are untidy, they lose things, they can't copy out right, and they
dither between theory and cribbing. . . . Of the cribs we supply,
some are tried out partially, some not at all, and one, at least, was
copied out wrong before being put on the machine. Sometimes we
produce a crib of 90% certainty. Turing and Twinn insist on adding
another word of less than 50% probability, because that reduces the
number of answers and makes the results quicker. Quicker, my
foot! It hasn't produced any result at all so far. The "slower"
method might have won the war by now.

   Adding to the friction was what the postwar history of Hut 8 referred to
as Turing's "almost total inability to make himself understood." For a crib to
work it was not enough for it to contain words that actually appeared in an
Enigma message; it also had to generate the right mathematical characteristics to yield an efficient bombe menu. That was what Birch failed to grasp,
even as he correctly railed against the way Turing and Twinn tinkered with
cribs that were based on a sound understanding of German naval procedure
and usage. Hut 8 did manage to explain one thing to the Naval Section: If
they could get hold of a month's worth of daily settings plus the bigram tables, that would get them over the hump once and for all. They could then
build up a good supply of reliable cribs and regularly recover the daily naval
Enigma settings thereafter. All sorts of plots began to be hatched. In early
September Knox suggested sending out a bogus German Enigma message
asking for the daily key to be transmitted. That idea was rejected as too
risky. But then, displaying the talents he would later exercise so brilliantly as
the author of the James Bond novels, Lieutenant Commander Ian Fleming
of the Naval Intelligence Division sent a note to Admiral Godfrey:

   D.N.I.

      I suggest we obtain the loot by the following means:

   1. Obtain from Air Ministry an air-worthy German bomber.
   2. Pick a tough crew of five, including a pilot, W/T operator and
      word-perfect German speaker. Dress them in German Air Force
      uniform, add blood and bandages to suit.

3. Crash plane in the Channel after making S.O.S. to rescue service in P/L.
4. Once aboard rescue boat, shoot German crew, dump overboard, bring rescue boat back to English port.

F. 12.9.40

The scheme was totally hare-brained, but desperate circumstances make for desperate action. Fleming was told to draw up a more detailed plan, which he did with obvious relish: The bomber would take off shortly before dawn on the tail of one of the big Luftwaffe raids against London. After picking a suitable rescue boat, the bomber crew would cut one engine, light a smoke candle in the tail, and ditch in the Channel. The pilot would need to be "tough, bachelor, able to swim." The requirements for the German speaker were "as for pilot." Fleming nominated himself for that part. In case the crew was captured and shot as francs-tireurs, the story would be put out that "it was done for a lark by a group of young hot-heads who thought the war was too tame and wanted to have a go at the Germans. They had stolen the plane and equipment and had expected to get in trouble when they got back. This will prevent suspicions that the party was after more valuable booty than a rescue boat."

Birch considered it a "very ingenious plot" and especially liked Fleming's cover story, which offered "the *enormous* advantage of not giving anything away if it fails." As unlikely as it seemed, the plan was approved and a bomber actually furnished, and Fleming recruited a crew and went to Dover to wait his chance. None came. Reconnaissance flights and radio monitoring failed to turn up a suitable rescue boat. On October 16 the mission, which had been theatrically dubbed "Operation RUTHLESS," was canceled. Birch reported that "Turing and Twinn came to me like undertakers cheated of a nice corpse" when they heard the news.

Nineteen forty came and went, and the naval Enigma remained unsolved.

The impossible problem that the U.S. Army's Signal Intelligence Service faced in 1940 was the Japanese Purple machine. After a year of work they had little to show for it.

Friedman was still operating on a shoestring, but at least it was a bigger shoestring. On September 8, 1939, Roosevelt had responded to the outbreak of war in Europe with a declaration of a "limited national emergency." It was mainly a rhetorical gesture: Most Americans felt no sense of emergency, unless it was to make sure that America did not get dragged into another war in Europe. In a 1937 opinion poll, 95 percent of those asked had said that if another world war broke out America should stay out. After the German attack on Poland, opinion shifted ever so slightly; a third of Americans now favored some "help" to Britain, France, and Poland. But an equal number were flatly opposed. And even among those who wanted to help the victims of the Nazi aggression, a large majority ruled out military assistance, then or ever. Roosevelt knew that budging America from her traditional isolationism would be a long and slow process of education, and he inched forward with a caution that maddened many in his inner circle but reflected his keener grasp of politics. Fewer than 10 percent of Americans said they would fight if the United States were not herself invaded by foreign troops; to press for more now would set off a backlash, if not a revolt. FDR's plan was to sway the middle third of voters whose sympathies lay with the British and French but who feared American involvement; to amend the 1935 Neutrality Act, which required the United States to maintain an arms embargo against both sides in any conflict; and to do *something* about the nation's still-comical state of military preparedness. "This nation will remain a neutral nation, but I cannot ask that every American remain neutral in thought as well," FDR said in a radio address on the evening of September 1, 1939. "Even a neutral has a right to take account of facts. Even a neutral cannot be asked to close his mind or his conscience." One fact that Roosevelt sought to plant in the minds of the middle third whose support he was courting was that isolationism was simply not possible in the twentieth century. America's first line of defense lay in Europe: "You must master at the outset the simple but unalterable fact in modern foreign relations between nations. When peace has been broken anywhere, the peace of all countries is in danger. It is easy for you and me to shrug our shoulders and to say that conflicts taking place thousands of miles from the continental United States, and, indeed, thousands of miles from the whole American Hemisphere, do not seriously affect the Americas—and that all the United States has to do is ignore them and go about its own business. Passionately though we may desire detachment, we are forced to realize that every word that comes through the air, every ship that sails the sea, every battle that is fought, does affect the American future."

Roosevelt had tried and failed to have the Neutrality Act's arms embargo provision revoked during the summer. Vice President John Nance Garner had polled the Senate leadership in July and given Roosevelt the results: "Well, Captain, we may as well face the facts. You haven't got the votes, and that's all there is to it." The House approved a small increase in the Army's budget, which would allow the purchase of five hundred additional airplanes. What the Army really needed was tens of thousands of airplanes, plus tanks, guns, and a few million men. What it had was 227,000 soldiers and rifles for 75,000. *Time* magazine reported that the U.S. Army looked like "a few nice boys with BB guns." Mobilization or a draft was out of the question; direct military aid to Hitler's victims was out of the question, too. But now that war had come there was hope at least of lifting the embargo, and so allowing Britain to *buy* goods from the United States and carry them across the Atlantic on British ships. Congress now was also willing to put some more money into modernizing and equipping the small professional Army and Navy.

Three days after the "limited national emergency" went into effect, the Army's Signal Intelligence Service put in its bid for some of the new funds that were soon expected. Friedman asked for a budget increase of $58,950 to hire twenty-six additional civilian employees and for eight thousand square feet of office space on the third floor of Wing 3 of the Munitions Building. Within three weeks the plan was approved. But even a "limited national emergency" could not budge Civil Service rules. Given the urgency and the secrecy of the work, Friedman had asked that he be allowed to hire "qualified civilian specialists" directly, without regard to Civil Service eligibility ("for example, members of the American Cryptogram Association"), but the request was turned down. That meant that only someone already employed by the U.S. government, or someone who had passed a Civil Service examination and was on a list of approved job candidates, could be considered. That was one reason that SIS never was able to emulate GC&CS in hiring "men of the professor type," at least until America was actually in the war. Even if the money had been available to hire a lot of top-rank professionals, which it wasn't, SIS simply had no legal authority to call up an MIT mathematician and offer him a job.

The SIS had grown at a crawl over the previous decade. Friedman had worked out a whole rank system; after six years on the job, the three "junior cryptanalysts" Sinkov, Rowlett, and Kullback were promoted in June 1936 to "assistant cryptanalyst" at a salary of $2,600; in July 1938 he had been able to give the three another raise to the rank of "associate cryptanalyst" at $3,200 a year. From there one could in theory ascend to cryptanalyst, senior cryptanalyst, principal cryptanalyst, and head cryptanalyst. But it was mostly a bureaucratic exercise; the new hires Friedman had been able to make since 1930 were virtually all "cryptographic clerks" at the lowest entry level on the Civil Service schedule, with a salary of $1,440. And even those new hires were precious few. When newly hired cryptographic clerk Samuel Snyder arrived at the Munitions Building in 1936 he recalled signing a log sheet that identified him as employee No. 10 of the Signal Intelligence Service.

But Friedman's ability to nurture talent, and the more or less formal training program he had begun to establish out of sheer necessity, continued to pay off even in the lean years. One of the more remarkable hires from the early period was a young man named John B. Hurt, who came from the same corner of Virginia as Frank Rowlett. Hurt was the nephew of a congressman and got the job more or less by pull. But he had precisely the sort of surprising mind suited to the job. He had taught himself Japanese without ever having set foot in Japan and without ever having formally taken a course in the language, picking it up first from a neighbor who had been a missionary in Japan and then from Japanese roommates at the University of Virginia. (Hurt had attended several other colleges too, but never quite managed to graduate from any of them.) Friedman, who was sold on the idea of hiring mathematicians, had hoped to find someone who knew both Japanese and mathematics, but decided to give Hurt a chance. Hurt in fact "had a problem with mathematics," as his colleagues later recalled. He had no patience with the formal analytical procedures such as frequency distribution counts, and would instead just plunge right into a cipher problem. Yet he obviously had something of the cryptanalyst's mind and he had a phenomenal ability to recognize linguistic patterns. Once, SIS received a coded message that had been intercepted from a prisoner in Columbus, Ohio. It had been written in made-up symbols and Rowlett, Sinkov, and Kullback all sat down and dutifully began making frequency counts of the letters. Hurt wandered over, peered over their shoulders, and almost at once began reading the text of the message: *Dearest Sweetheart Sarah. . . .* "And of course that's what it was," Kullback recalled. It was a simple monalphabetic cipher and to Hurt the pattern of words had leapt right off the page. (The message, rather anticlimactically, turned out to be nothing but a letter to the prisoner's girlfriend.) Hurt, rather like Dilly Knox, was also defiantly oblivious to worldly considerations such as traffic. "He had a theory that when you crossed a street, you just ignored traffic rules and you don't worry about cars," Kullback said. "You stared at them, as you would an animal." Once Hurt was actually knocked down by a taxi as a result of this practice. The cabman ran over and said, "Are you hurt?" Apparently the temptation to get off a Groucho Marx line was greater than his injury: He got up, brushed himself off, said, "Yes, John B.," and walked away.

The official SIS plan adopted immediately after the onset of the war in Europe added Germany and Italy to the list of countries whose radio messages were to be studied. Friedman was well aware how behind his team was in these areas. They had focused almost exclusively on diplomatic traffic, and mainly Japanese; the military signals of Japan, Italy, and Germany for the most part could not even be picked up from U.S. monitoring stations. German diplomatic traffic seemed impenetrable and unpromising. A huge increase in cryptanalysts, linguists, and intercept operators was obviously needed.

Among Friedman's first hires with his $58,950 windfall were four senior civil servants from other agencies. Strikingly, they were two men and two women. At a time when GC&CS was hiring even university-trained "girls" as mathematicians and linguists at £2 a week—about $400 a year—Friedman was offering jobs at the "PS-1" rank to women. This was the first rung of the professional category of the Civil Service, and it paid $2,000 a year, five times what the British thought suitable for women to earn. Friedman's enlightened views about the capabilities of women were no doubt influenced in part by his own household. He had met his future wife, Elizebeth Smith, when the two worked together at George Fabyan's "Department of Ciphers." They were married in 1917, and Elizebeth Friedman went on to become a senior cryptanalyst in the U.S. Treasury Department, where she broke codes used by drug runners and bootleggers. Friedman found his four top-rank hires—Genevieve Grotjan, Delia Taylor (who later married Abe Sinkov), Frank Lewis, and Al Small—by putting out an urgent call to the Civil Service offices. It was as scattershot a process as ever, and it was pure chance that brought each of these men and women, all of whom were destined to make significant contributions to American cryptanalytical successes, to SIS.

Lewis, who had held twelve different government jobs over the previous five years—he had been private secretary to officials at several New Deal public works and housing agencies, headed an office that received designs for new post offices, and handled payroll for the night crew that was building the planned community of Greenbelt, Maryland—was now at the main Civil Service office in Washington handling "dead claims," payments to spouses of retirees who had died. He was bored to tears. He had reorganized the office files and done everything else he could think of, and finally one day he stormed into the Civil Service's personnel office and began hollering at the man behind the desk: "Do you know what you're doing? I'm being paid a good salary and don't really have a job!" The next day the man behind the desk summoned him back. Several interviewers were present. Did he play bridge? Yes. Chess? Kind of. Did he like crossword puzzles? Yes, especially the "cryptic" kind—even invented his own. Would he like to go down to the Munitions Building and meet a Colonel Friedman? Why not.

Friedman gave Lewis a quick trial by fire, a week's worth of problems from his course to work on at home. At the end of the week he was offered the job and was assigned to the German section that was being formed under Kullback. For most of the next two decades Lewis would serve as Kullback's deputy.

Even as SIS scrambled to build up its German and Italian sections, its number-one priority remained the unsolved Purple machine. The Red machine continued to be used on some Japanese diplomatic circuits even after Purple made its appearance on February 20, 1939, and that provided a wealth of very long cribs when identical texts were sent in both systems. Additional cribs were provided by the Japanese practice of spelling out a message number at the start of each dispatch; by keeping careful logs of every message sent by each station, it was possible to predict the first few words of each message this way. It was immediately apparent that while vowels no longer stood for vowels and consonants for consonants, as had been the case with the first Red traffic, Purple still treated six letters differently from the other twenty. This was evident from the fact that six letters in each message appeared with a different frequency from the others. (In a good polyalphabetic encipherment each cipher letter appears with about equal frequency in the cipher text. The separation of the alphabet into two separate sets meant that the letters of one set appeared with a frequency equal to the average plain-text frequency of all the letters in that set, which by chance was likely to differ from that of the letters in the other set. Appendix C contains a fuller explanation of the cryptanalysis of Purple.)

Using the abundant cribs, Rowlett and his coworkers soon cracked the "sixes." The result was a 6 × 25 substitution tableau. One thing was certain: The Purple machine did not use rotors. The twenty-five different ways that the six letters were sequentially shuffled among themselves followed no systematic pattern the way the Red machine had. Still, recovering the shuffling pattern was a great step forward; a pencil-and-paper model was quickly built, and just as for Red, it was a fairly easy matter, by using frequency counts of the cipher text letters, to figure out which six letters of the alphabet were each plugged into which of the "sixes" each day.

The paper-and-pencil method was clumsy and slow, and Rowlett was hoping to develop some way to automate the task using IBM tabulating machines and punch cards. The catch was that while it was possible to diagram the shuffling sequence on paper, it was all far too complex for the IBM machines to handle with punch-card methods alone. What was needed was a piece of actual cryptological hardware that could be plugged into the IBM machines. This black box would connect each of the six input letters to six output letters in the twenty-five different patterns required. But even with standard relays and switches there was no obvious way to reproduce the Purple machine's complex scrambling pattern. Various schemes were tried and rejected, and finally the task was handed over to Leo Rosen, an MIT electrical engineering major who had joined ROTC in college and had just been called up for active duty. In early September 1939 Rosen was flipping through an electrical supply catalogue when his eye fell on a gizmo that he thought would do the trick. It was a stepping switch, or "uniselector," designed for automatic telephone exchanges, and Rosen saw that it almost seemed made for the job: It was six switches ganged together, each of which rotated through twenty-five contacts. It also ran on 110 volts DC, which just by chance happened to be what the Munitions Building had as its power supply—the building shared its power with the District of Columbia streetcars, which ran on direct current, not the alternating current more commonly provided by commercial utilities. For once urgency overcame bureaucracy; Colonel Spencer B. Akin, Friedman's superior in the Signal Corps, picked up the phone, called the manufacturer in Chicago, and arranged for two of the switches to be delivered by air immediately.

There was a simple reason that the stepping switches seemed so admirably suited to the job: They were precisely what the Japanese had used in the Purple machine, though neither Rowlett nor Rosen had any way to know it at the time, nor would they know for sure until the war was over and a fragment of an actual Purple machine was captured with a bank of its stepping switches intact.

The "six-buster," as they dubbed Rosen's device, worked admirably. But meanwhile no progress was being made on the much more difficult problem of the "twenties." Samuel Snyder, Al Small, and Genevieve Grotjan had joined the work on the problem, and the Purple team spent a year poring over endless streams of traffic. On September 20, 1940, almost exactly a year after Rosen's chance discovery of telephone switches in an electrical catalogue, Genevieve Grotjan discovered something, too. With the large quantity of matching plain and cipher text they had to work from, the Purple team had already been able to reconstruct the cipher alphabets for the twenties—that is, the patterns in which the letters at successive key positions were shuffled with one another. But there were thousands upon thousands of these patterns, and nothing seemed to link one to another. Grotjan was poring over a worksheet of these cipher alphabets one day when suddenly she noticed there *was* a pattern to the patterns. They formed cycles twenty-five key positions long that were related to one another in a systematic fashion that unmistakably showed that they had been generated by a cascade of scrambling units. (Appendix C more fully describes Grotjan's discovery.) Instead of the single scrambler that shuffled the alphabets of the sixes, a series of three scramblers, each of which fed into the next much like the sequence of rotors in the Enigma, was being used. The pattern of cipher alphabets also showed that any of the three scramblers could be selected to act as the "fast" rotor. But though a complication in one sense, this shifting of the fast rotor also meant that far more data was available to tease apart the wiring of each scrambler.

Grotjan rushed in to show her results to Rowlett, and soon the commotion and shouting drew everyone in the section. The more cosmopolitan scientists who came from around the world to work on America's Manhattan Project celebrated at times like this with a drink of the distinctly alcoholic variety. Friedman's band of very middle Americans sent out for a round of Cokes.

———

There were still weeks of work ahead to recover the actual substitution tables for each of the three scrambler stages. The un-air-conditioned Munitions Building was never a pleasant place to work in a late Washington summer, but now it became sheer torment. The War Department had nowhere in Washington to expand but up, and workmen were adding a fourth floor to the Munitions Building, one wing at a time. They chose this moment to begin work on the wing where the Purple team occupied the third floor. From morning to night the offices shook with the banging of hammers; when the cranes dropped heavy loads on the roof directly overhead, the whole wing would tremble. Worst of all were the jackhammers, which were so loud that the code breakers had to shout to one another to make themselves heard. Closing the windows cut down the noise but turned the offices into a steam bath. And the work required absolute concentration; even a single mistake in the substitution tables could throw off the entire results.

But after three weeks the job was done. Rosen wired up additional uniselectors according to the tables and incorporated a keyboard raided from a prototype SIGABA code machine that had been lying on a bench in the section. Wiring the uniselectors together was a huge mechanical task that again required absolute accuracy. Rosen and Rowlett took turns with the Signal Corps enlisted men who had been sent over to do the soldering. Each of the "twenties" banks was connected to the next with 500 wires; altogether there were thousands of soldered connections to be made.

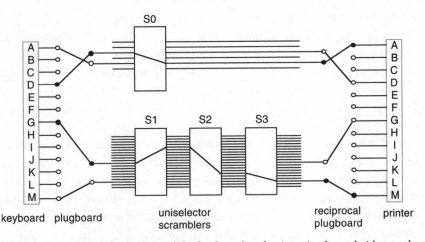

The Purple machine. Each letter of the keyboard and printer is plugged either to the "sixes" circuit (uniselector SO) or the "twenties" (the cascade of three uniselectors, S1, S2, and S3). The schematic traces the paths followed by one letter through each channel, at one position of the machine.

It was one thing to break a code in theory, but there came a terrible moment of truth when, late one night, Rosen and Rowlett soldered the last connection, plugged in the power supply, and flipped the main switch. Rowlett sat down at the keyboard and began to type in the cipher text of a Purple message he had selected earlier that day for the test. The two cryptanalysts watched in awe as deciphered Japanese text began to emerge from the printer, then went home, exhausted and triumphant.

Finding the correct starting point of the four uniselectors for each message was made easier by the discovery that the Japanese apparently used a key list that contained only 240 combinations (out of the $25 \times 25 \times 25 \times 25 = 390,625$ possible positions). Cribbing stereotyped serial numbers and punctuation into the first twenty-five letters of message text helped to recover the keys. Likewise, of the $26 \times 25 \times 24 \times 23 \times \ldots \times 3 \times 2 \times 1 = 400$ million million million million possible pluggings, it was found that only about 1,000 were ever used. Once these lists had been reconstructed, the task of daily decryption became far easier.

Almost immediately the flow of decrypts, code named MAGIC, began. Success reawakened the old bureaucratic war with the Navy almost immediately, too. The two services had just spent the last two months, July and August 1940, wrangling inconclusively over the division of diplomatic traffic. Safford, at OP-20-G, proposed that the Navy should take charge of intercepting and decrypting all Japanese and Russian diplomatic communications, and the Army take all other countries. But Safford also quite sensibly said that it ought to be one or the other; if the Army would not agree to turning all Japanese diplomatic work over to the Navy, then the Navy should turn it over to the Army. But that solution was not acceptable to the veteran turf warriors in the higher ranks. The break of Purple only compounded those feelings. It was now inconceivable for the Army to bow out; likewise the Navy was not going to give up a chance to have so important a piece of the action.

But there were sound practical reasons, as well, that made it difficult to effect a simple division of labor. The two services had each built their own networks of monitoring stations without any coordination with the other; that was a historical fact by 1940. The choice of where to build a station was dictated as much by where Army and Navy bases happened to be located as by any rational scheme to maximize the coverage of the airwaves. Yet no one location could pick up all Japanese traffic anyway. The Army's station at the Presidio in San Francisco and the Navy's at Bainbridge Island, Washington, were both able to intercept the signals between Tokyo and the United States. But the best sites for listening in on other Tokyo circuits were almost evenly divided between the Army stations at Fort Monmouth, New Jersey; Fort Sam Houston, Texas; Fort Shafter, Hawaii; and Corozel in the Canal Zone and the Navy stations at Winter Harbor, Maine, and Heeia, Hawaii.

The final deal the services cut on MAGIC was the oddest of compromises. Although a formal plan on divvying up intercept to avoid unnecessary duplication was signed in October, the crux of the dispute—how to handle decryption and distribution of the results—was left to an unwritten understanding: The Navy would decipher Purple on odd-numbered days, the Army on even days. The two services would alternate by month the responsibility for circulating important intelligence from the decrypts to other departments, including the White House and the State Department. Neither OP-20-G nor SIS was happy with the arrangement, which was evidently forced on them by higher-ups. The cryptanalysts from both services would share results and occasionally phone one another, but rarely met.

One bureaucratic revolution had occurred, though. In late June 1940, as Roosevelt was still coyly deprecating the idea that he would seek an unprecedented third term as President, he had pulled off a political master stroke. On the eve of the Republican convention he announced the appointment of two eminent Republicans to his Cabinet. The new Secretary of the Navy would be Frank Knox, who had actually been the Republican vice presidential nominee just four years before. The new Secretary of War would be Henry L. Stimson, seventy-three years old, a strong opponent of isolationism—and the very man who as Secretary of State had shut down the Black Chamber a decade earlier. The Republicans denounced FDR's appointments as "dirty politics." The SIS liked it even less, and General Mauborgne at once went to see FDR's military aide, the garrulous Major General Edwin M. "Pa" Watson, to explain his concerns about the new secretary; if necessary he wished to speak to the President directly to make sure that SIS would be allowed to continue its work. Reassuring word soon came back: Stimson need not be told about the new Black Chamber that was now under his direct authority.

As it happened Mauborgne needn't have worried. A parade of brass came by to see the Purple machine that the wizards of SIS had built, the new Chief of Staff George C. Marshall among them. Two weeks later Marshall asked if he could bring the Secretary of War over to see the show. Stimson came, was duly impressed, and expressed nothing but satisfaction with the work. As Stimson later explained, he believed the world was in an entirely different situation in 1940 from where it had been in 1929. Europe was at war, America on the verge. In any event, Stimson insisted, his opposition to the Black Chamber had been a narrow and principled one; he understood the importance of intelligence gathering, but simply saw it as fundamentally dishonest for the State Department to be the ones to do it, for diplomats to be spying directly on the diplomats of other nations at the very time they were attempting to negotiate with them in good faith.

There was one other telling footnote. One of the first things that both Mauborgne and Marshall had asked when shown how the Purple replica could crack the most secure Japanese diplomatic messages was, If we can do this to the Japanese, can the Japanese or other countries do this to us? It underscored a striking difference in attitude from what America's soon-to-be enemies would assume about the security of their own codes.

The legality of intercepting foreign communications had never really been resolved, but there was no doubt at all about the complete illegality of one thing the SIS had begun to do under the greatest of secrecy earlier that year. On January 24, 1940, David Sarnoff, president of the Radio Corporation of America, received a letter from the Adjutant General of the U.S. Army expressing the War Department's desire to have a "technically qualified officer" of the Signal Corps pursue a course of study at RCA for about six months. It was a common arrangement for the technical branches of the military services to send officers on such visits to the companies they did business with, but in this case it was nothing but a convenient cover story. Lieutenant Earle F. Cook, the officer chosen for the assignment, was going to RCA to "study" one thing only: copies of cable messages that passed through the company's New York office. Cook soon realized that it would be more efficient for him to take up his post at RCA's Washington office, which was downtown on Connecticut Avenue just opposite the site of the

Mayflower Hotel. RCA obligingly provided Cook with a secure room, complete with photographic equipment and a separate entrance to the corridor so he could come and go with no one in the office being aware of the fact. Each morning Cook would stroll over and photograph the messages that had passed through the day before, then deliver the film to the Munitions Building. Frank Rowlett remembered seeing Cook's thumbs in the photographs of many messages that came to SIS this way.

Having the actual cable forms eliminated any possibility of the garbling that occurred all too often as messages were copied by radio operators sometimes straining to hear Morse code beeps through a hail of static or "atmospherics." And a certain fraction of commercial telegram traffic actually did travel by cable, under the oceans to Asia or Europe; there was no chance of picking that up by radio. With the coming of war in Europe, that was no longer the case for German and Italian traffic, most all of which now went by commercial radiogram. Most if not all of Japanese traffic went from San Francisco by the radio transmitters operated by RCA and Mackay. But SIS may also have wanted to gather diplomatic traffic from some other, neutral countries for eventual study—including Vichy France, Mexico, Spain, Portugal, and the South American countries. The flagrant illegality of the operation would end only with America's entry into war and the imposition of official cable censorship.

Slowly through the terrible events of 1940 American sympathies began swinging more to the side of Britain, though not without months of confusion, of wanting to have it both ways. The shock of the German attack in the west did clarify most people's view about one thing, though: America needed to be a lot better prepared; it needed to trade in its pop-gun Army for something that would at least command some respect in a world where the only voices that were being heard were the ones that had guns to back them up. On May 16, Roosevelt drove to Capitol Hill and asked for urgently needed improvements to the nation's defense, including fifty thousand new aircraft a year, a staggering number. A month later, as Pétain was capitulating to Hitler, Congress approved a billion-dollar tax increase to pay for it, and a month after that passed a $37 billion defense appropriation to rebuild the Army and Navy with new ships and tanks. Roosevelt kept promising that he would not send American boys to war, but sensing the nation's new mood he began to call more openly for aiding the victims of Axis aggression. In an age when statesmen were expected to speak of even the most perfidious acts in international relations in circumspect euphemisms, Roosevelt began to use direct, blunt language, and Americans responded. "Discretion" told him not to use the phrase "stab in the back" to refer to Italy's belated declaration of war against a defeated France on June 10, but "the old red blood" told him to use it, and use it he did at a commencement speech that afternoon at the University of Virginia.

By the time of the fall of France, two-thirds of Americans were now in favor of the unprecedented step of a peacetime draft. Twenty years of isolationism was eroding quickly under the tidal wave of Nazi force. England's lone stand against Hitler and its desperate fight against the Luftwaffe during the Battle of Britain that summer weakened the isolationist forces still more. Churchill's inspiring words were felt on both sides of the Atlantic. "Even though large tracts of Europe and many old and famous States have fallen or may fall into the grip of the Gestapo and all the odious apparatus of Nazi rule, we shall not flag or fail," Churchill declared to the House of Commons on June 4. "We shall go on to the end . . . we shall defend our island, whatever the cost may be, we shall fight on the beaches, we shall fight on the landing grounds, we shall fight in the fields and in the streets, we shall fight in the hills; we shall never surrender, and even if, which I do not for a moment believe, this island or a large part of it were subjugated and starving, then our Empire beyond the seas, armed and guarded by the British Fleet, would carry on the struggle, until, in God's good time, the New World, with all its power and might, steps forth to the rescue and the liberation of the Old."

As they lost ground in public opinion, the isolationists became more strident than ever. Charles Lindbergh, a national hero and the leading voice in the America First movement, began to blame Jewish ownership of newspapers, radio, and Hollywood studios for pushing America into war. Senator Burton K. Wheeler of Montana said the draft would be Hitler's "greatest and cheapest victory" and would lead to an American dictatorship. America's ambassador to Britain, Joseph P. Kennedy, nurtured an Irish-American dislike of the English and sourly predicted that Britain would shortly fall.

The draft bill came out of committee in September on the very day that newspaper front pages were emblazoned with pictures of London in flames from the bombs dropped by an armada of a thousand German planes. The bill called for training 1.2 million soldiers and 800,000 reserves, and required all men age twenty-one to thirty-five to register, and it was passed at the end of the month. The New World had made a few little steps; Churchill was now determined to do everything he could to make those steps irreversible. Throughout the summer he beseeched Roosevelt in personal telegrams to go beyond "cash and carry" neutrality and supply Britain with urgently needed military assistance. The amendment to the Neutrality Act passed the previous year permitted Britain to buy arms, including surplus stocks in the

American military inventory, but Churchill hoped to get America committed beyond a mere business proposition. For one thing Britain was simply running out of cash: Britain had begun the war with $4.5 billion in dollars, gold, and U.S. investments. A vigorous effort to export Scotch whisky and other luxury goods to the United States had generated another $2 billion. But by the fall of 1940 Britain had already paid out $4.5 billion to American factories for the tanks, planes, and merchant ships it had ordered, and its remaining dollar reserves were mostly locked up in illiquid investments.

Churchill's larger aim, as he later candidly told the House of Commons, was to see the United States and Britain "somewhat mixed up together." Once America committed herself to aiding Britain, the process would become unstoppable: "Like the Mississippi, it just keeps rolling along," he declared. "Let it roll on—full flood, inexorable, irresistible, benignant, to broader lands and better days." A precedent for direct American aid would be the first trickle of that inexorable flood.

The crux of Churchill's campaign was his plea for a loan of fifty First World War–era destroyers. Again, it was both a tactical and a strategic move. Britain needed destroyers; it had lost almost a dozen in one ten-day period that summer, and the threat of a German seaborne invasion loomed menacingly across the Channel. But it needed American commitment still more. Roosevelt, delicately balancing public mood, election-year politics, and a belligerent Senate, knew that he could not get a destroyer deal through Congress. Bypassing Congress under existing law would require a certification that the destroyers were "surplus" and unneeded for the nation's defense, a stretch; the law also was clear that they could not be given or loaned, only sold. But at the end of August, Roosevelt took a leap: The United States announced it was not giving or loaning the ships to Britain, but it would sell them for an in-kind payment, a lease on British naval bases in Canada and the Caribbean that were vital to the defense of the United States. Isolationists screamed again about dictatorship, but a majority of the public thought it was a good deal.

Throughout the summer and fall of 1940 the British tried to come up with more ideas of how to get America further "mixed up," with cashless exchanges and direct contacts between the British and American military staffs. In June, Lord Lothian, the British ambassador to Washington, began proposing confidential military and naval staff discussions, and offered to provide briefings on the effectiveness of RAF bombing and on other weaponry. Roosevelt overrode Secretary of State Cordell Hull's initial rejection of the idea ("Of course, we could not be connected with any exchange of information of that nature," Hull had said at first), and on August 6 a U.S. mission headed by Brigadier General George V. Strong (who would later direct Army intelligence, including the Army's code breaking work), Major General Delos C. Emmons (who commanded the Army Air Corps headquarters in Washington, and would be the Army commander in Hawaii during the Battle of Midway), and Rear Admiral Robert L. Ghormley (the Assistant Chief of Naval Operations and former head of the Navy's planning staff) sailed for England. They were under strict orders to keep their discussions absolutely secret. And while they should seek to learn all they can, they were told: "You must not be understood to commit your government in any manner or to any degree—whatsoever."

The British lavished unexpected gifts of information on their visitors. On August 23 General Strong cabled Marshall that England was "a gold mine" of technical information. The Americans were offered full briefings on the actual performance of weapons and tactics in wartime, including full details of Britain's super-secret ASV, or "anti-surface vessel" radar, used to spot surfaced submarines at night. At one point during the visit Ghormley received a request from the Navy Department: Probe the Admiralty for information about the Japanese Mandated Islands in the Pacific. Ghormley inquired; Admiral Godfrey was noncommittal and said he would have to seek approval of the Admiralty. That afternoon a motorcade pulled up with sirens screaming at the American Embassy in Grosvenor Square. Messengers accompanied by guards marched in and "plunked down on my desk the entire portfolio of the Far East from the British Admiralty and asked me to sign a receipt for it," recalled an aide to Ghormley. The charm offensive was on.

———

The U.S. Navy certainly needed to be charmed if the British were going to get anywhere. By 1940 American sailors might have forgotten the War of 1812, but they still harbored a mistrust and suspicion of the Royal Navy, which could easily become a downright dislike. Admiral Ernest King, who commanded the U.S. Navy's Atlantic "Neutrality Patrol" and would be named Commander in Chief of the U.S. fleet and Chief of Naval Operations following Pearl Harbor, was widely viewed as a dyed-in-the-wool anglophobe. Anything he took to be an air of class superiority would send him into a rage, and sometimes all it took was an English accent. He would pointedly tell British naval officers that Britannia no longer ruled the waves, that it was America that now commanded the most powerful navy in the world. Some of this ill will was a legacy of the First World War, when the Royal Navy had airily assumed a manner of dictating to a junior partner when America had entered the war, even suggesting that the U.S. Navy adopt British standards for guns and ammunition. Some of this, however, was simply King's basic cussedness. Roosevelt remarked that King shaved with a blowtorch and trimmed his toenails with torpedo net cutters; he was tactless with everyone.

The Army had fewer such feelings of mistrust toward the British. But suspicions were almost impossible to avoid in the area of signals intelligence. This was territory where suspicions bordering on paranoia were endemic, and there was little doubt in the minds of the American code breakers that the British were trying to read American codes. So when General Strong urgently cabled from London on September 5, 1940, asking if Washington would approve a technical exchange of information on cryptography with the British, both SIS and OP-20-G reacted cautiously at first. Strong had apparently raised the idea himself with a British staff group, suggesting that the United States might be prepared to offer full information on Japanese diplomatic systems. The British were astonished, but immediately accepted.

Friedman at once saw the advantage, especially if an exchange of intercepts would provide the United States with the actual German, Italian, and Japanese military traffic it currently lacked. He was more careful about the idea of exchanging cryptanalytic research or results, but thought that a strict item-by-item quid pro quo could be beneficial to both sides. Where Friedman drew an absolute line was on divulging anything about the American SIGABA cipher machine; even the existence of the machine was a fact that was off limits. Trust only went so far.

The Navy agreed to almost nothing. Safford opposed any exchange of cryptanalytical results, training materials, or research on general cryptanalytical theory; he would only agree to an exchange of intercept traffic. But after further tugging the Navy at last gave in and in late October agreed that a technical delegation could be sent to Bletchley Park for discussions. President Roosevelt gave the exchange his blessing, and the next day, October 25, General Mauborgne forwarded to the U.S. Army Staff a list of what the SIS proposed to provide the British, including its solution of the Japanese Purple machine. OP-20-G subsequently tried to back out again, but the War Department insisted that a deal was a deal.

It was then GC&CS's turn to begin hemming and hawing. The British signals intelligence service existed in a state of near perfect bureaucratic limbo; it officially answered to the Foreign Office but in reality was a force unto itself, and it was used to viewing its secrets as existing in a class by themselves, too. Denniston and the armed-services intelligence chiefs at once began plotting ways they could make polite noises at the Americans— and at their British masters who were trying so hard to romance the Americans—while keeping their visitors from actually learning too much. Cooperating on Russian and Japanese codes would be possible, Denniston wrote "C" on November 15, but, "As regards German and Italian, any progress we may have made is of such vital importance to us that we cannot agree at once to hand it over unreservedly. We should require to be informed in detail as to the security of such information, after it left our hands. (In this respect we are entitled to recall that the Americans sent over at the end of the last war the now notorious Colonel Yardley for purposes of co-operation. He went so far as to publish the story of his co-operation in book form)."

The heads of military, air, and naval intelligence agreed. When the American "expert" arrives, "steps will be taken to steer him away from our most secret subjects":

> Should this expert make a favourable impression, we could consider opening out on the Italian material, and possibly discuss generally 'Y' work problems as regards Germany, upon which subject their assistance might be valuable. . . . I would add that the matter has been discussed with Sir A[lexander] Cadogan [the Permanent Undersecretary of the Foreign Office], who concurs that we cannot possibly divulge our innermost secrets at this stage, but that if the Americans return to the charge, it might become necessary to refer the question of policy to the Prime Minister.

"Y" referred to the interception and direction finding of enemy signals and to the decryption of low-level tactical and field codes, and most definitely did not include the "innermost secret" of the Enigma.

Friedman was originally slated to make the trip. But in early January the strain of the past two years suddenly overwhelmed him. On January 4, 1941, he was admitted to Walter Reed Army Hospital in Washington suffering from what was termed a nervous breakdown. He spent almost three months there; upon his release he was forced to take an honorable discharge from the Signal Corps reserve, and while he resumed his duties at SIS, and was eventually promoted to director of communications research, the day-to-day management of the organization was no longer his. His role became much more that of a technical adviser, and his superiors were careful not to let him work more than a few hours a day.

In Friedman's place, Abe Sinkov, who had just been called to active duty as a captain in the Signal Corps, and Leo Rosen, then a lieutenant, were given orders to make the trip to England. From the Navy, Lieutenants Robert Weeks and Prescott Currier were assigned to the mission. Secrecy was paramount. They were to wear plain clothes, and even at Bletchley were to be passed off as a visiting Canadian delegation. The Americans insisted that for the safety of the delegation the British provide passage in a warship. On January 24, 1941, the battleship *King George V,* on its maiden voyage, anchored at Annapolis Roads. The ship had brought Britain's new ambassador to the United States, Lord Halifax; the four American cryptanalysts were told to be ready to board her for the return voyage. They arrived at Annapolis in four government cars, their precious cargo—a Purple machine, a kana typewriter, stacks of documents—crated up in four large wooden boxes. The gear was loaded into a motorboat; then they bounced around for several hours in a cold and bumpy sea waiting, as protocol demanded, for Lord Halifax to disembark before they could go aboard. Arriving at Scapa Flow, they were met by two flying boats that had been dispatched to take them south. Unfortunately the crates could not fit through the planes' hatches, and the Americans and their cargo were accordingly shifted to a cruiser, *Neptune,* that was heading for Sheerness. The following noon *Neptune* overtook a coastal convoy; just at that moment a German reconnaissance plane appeared overhead, and twenty minutes later the convoy was attacked. Bombs fell on either side of the ship, and the deck—where the wooden crates had been stowed—was strafed. After the smoke cleared the Americans hastened atop to see what was left of their gear. The wooden crates were pockmarked but otherwise undamaged: The Germans had used antipersonnel ammunition that exploded on contact rather than penetrating.

The delegation was met by two staff cars and a truck, and from that point things proceeded more smoothly. Shenley Park, a huge country house a few miles from Bletchley that was owned by the chairman of the Anglo-Persian Oil Company, was placed at the disposal of the four visiting Americans; it came complete with unimaginable wartime luxuries, including a butler, a cook, three upstairs maids, and a full larder. The Americans were warmly welcomed at Bletchley Park, plied with Denniston's sherry, which he bought in huge casks from the Army & Navy Stores, and shown around the Park. They were also driven to various sites of interest around the country, including intercept stations and radio factories, and then taken to London, put up at the Savoy, and ushered in to see "C" himself, who impressed the Americans by breezily discussing the difficulties of running agents behind enemy lines. On one of these expeditions Currier and Weeks were stopped in a small village by a zealous constable whose suspicions were aroused by the sight of two obvious foreigners in civilian clothes being driven in a War Office staff car. The policeman asked the passengers if they would mind stepping out and accompanying him to the station. At this point their Scottish driver leapt out in a fury to remonstrate. "Ye canna do this!" he shouted. "They're Americans on a secret mission!"

As it happened, the secrecy of their mission was particularly important at that very moment. American isolationists were fighting their final battle in the United States Senate, and like all final battles it was a furious and brutal one. A month after Roosevelt's reelection, in November 1940, Churchill sat down to write what historian James MacGregor Burns called "perhaps the most important letter of his life." "The decision for 1941 lies upon the seas," he told the President. Britain had lost four hundred thousand tons of shipping during the five weeks ending November 3. At any point Vichy France might surrender the French fleet to Hitler, despite its promises to the contrary; that would tip the balance cataclysmically against the Royal Navy. The only hope, Churchill argued, was for American merchant ships to begin carrying war supplies to Britain, guarded by American warships. Britain had also run out of money to pay for what it desperately needed; cash and carry had run its course: "I believe you will agree that it would be wrong in principle and mutually disadvantageous in effect if at the height of this struggle Great Britain were to be divested of all saleable assets, so that after the victory was won with our blood, civilisation saved, and the time gained for the United States to be fully armed against all eventualities, we should stand stripped to the bone."

Churchill's letter arrived in December while the President was relaxing on a ten-day postelection vacation, a Caribbean cruise aboard the cruiser *Tuscaloosa*. A seaplane delivered the letter to the ship; Roosevelt read it with no comment and no obvious reaction. But Harry Hopkins, the President's close aide and confidant, could see that Roosevelt was turning it over in his mind. "Then, one evening, he suddenly came out with it—the whole program," Hopkins later wrote. On December 17 FDR casually told the reporters gathered in his office for their regular press conference that there wasn't "any particular news." He then proceeded to lay out "the whole program." It was what would soon be known to the world as Lend–Lease. America, he explained two weeks later in one of his fireside chats, "must be the great arsenal of democracy." While still insisting that his "sole purpose" was to keep America *out* of war, never before, he said, had American civilization been in such danger. He called upon Congress to provide Britain with whatever it needed for the struggle against Nazism; the British could pay America back when the war was over, and Hitler defeated.

The speech electrified Britain, and most Americans, too. It sent the isolationists into a frenzy. The Women's Neutrality League picketed the British Embassy carrying signs declaring BENEDICT ARNOLD HELPED ENGLAND TOO. Members of a "Mothers Crusade" blockaded the offices of Senator Carter Glass of Virginia, who supported the President. (Glass shot back, "It would be pertinent to inquire whether they are mothers. For the sake of the race, I devoutly hope not.") But the ugliest rhetoric came from Senator Wheeler. Referring to the New Deal agency that had paid farmers to plow under their crops, he said, "The lend–lease–give program is the New Deal's AAA foreign policy. It will plow under every fourth American boy." Wheeler had picked up the line without apparently realizing that it had been penned by George Sylvester Viereck, a U.S. citizen working as a Nazi propagandist for the German Embassy in Washington. Roosevelt called it the rottenest, most dastardly, and most unpatriotic thing that had been said in public life in his generation. Opponent Senator Robert Taft was lower in key and closer to the mark when he observed that "lending" arms was like lending chewing gum: You didn't want it back. The truth was that Lend–Lease was something of a pious fraud. It was a decidedly unneutral act masquerading as neutrality, and that was its brilliance. The bill passed on March 11, 1941; a few hours later the White House sent a list of available items to British officials and asked Congress for $7 billion to pay for them.

The American delegation to B.P. had meanwhile arrived and turned over their gifts to their hosts with still no official decision on the part of the British to reveal anything about the Enigma in return. But on February 26, "C" reported to Churchill that the Chiefs of Staff were now "on balance" in favor of "revealing to our American colleagues the progress which we have made in probing the German Armed Force cryptography." "C" assured the Prime Minister that the discussions would be limited to technical aspects of cryptanalysis only, however, and would not extend to "the results"—that is, actual intelligence derived from Enigma traffic. A handwritten scrawl at the bottom of the memo recorded the Prime Minister's approval: "As proposed. WSC. 27.2."

The agreement to show the Americans the Enigma work came with other strings attached. Weeks wrote out and signed a handwritten commitment to preserve the secrecy of the information, informing "by word of mouth only the head of our section, Commander L. F. Safford, USN." The Army representatives gave a similar pledge to inform only Friedman and Colonel Akin. Although the point was later the subject of considerable confusion and contradictory statements, the Americans apparently were not permitted to make written notes about the Enigma work. The British agenda was clear: They did not fully trust the ability of Americans to keep a secret; they were willing to share technical know-how; but they were the ones fighting the war and Enigma intelligence was their baby.

The Army's interest in the Enigma at this stage was in any event mainly a matter of preparedness. Knowing how to crack the Enigma was a prudent precaution exactly akin to building tanks and buying airplanes. The only immediate operational value as far as the Army was concerned was in possibly being able to gather intelligence about the activities of German agents in the United States and Latin America.

The Navy was in a slightly different position. It was clear that it would soon be convoying ships in the Atlantic. The Neutrality Patrol was about to become the Atlantic Fleet, with Admiral King as its commander. Its need for operational intelligence was soon to be very real. The Navy accordingly pressed further, asking if Bletchley Park could supply it with the wiring for the Enigma rotors. The answer was a bit vague. What GC&CS would later call a gentleman's agreement was made: The Americans were given a "paper" Enigma that included the rotor wirings, but Weeks and company had to pledge that they would "divulge" the information—presumably to anyone else at OP-20-G or SIS —"only when it is decided to work on the problem," a bit of ambiguity that papered over the fact that the British intended to be the ones to decide when anyone else might be permitted to "work on the problem" in an actual operational sense. Meanwhile, when the United States Navy had a direct "interest" in operational intelligence derived from Enigma traffic, the Admiralty would provide it. At every turn over the months ahead the British would try to forestall American demands for more direct involvement in the Enigma work by emphasizing the need to "avoid duplication."

Where GC&CS *really* wanted to cooperate with the Americans was in the area they desperately needed help, and that was Japan. GC&CS had earlier tried and given up on solving the Purple machine. The British had begun to intercept some Japanese naval traffic at a listening post in Hong Kong in the 1930s, and continuing the work after moving to Singapore in September 1939 the Far East bureau had made some progress on JN-25, the main Japanese naval code. But the British had very few Japanese linguists, and indeed suggested they would hand over all of their cryptanalytic work in the Far East to the United States if the Americans did no more than supply the needed translators. The British gratitude for American help on Japanese codes and the gift of the Purple machine was genuine, and for the moment at least that helped to smooth over the British reticence on the Enigma front.

The four Americans returned after ten weeks bearing crates of documents, an entire naval radio direction-finding unit, and a glow of general good fellowship that would begin to fade almost immediately.

CHAPTER 6

Success Breeds Success

FROM the fall of 1940 to the spring of 1941 the Luftwaffe pounded British cities in nightly raids. In one way it was actually a relief, for it signaled that the invasion threat was waning. The attacks on airfields and radar installations in August and September 1940 that had come within days of crippling the RAF and leaving the island defenseless had abruptly ceased when an errant German bomber accidentally hit London. The RAF had responded by bombing Berlin, and Hitler in a rage had ordered an all-out reprisal against British cities.

One errant bomber dropped its load over Bletchley the night of November 20, 1940; Elmers School took a direct hit, destroying the typists' room and the telephone exchange, and another bomb landed near Hut 4, blowing out a half-dozen panes of glass. Bletchley Park had drawn up a not entirely convincing evacuation plan in the event of a more serious bombing attack or invasion. If German armies landed in England, a mobile unit dubbed the "B.Q. Party" and staffed by members of Huts 8, 6, 4, and 3 would attempt to keep up the work during an actual invasion. Bletchley was provided some decrepit buses that were to be kept at the ready, but they were in such a state of disrepair and were such gas guzzlers that no one believed they would go more than a few miles before giving up for good. Vaguer plans envisioned carrying on the work of Bletchley in Canada or even the United States if Britain were occupied.

The second bombe, which incorporated some important improvements, had arrived on August 8, 1940, and at last Hut 6 had its chance to start solving Air and Army Enigma keys on the machines. But it was still the cillis that were carrying the day. The Luftwaffe was maintaining its bad habits and the Red key had been read currently, almost without interruption, since the new indicator system was overcome with the use of cillis on May 20, 1940.

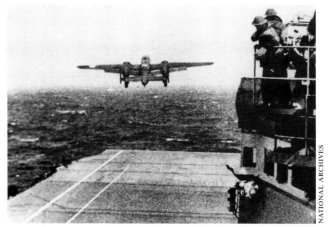

The Doolittle raid: A B-25 lumbers off the deck of USS *Hornet* on the morning of April 18, 1942, headed for Tokyo.

Admiral Isoroku Yamamoto

Midway, under Japanese attack the morning of June 4, 1942. Two of the island's ubiquitous gooney birds are in the foreground.

Station Hypo's chief, Joseph Rochefort

With the closing of the Black Chamber, the War Department was left with a cryptanalytic staff of one: William F. Friedman, shown here in a 1945 photograph.

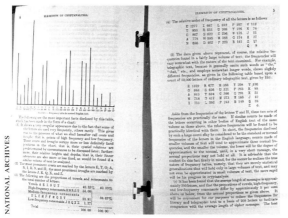

A page from Friedman's training manual, *Elements of Cryptanalysis*

An analogue of the Japanese Purple machine, constructed by the U.S. Army's Signal Intelligence Service

The three-rotor Enigma machine, with a box containing its five interchangeable rotors. The stecker board (with plugged cables) is at the front.

The four-rotor naval Enigma machine. The lid can be swung open to change the rotors; the rotors are then turned so that the letters corresponding to a selected starting position show through the window in the lid.

Alan Turing

One of the "standard" British three-wheel bombes. Ten of the machines, oper-ated by a U.S. Army contingent that arrived in Britain in the summer of 1943, were given the names of American cities. The large plugged cables were used to link the machine's thirty-six Enigma analogues together according to a "menu" derived from matching plain and cipher text.

The mansion at Bletchley Park, before the war

The Colossus computer, developed by mathematician Max Newman's group at Bletchley Park to break the German teletype ciphers.

Antiquated Swordfish torpedo bombers crowd the deck of HMS *Victorious* just before their successful attack on the German battleship *Bismarck* on May 24, 1941.

Admiral Andrew Cunningham, Commander in Chief Mediterranean. A fighting admiral of the anti-intellectual "blue water" school, Cunningham defied the stereotype to become the first commander of the war to score a clear-cut victory through the timely and imaginative use of signals intelligence.

A. G. Denniston, head of the Government Code & Cypher School between the wars. Cryptanalyst William "Nobby" Clarke complained that Denniston was "possibly fit to manage a small sweet shop in the East End," but others appreciated the trust he placed in subordinates.

Sir Edward Travis, who replaced A. G. Denniston as head of Bletchley Park in 1942. Complaints from the British cryptanalysts about chronic shortages of manpower and resources at Bletchley Park prompted the change.

John Tiltman, chief of Bletchley Park's military section, was a master cryptanalyst who achieved crucial breaks in German police ciphers, the Japanese military attaché code, the German teletype ciphers, and Japanese army codes.

British troops burn their supplies as they fall back toward El Alamein, July 1942. General Bernard Montgomery's brilliant use of decrypted German signals at the Battle of Alam al Halfa a month later marked the final turning point of Allied fortunes in the desert campaign.

A tanker explodes after being torpedoed by a U-boat in the Caribbean.

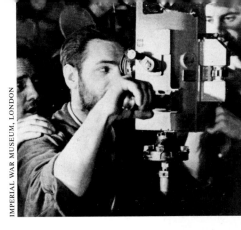

Conditions aboard U-boats were claustrophobic and dangerous. Seventy percent of the men who served aboard German submarines were killed in action.

Grossadmiral Karl Dönitz greets members of a returning U-boat crew.

The Coast Guard cutter *Spencer* drops depth charges on a German U-boat that attempts to break into a large convoy, April 17, 1943.

Arlington Hall, wartime home of the U.S. Army's code-breaking bureau

The staff at Vint Hill Farms, the Army intercept station thirty-five miles west of Washington, near Warrenton, Virginia

Bombe operations in Building 4 of the Naval Communications Annex

The U.S. Navy's four-wheel bombe

SECRET   N-530 BOMBE
SECOND DECK BUILDING 4   MAY 25,194

Women at Arlington Hall, B Building

The CAMEL machine, an IBM attachment used to search for characteristic sequences of code groups in Japanese signals that carried instructions for new code tables

The SC-1, one of the more nightmarish experiments of Arlington Hall's Machine Branch

Racks upon racks of relays that made up the U.S. Army's 003 bombe, also known as "Madame X"

Copperhead was an optical comparator that searched for repeated cipher groups in messages that had been punched onto special paper tape.

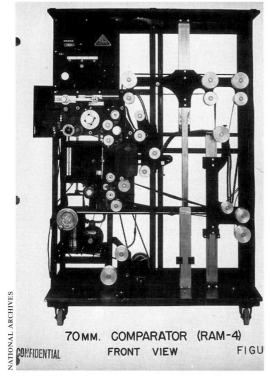

70MM. COMPARATOR (RAM-4)
CONFIDENTIAL   FRONT VIEW   FIGU

The 70mm comparator, one of the optical computers developed by OP-20-G to scan large quantities of messages for repeats

An intercept room at the U.S. Army's monitoring station at Avon Tyrell, England

Direction-finding unit at Two Rock Ranch. The station, forty miles north of San Francisco, became the Army's primary West Coast monitoring site in January 1943.

Arlington Hall, 1944. In the back row are Solomon Kullback (second from left), John Hurt (center, in civilian clothes), and Frank Rowlett and Abe Sinkov (far right); in the front are William Friedman (right) and Preston Corderman (center).

D-Day: American troops come under heavy German machine gun fire as they leave the ramp of their landing craft.

2 MAY 1945 TOP SECRET ULTRA GP W

Ø2Ø1/2/D53   64   FPEX XCBA

TO: ALL U/B'S

ON 3Ø/4 THE FUEHRER DIED IN THE HEROIC BATTLE FOR
GERMANY IN BERLIN WITH HIS MEN.   ACCORDING TO HIS
ORDERS GRAND ADMIRAL DOENITZ HAS BECOME HIS SUCCESSOR.
THE GRAND ADMIRALS' ORDER OF THE DAY WILL FOLLOW.
THE BATTLE FOR OUR PEOPLE IS BEING CARRIED ON.

        ØØ32 Z   42ØØ KCS   Ø5Ø5/4

OP-20-G's decrypt of a German Enigma message sent to U-boats on May 2, 1945, announcing Hitler's death

General Douglas MacArthur signs the Japanese surrender aboard USS *Missouri* in Tokyo Bay, September 2, 1945.

The German Army ran a much tighter operation, however, and even with the arrival of additional bombes their keys were resisting solution. It was a maddening situation that the cryptanalysts of Hut 6 found themselves in. They had achieved a cryptanalytical triumph of the first order; Turing's bombes had actually been built and were actually running; and still they were stuck. A memorandum from Welchman summed up the situation:

> In general the only method which is likely to lead to regular break-
> ing is cribbery, but this method can only be applied after a colour
> [i.e., a key] has been broken by some other methods, and depends
> on the existence of routine messages.
>
> Cribbery is seldom an easy or reliable method. Even when regu-
> lar breaking of a colour has been achieved, the form of the cribs is
> seldom static, and failure to break a few days may result in a com-
> plete stoppage.

It was the same problem that beset naval Enigma: In order to find suit-able cribs one first had to read the traffic to know what words the messages were likely to contain. Another problem was that absolute accuracy in tran-scribing the intercepted messages was essential for the system to work, even with a correct crib. A single error in the cipher text would make the whole bombe menu worthless. A crib could "only be run safely," Welchman con-cluded, "when at least two good independent interceptions of the crib mes-sage are available." The new RAF intercept unit established at Chicksands Priory in Bedfordshire lacked the experience of the Army operators at Chatham, and this did not help the situation.

The traffic that was being broken—the Luftwaffe Red key—dealt mainly with strategic and organizational matters and only occasionally provided in-telligence of operational importance. It provided only the most general in-telligence during the Battle of Britain, such as information on the Luftwaffe order of battle. Another Luftwaffe key, Brown, was broken beginning Sep-tember 2, 1940, and proved to contain information about the navigational beams being used to direct bombing raids against England. But the direc-tion in which the navigational beams themselves were aimed provided as much or more information about the intended targets of the nightly Ger-man raids, and the Brown decrypts were again only of limited value. The false but still repeated story that Churchill knew in advance of the devas-tating raid on Coventry planned for the night of November 14, 1940— and chose to take no countermeasures, such as evacuating the city or deploying antiaircraft batteries, lest the Germans guess the Enigma was compromised—credits Enigma intelligence with far more than it was providing at the time. The German attack, in which 449 bombers dropped 56 tons of incendiaries, 394 tons of high explosives, and 127 land mines, killed 550 people and destroyed 50,000 houses and the fourteenth-century St. Michaels cathedral. Enigma decrypts did give advance warning of *a* planned attack, but the targets mentioned were London and the Farnborough–Maidenhead–Reading area about forty miles to the west of the capital, and the date of the operation was given as November 15, continuing for three successive nights.

The reason for the discrepancy was that Brown was not really an operational or tactical circuit; it linked a Luftwaffe base in France that operated the beams to the radio research center in Germany that was developing the equipment, and the information passed over the circuit about upcoming plans did not always reflect the latest tactical orders. The intelligence that was available to Churchill on the fourteenth, particularly beam bearings detected late in the day, convinced him that London was indeed the target, and he even canceled a planned trip out of town, explaining to an aide when he arrived back at Downing Street that he was not going to spend a peaceful evening in the country while the metropolis was under heavy attack. Churchill was actually scanning the skies impatiently from the roof of the Air Ministry, waiting for the German bombers to appear, when the attack on Coventry began a hundred miles away.

––––––

In September 1940, Mussolini, craving an Italian feat of arms of his own to set beside the Führer's, ordered a reluctant Marshal Rudolfo Graziani to advance with his army of 250,000 men from Libya into Egypt. Graziani's force outnumbered all British forces in the region by better than three to one. But as Churchill concluded after reading inside accounts of the Italian high command that became available after the war, British anxiety over the Italian attack was far surpassed by the Italian marshal's. (The Duce had concluded: "One should only give jobs to people who are looking for at least one promotion. Graziani's only anxiety is to remain a Marshal.") Graziani advanced sixty miles, called a halt, and spent three months building fortifications. The British counterattack that began December 9 was devastating. By February 8, 1941, the British army of 30,000 men had taken half the Italian army prisoner and had reached El Agheila.

The Red traffic had given hints of what would come next. Messages broken around the New Year gave warning of the transfer from Norway to Sicily of the Luftwaffe's Fliegerkorps X, a battle-tested unit of three hundred dive bombers and fighters that had been trained especially to attack shipping. Then a new key, which Hut 6 dubbed Light Blue, appeared. It was broken on February 28, and it immediately revealed that a sizable German air and ground force had been landed at Tripoli to help the battered Italians. It was a whole new war.

Hints were all that the Enigma traffic was yet providing, however. Light Blue was another Luftwaffe key; the much more important Army keys remained unbroken. A year and a half into the war, a year after GC&CS made its first break into the Enigma, signals intelligence had yet to deliver its first solid victory. (Hut 3 had at least scored one important bureaucratic victory: On March 13, GC&CS was suddenly empowered to begin transmitting important intelligence directly to the British Army command in Cairo.)

When it did at last come, GC&CS's first real triumph was, from a cryptanalytic point of view, almost an anticlimax after the heroic cryptanalytic battle against the German Enigma. The Italian Navy had begun using an unsteckered Enigma, with rewired rotors, during the Spanish Civil War; when Italy entered the war against Britain and France in June 1940, Italian Enigma traffic again appeared and by September, Dilly Knox had it completely solved, using the hand cribbing methods he had worked out years before. Even with its reappearance, though, the Italian system carried relatively little traffic, and the Admiralty viewed it with its customary lack of interest. In fact, given how easy it was to break, Nobby Clarke observed on March 24, 1941, "the Admiralty are beginning to think that it is only intended to deceive."

A message transmitted the next day made them eat those words. It was broken by Bletchley Park and teleprinted to the Admiralty, and at eight-twenty on the morning of the twenty-sixth a secret signal was on its way to Admiral Andrew B. Cunningham, Commander in Chief Mediterranean, using the British naval officers' code, enciphered with a special flag officers' additive table for extra security. The Italian message was a warning forwarded from Rome to Rhodes that the Italian fleet, under the command of Admiral Angelo Iachino, was preparing to move in three days. Their intended target was without a doubt the lightly escorted British convoys carrying troops and supplies from Alexandria to Greece.

The British convoys were an extremely risky part of an extremely risky operation. Churchill, motivated by pro-Greek British sentiment and his notorious obsession with attacking on the enemy's flanks—in the broadest geographical terms—had called off the pursuit of the defeated Italian army; fifty-eight thousand troops were ordered to depart from Egypt to shore up Greece's defenses against an impending German attack. German dive bombers and the Italian fleet posed a constant threat to the British ships. Cunningham was eager to draw the Italians into a fight and eliminate the threat once and for all, but Iachino had no intention of taking on the British fleet directly; if Cunningham stationed his battle fleet to the west of Crete to protect the convoys, the Italians could just wait until the British ran out of fuel and returned to Alexandria, and then come out. The only hope was to surprise the Italians at sea.

The intercepted signal was precisely what Cunningham had been waiting for. Cunningham was a fighting admiral from the word go. Throughout his career he had steadfastly refused staff appointments and considered the normal tours of duty to gain an expertise in gunnery, torpedoes, or signals a waste of time since they did not involve the direct command of ships at sea. He was, as the historian Ralph Bennett put it, precisely the sort of anti-intellectual "blue water" admiral who could be counted on to scorn intelligence and prefer action to thought. But Cunningham defied the stereotypes, for he rose to the occasion with aplomb and an agility of mind that was notably lacking in the Admiralty itself at the time.

On March 27, the day Cunningham prepared to act, there was but one British convoy at sea; it was carrying troops to the Piraeus, and was currently south of Crete. The convoy was ordered to continue steaming on until sunset, then reverse course under cover of darkness. A returning convoy scheduled to depart from Greece was ordered not to sail. Not to tempt fate, Cunningham bet his operations officer ten shillings that they would see nothing of the enemy. "I also arranged a little private cover plan on my own," Cunningham later recounted. The Japanese consul at Alexandria played golf every afternoon. He was also suspected of reporting British fleet movements to the Germans and Italians. So Cunningham went ashore to play a leisurely round of golf himself that afternoon, carrying an "obvious suitcase" with him as if he intended to stay overnight. (The consul, Cunningham observed, was a short and squat man who was a "remarkable sight" when he bent over to putt; Cunningham's chief of staff had nicknamed him "the blunt end of the Axis.") After finishing his round of golf, Cunningham retrieved his suitcase, slipped back aboard his flagship *Warspite* after dark, and the fleet sailed at 7:00 P.M.

Warspite was accompanied by two other battleships, *Barham* and *Valiant*, the aircraft carrier *Formidable*, and nine destroyers. A separate squadron of four cruisers and four destroyers operating in the Aegean under Vice Admiral Henry D. Pridham-Wippell was ordered to rendezvous with Cunningham the next morning. Before the rendezvous could take place, the Italian and British cruiser squadrons discovered each other at dawn south of Crete. Pridham-Wippell knew that his six-inch guns were no match for the Italians' eight-inch guns and tried to lure the enemy back toward the British battleships, now ninety miles to the south. But after forty-five minutes of raining near misses down on the British, the Italians broke off action. The crews aboard the British cruisers stood down from their battle stations. Bully beef sandwiches were handed around, and the squadron's operations officer was standing on the bridge of the cruiser *Orion* when another staff officer, his mouth full of sandwich, nudged him and said, "What battleship is that over on the starboard beam? I thought ours were miles to the east." A second later he received his answer when a fifteen-inch shell exploded in the sea among the British cruisers. The battleship was the *Vittorio Veneto,* flying Admiral Iachino's flag. Pridham-Wippell immediately ordered the squadron to turn together 180 degrees, make smoke, and "proceed at your utmost speed." As Cunningham recalled, one of the cruisers, *Gloucester,* had reported engine trouble the night before and had been unable to do more than twenty-four knots; "However, the sight of an enemy battleship had somehow increased *Gloucester*'s speed to thirty knots."

Cunningham's aging British battleships could do no better than twenty-two knots, however, whether chased or chasing. Still scores of miles away, the admiral paced the bridge like "a caged tiger," one officer said, always on the side of the ship closest to the enemy, as if willing the ship forward with all his might—and in a barely contained fury that the enemy might slip away. His plan had been to strike with aircraft only when the battle fleet was in position to finish off any enemy ships crippled by a torpedo attack, but to relieve the pressure on Pridham-Wippell's cruisers Cunningham reluctantly ordered the *Formidable*'s antiquated Albacore and Swordfish biplane torpedo bombers into action at once. Their torpedoes all missed; the Italians realized that a British aircraft carrier, and thus British battleships, were nearby, and turned tail. The chase was on in earnest now. A second air strike a little after 3:00 P.M. found its mark, jamming *Vittorio Veneto*'s rudder and bringing her to a dead halt, but her engineers soon had the ship under way again at seventeen knots, and now tightly protected by a screen of cruisers and destroyers. A third strike at 7:25 P.M. again missed the Italian battleship but scored a direct hit amidships of the heavy cruiser *Pola.* Iachino dispatched two cruisers and four destroyers to go to her aid.

Night fell, and Cunningham was mad to continue the pursuit. His staff counseled caution, pointing out that they would be charging blindly into the range of the enemy's land-based dive bombers by daybreak. In his memoirs Cunningham wrote that he paid "respectful attention to this opinion" and told his staff he would have his evening meal and see how his state of "morale" was when he returned. One of his officers recalled a slightly different version: Cunningham said, "You're a pack of yellow-livered skunks. I'll go and have my supper now and see after supper if my morale isn't higher than yours." It was; Cunningham ordered the pursuit to continue. Although the British could not know it until many hours later, *Vittorio Veneto* had made good her escape. But the cruisers would prove to be sitting ducks. Shortly after 10:00 P.M. *Valiant* picked up on her radar a large ship, stopped in the water, six miles ahead on the port bow. A few minutes later as the officers on the bridge of the flagship swept the horizon with their night glasses, the silhouettes of three cruisers loomed into view. The British battle fleet swung into line ahead, and an absolute silence fell over *Warspite* as her fifteen-inch guns fixed on their targets, now at point blank range, less than thirty-eight hundred yards. The firing gongs sounded and at that same instant one of the British destroyers switched on her searchlight, capturing the ghastly sight of six huge shells hurtling toward one of the cruisers, then smashing into its upper deck in a huge fireball. A voice of wondering astonishment rose through the din. "Good Lord," said *Warspite*'s captain, a noted gunnery officer, "we've hit her."

In a few minutes two of the Italian cruisers were literally blown to pieces by the shells from the three British battleships. *Pola* was sunk the following morning by torpedoes fired from destroyers. Two Italian destroyers were also sunk. In all, twenty-four hundred Italian officers and men were killed. The Battle of Cape Matapan, as it came to be called, took place a year before the Battle of Midway; it was the first naval battle in which carrier-based aircraft played a decisive role, and the first battle of any kind in the Second World War in which the timely use of signals intelligence played *the* decisive role. Dilly Knox penned a few lines of celebratory doggerel in praise of his staff to mark the occasion, to which Nobby Clarke added his own verse in praise of Knox himself:

> *When Cunningham won at Matapan*
> *By the grace of God and Dilly,*
> *He was the brains behind them all*
> *And should ne'er be forgotten. Will he?*

———

Another source that had at first seemed mundane, from both a cryptographic and intelligence view, provided a second intelligence coup that month of March 1941. The German railways in Eastern Europe used their own version of the Enigma machine; like the Italian Enigma it was unsteckered, though it did offer one cryptanalytic complication in that the reflector could be set to any of twenty-six positions. But it was still a fairly simple proposition, and within a few weeks of intercepting the first railway Enigma signals in July 1940, GC&CS had broken it. (Once again, it was the sharp eye and mind of John Tiltman at work: He spotted two messages in depth, which provided a wedge into the system.) The railway network shortly thereafter went silent for six months, reappearing in late January 1941 with long and tedious lists of timetables. But toward the end of March the railway schedules suddenly began to tell a gripping story. Over the course of several weeks, German armor had been shuttled from Bucharest and Krakow following the capitulation by Yugoslavia's Prince Paul. The panzers were then hurried back to the south when anti-Nazi officers in Belgrade seized power in a coup. Churchill was electrified by this intelligence; he realized at once that it meant Hitler intended to invade Russia in May, a move that was now postponed by events in Yugoslavia—but surely not for long. Churchill had had no contact with Stalin since the previous summer. But he immediately sent him a "short and cryptic" message of warning:

> I have sure information from a trusted agent that when the Germans thought they had got Yugoslavia in the net—that is to say, after March 20—they began to move three out of five Panzer divisions from Rumania to Southern Poland. The moment they heard of the Serbian revolution this movement was countermanded. Your Excellency will readily appreciate the significance of these facts.

Like Churchill, Stalin believed he was his own best intelligence officer; unlike Churchill, his view of the world was so clouded by visions of conspiracy and intrigue that he refused to accept findings at odds with his preconceptions. Stalin was convinced it was irrational for the Germans to attack Russia, especially when Britain lay so close to defeat, and dismissed Churchill's news as a bluff—either by Churchill or by Hitler. Then Rudolf Hess's bizarre flight to Scotland in May evoked in Stalin's paranoid mind thoughts of a secret British–German conspiracy to join forces against the Soviet Union, which made him still more suspicious of the warnings Britain was sending.

Stalin so refused to heed the signs of massing German troops on the frontier that it is unlikely he would have responded differently even had Churchill been totally frank about the source of his information. But it did not help that Churchill was so cryptic, and that pointed up a dilemma that would recur again and again now that signals intelligence was actually beginning to deliver the goods. There was no point in going to all the trouble of breaking enemy codes unless the information derived from them could be put to use. But every time it was put to use risked giving away to the enemy the truth, that its codes were being broken. Circulating the information while concealing the source seemed like a sensible compromise, but in practice the cover stories in which signals intelligence was wrapped might end up fooling the wrong side: The enemy, who knew what was in his signals, would not be deceived about how the information was obtained, while friendly forces might simply discount information that was chalked up to a "reliable source" as the usual cocktail-party drivel from self-dramatizing secret agents. The fact that signals intelligence came directly from the enemy's own words gave it an authority that was unmatched, yet that fact also was precisely what needed to be concealed for the sake of security.

At first Bletchley Park referred to what would become known as ULTRA or Special Intelligence with the code name "Boniface," a purported secret agent. (Churchill was especially taken with "Boniface" and continued using that as his cover name for ULTRA throughout the war.) Intelligence reports sent from Bletchley Park bore the code letters CX/FJ, where CX meant an agent's report and FJ was the "agent's" initials. FJ was then killed off and replaced by JQ, who was later followed by MSS.

There was also the question of how much information actually to include in the signals, even when they went to good British commanders. Disguising the source was arguably not enough. A message could always be captured in the field, and if its contents could be matched up with one of the enemy's own signals, no amount of window dressing about POW reports or wastebasket scroungings was likely to conceal the truth of where it had come from. The Typex machines used by the British Army and the RAF for transmitting such intelligence reports were justifiably considered to be quite secure, as were the one-time pads that the Admiralty somewhat belatedly put into service in June 1941 for the transmission of ULTRA, but even the length and timing of signals that relayed the contents of decrypted enemy messages might arouse suspicions if an alert enemy started analyzing their patterns.

More generally, "C" for both bureaucratic and security reasons insisted on tightly limiting the circle of initiates who knew the truth about Boniface. Even many cabinet ministers were kept in the dark; as fantastic as it seems, "C" even refused at first to give the First Lord of the Admiralty ULTRA clearance, and in one farcical episode early in the war the First Lord was hustled out of the Admiralty's map room. Bletchley Park was never permitted to send ULTRA directly to the Foreign Office, to whom "C" in theory reported. When Churchill asked that certain officials—including the Commander in Chief of Bomber Command—be included on the ULTRA distribution list, "C" simply decided otherwise, and quietly vetoed the Prime Minister's requests.

Heavily paraphrasing the information revealed by ULTRA was one way to cope with these security restrictions, but that posed the same problem the fictional-agent cover story did, of watering down its potency. Churchill continually pestered "C" and the service ministries to provide commanders in the field with full, verbatim texts, insisting that paraphrases could never capture the flavor and force of the original, and at times calling attention to specific wordings of Hut 3 paraphrases that he believed departed in meaning from those of the actual decrypts. Bletchley Park held its ground. Verbatim copies of translated decrypts were teleprinted to the Admiralty, but all signals that went to commanders were paraphrased and their source disguised.

In addition to protecting security during transmission, paraphrases prevented those not in the know who happened to see the message from learning of the source. Gossip being what it is in wartime, Bletchley Park was keenly aware of the dangers if rumors about British breaking of German codes became widespread in the ranks. But the most difficult security dilemma ULTRA posed was how it actually could be employed operationally without giving away the store. At Knox's insistence, Cunningham had sent up an aerial reconnaissance patrol—and made sure the Italians spotted it— in order to establish a convincing cover story for his knowledge of the Italian fleet movement, and this became a standard drill in the Mediterranean whenever ULTRA intelligence was to be acted upon. (Knox's doggerel about Cunningham's victory had referred facetiously to members of his cryptanalytic team in the Cottage piloting "the aeroplane that spotted their fleet.")

The Admiralty felt its way with extra caution at first, exercising "a more direct control over operations than was the usual practice," according to an Admiralty history. Still, almost immediately there were a series of scares. In spring 1941, still lacking one-time pads or other secure channels, the Admiralty had sent rather elliptically worded intelligence reports to its naval commanders during the chase of the German battleship *Bismarck*. The first signals intelligence about *Bismarck*, though broken a month late, was still timely: An Enigma message revealed that prize crews and special charts had been placed aboard her at the Baltic base of Gdynia, signaling that the much-feared breakout of the German surface raider to the Atlantic would soon come. The Admiralty decided that given the time lapse, the information could plausibly have come from an agent's report and so went ahead and passed a warning to the fleet.

On May 18, *Bismarck* sailed, setting off an epic chase. Hut 8 was still not reading the naval Enigma currently, or anything close to it, but direction finding and traffic analysis at a crucial moment in the ensuing battle proved decisive. In the early morning hours of May 24 *Bismarck* destroyed the battle cruiser *Hood*, the most famous ship in the British fleet. There were three survivors. *Bismarck*, hit by three shells from the battleship *Prince of Wales* during the action, steaming at reduced speed and trailing an oil slick, broke off and vanished into the mists of the Atlantic. The next morning a signal went out from the Admiralty: INFORMATION RECEIVED GRADED A1 THAT INTEN-TION OF 'BISMARCK' IS TO MAKE FOR THE WEST COAST OF FRANCE. At Bletchley Park, Harry Hinsley had discovered that radio control of *Bismarck* had shifted from Wilhelmshaven to Paris following the engagement, a sign that she was heading for one of the French ports. Hinsley had several heated discussions with the Admiralty to convince them of the fact, but they had accepted his conclusion and informed the fleet by the time an unusual confirmation came in. General Hans Jeschonnek, the Luftwaffe's deputy chief of staff, was in Athens for the German invasion of Crete; his son was a midshipman aboard *Bismarck*. Jeschonnek sent a signal—in the Luftwaffe Enigma Red key— inquiring where *Bismarck* was heading. The reply came: Brest. The messages were promptly read by GC&CS. *Bismarck* was finally spotted by a Coastal Command Catalina on the morning of May 26, 690 miles from Brest; the Swordfish biplanes of the carrier *Ark Royal* armed with their single torpedo apiece attacked through murderous antiaircraft fire and rain squalls that obscured visibility. In the fading light one torpedo finally found its mark in the ship's vulnerable stern, jamming both rudders and leaving the mighty battleship circling helplessly. It was the end; early next morning British destroyers and the guns of the battleships *King George V* and *Rodney* finished her off.

Naval Enigma decrypts came too late to play a direct role in the battle; only later did Bletchley Park read the last words of Admiral Günther Lütjens, commanding the German fleet from *Bismarck*: "Ship unmaneuverable. We fight to the last shell. Long live the Führer." But Turing's first break into the naval Enigma did come in time to help track down eight supply and weather ships that had been at sea to aid *Bismarck* in its dash to the Atlantic. The Admiralty prudently decided that sinking all eight might tip the Germans about the source of their intelligence, and so ordered that two be deliberately spared, *Gedania* and *Gonzenheim*. Fate was against such subtleties. By pure chance both were intercepted the same day, June 4, by British ships that happened to cross their paths. One was captured and the other scuttled by her crew.

The Admiralty and GC&CS's head reacted to the scare by becoming more cautious than ever in disguising ULTRA intelligence. Direction-finding was a good cover, but it could go only so far. Some information could not be disguised. When British captains were supplied with such details as the names, bearings, destinations, or times of arrival of enemy ships, it had to be obvious that such facts could only have come from reading enemy singals. So the simplest solution was not to send out such information at all. For Bletchley Park it was an odd and frustrating situation. For two years their work had been dismissed as too unimportant to be used. Now it was too important to be used.

————

It was certainly too important to pass on to the loose-lipped Americas. By the summer of 1941 U.S. Navy ships were committing themselves ever deeper in an undeclared war against U-boats in the Atlantic. In a carefully orchestrated arrangement, American troops and patrol planes took over the defenses of Iceland. American warships began convoying merchant ships as far as Iceland and Greenland. The U.S. Navy was under instructions not to attack U-boats but to report their location to the British, and the Germans were well aware of the fact. On June 20, *U-203* had shadowed the battleship *Texas* for 140 miles, repeatedly trying to get into position for a torpedo shot. This was certainly going a long way toward Churchill's aim of having America and Britain "somewhat mixed up together." Under the circumstances Churchill felt the Americans ought to be getting ULTRA intelligence about German U-boats chasing their warships, and following the *Texas* incident he pressed his secret service chief on the matter. But four days later "C" replied:

> I find myself unable to devise any safe means of wrapping up the information in a manner which would not imperil this source . . . it [is] well nigh impossible that the information could have been secured by an agent, and however much we insist that it came from a highly placed source, I greatly doubt the enemy being for a moment deceived, should there be any indiscretion in the U.S.A. That this might occur, cannot be ruled out, as the Americans are not in any sense as security minded as one would wish, and I need only draw your attention to the attached cutting from to-day's "Daily Express" on a matter which, in my opinion, should not have been made public if the two Secret Services are to work together.

The attached article that had caught "C"'s attention reported that Colonel William "Wild Bill" Donovan "has a new hush-hush mission—to supervise the United States Secret Service and ally it with the British Secret Service. . . . The American 'Mr. X,' as he is known privately, will report direct to the President." It wasn't exactly "hush-hush" by the time the *Daily Express* got hold of it, and "C" had a point.

Supplying naval ULTRA to the Americans had become an issue because Turing had at last achieved his breakthrough. In August 1940 the eighth and final Enigma rotor was recovered, possibly by divers combing through a

German submarine sunk earlier near the British coast. Turing and Twinn kept trying out cribs and running the bombe a week at a stretch on all 336 wheel orders; a few days of traffic were broken here and there, and the bombe's first success with a "straight crib" came in February 1941 when the mathematicians succeeded in breaking the naval Enigma setting for April 28, 1940, after running all wheel orders on a crib that Birch and the Naval Section had rejected, a small triumph. But this still was a very long way from continuous solution.

Operation RUTHLESS was dead, but its functional equivalent was still the thing that was needed. And where Ian Fleming's ingenuity had failed, the fog and fortunes of war stepped in to provide—twice. In early March 1941 the British destroyer *Somali* was sent to lead a commando raid to Norway to destroy cod oil factories and evacuate Norwegians who wanted to continue the fight against the Nazis. In the Vestfjord, near the Lofoten Islands, the expedition was fired upon by a German patrol boat. The British returned fire, the German crew jumped out, and their boat, *Krebs,* was left stranded on a small rocky islet. A British boarding party gathered up papers from the captain's cabin and carried them back to Scapa Flow. On March 12 the documents were at Bletchley Park. They contained the daily Enigma settings for February.

By late March, Hut 8 had worked out some of the bigram tables. Banburismus was working better, too, thanks to months of practice. The actual work of lining up the punched sheets was, as Jack Good recalled, "not easy enough to be trivial, but not difficult enough to cause a nervous breakdown." Turing had worked out a statistical method of determining how probable it was that two messages found by Banburismus to contain repeated bits of cipher text were truly in depth. A two-letter repeat between two messages could occur quite readily by chance, but longer runs were more likely to be "causal," a result of the two messages actually having been enciphered with the same stretch of key. It would require something close to a miracle for a sequence of nine or more letters to be the result of random coincidence. (The longest noncausal repeat the Banburists ever found was an octograph.) Turing dubbed the statistical unit he invented for this test the "Ban," and most scores were expressed in "decibans"—a tenth of a ban. Jack Good recalculated the whole system and came up with much more accurate statistics, which Hugh Alexander facetiously dubbed "ROMS," Resources of Modern Science. There certainly wasn't anything very scientific about the grunt labor of shifting sheets across one another over and over. On one night shift, Good became so sleepy he lay his head down on the table. When he woke up he shifted a sheet one place to right and found a twenty-two-letter repeat, a record.

With the settings for February and the partially reconstructed bigram tables, Hut 8 was able to read about 180 naval Enigma messages covering the period February 8–28. But March and April proved almost impenetrable; it was the old problem of cribs again, compounded by the incomplete bigram tables. Then on May 9 a second chance encounter between British and German naval vessels took place, and changed everything. The British destroyer *Bulldog* was escorting forty ships in a westbound convoy across the Atlantic; two days earlier a U-boat had gotten in among the columns of the convoy and sunk two merchantmen. Just after noon on the ninth, a sunny day under a bright blue sky, the air was rent by two enormous explosions. Columns of water shot upward. The captain of the *Bulldog,* Commander A. J. Baker-Cresswell, immediately swung the convoy forty-five degrees and from the head of the column sheered off with two corvettes to hunt down the intruder. The corvette *Aubretia* almost immediately made sonar contact and began firing depth charges. Her second attack scored a near direct hit on the submerged German—*U-110,* commanded by Kapitänleutnant Julius Lemp. Aboard the U-boat electric motors were knocked out, depth gauges cracked, oil tanks began leaking. An instant later the U-boat blew to the surface—"the dream of all escort vessels," said Sublieutenant David Edward Balme of the *Bulldog.* Baker-Cresswell ordered his crew to open fire with every available weapon, and two 4.7-inch guns, machine guns, even an anti-aircraft gun unleashed a deafening hail of ordnance bouncing off the U-boat's hull. The psychological shock had the desired effect. The U-boat crew forgot about everything except abandoning ship, and at once began spilling out on deck and into the water.

When it became apparent that the Germans in their haste had not even set scuttling charges, Baker-Cresswell turned to Balme and said: "Right, we will board her. Sub, you take this sea boat." Sublieutenant Balme, twenty years old, answered, "Right." The sea boat, one of two the ship carried, was a rowboat with five oarsmen, and the trick was to lower it exactly in time with a crest of the six-foot rollers, which the crew on the deck of the destroyer just managed to do. Meanwhile Baker-Cresswell had shouted across to the *Aubretia,* which was picking up survivors: "Get them down below quick"—so the Germans wouldn't see that a boarding party was heading over to their submarine.

To Balme fell the bracing but unnerving job of being the first to climb down the conning tower, "a nasty moment," he recalled, wondering if there were any Germans still down there. He "waved" his revolver around a bit and plunged down the ladder. The submarine was deserted. Balme then brought the rest of the crew down and formed them into a human chain to pass charts, books, and documents up. Among the haul was an intact Enigma machine spotted by *Bulldog*'s radio operator (he had pressed the keys and "finding the results peculiar" unbolted it and sent it up the hatch). Bletchley Park already had an Enigma machine. What they did not have, and which was worth more than a thousand Enigma machines, were some of the documents left undestroyed by the U-boat crew in their haste. Among the booty gathered up by Balme's boarding party were the complete bigram tables for the naval Enigma.

U-110 was taken under tow, but sank the next morning. That turned out to be "one of the greatest blessings in disguise," Balme later remarked, for *Bulldog* had been ordered to tow the captured U-boat to Iceland, which "was actually full of German spies, and if we had arrived with the U-boat it would have been reported back to Berlin probably quite soon. As it was we just appeared in the normal course of events to refuel."

Three days later, May 13, the haul of captured documents from *U-110* was being eagerly pored over by Lieutenant Allon Bacon, a Royal Navy intelligence officer who had been dispatched from Bletchley Park to Scapa Flow to meet the returning ship. Bacon photographed all the documents for safety's sake and hastened aboard a plane for London with the originals. He arrived at Bletchley that afternoon. That night at 9:37 the teleprinters that connected B.P. to the basement bunker of the Citadel, the Admiralty's ugly but bombproof addition in Whitehall, came alive for the first time in forty-eight hours. An uninterrupted stream of a hundred almost-current decrypts followed. For his actions Balme was awarded the DSC; the King, as he presented the award, apologized that because of security considerations he could not make it a higher honor.

Meanwhile two deliberate "pinches" set in motion in the spring of 1941 bore fruit, adding to Hut 8's windfall. The scattered traffic read from February had shown that the German Navy stationed isolated weather ships in the North Atlantic to transmit crucial meteorological information. Hinsley reasoned that since the ships used the naval Enigma, and since they were on station for months at a time, they must have been issued lists of daily settings far in advance. He proposed a slightly more grown-up version of Operation RUTHLESS. The Admiralty agreed, and taking no chances dispatched a huge force. The Enigma decrypts had pinned down one German weather ship to the area of a single grid square on the German charts, which was one degree of latitude on a side, sixty-nine miles. The British plan was to sweep the grid with three cruisers and four destroyers spaced at ten-mile intervals. In the fading light of the afternoon of May 7 a lookout on the cruiser *Edinburgh* spotted smoke on the starboard bow. After a short chase the weather ship *München* was brought to bay. A British communiqué casually referred to the incident and stated that "a German armed trawler" had been scuttled before it could be boarded. In fact she was captured intact, with a complete list of the daily Enigma settings for the month of June. A month later the Admiralty agreed to an encore, and on June 25 the trawler *Lauenburg* was picked off. (When the gun crew on the destroyer *Tartar* was ordered to fire at the German ship but not hit her, the gunner's mate replied, "Christ, that should be easy.") The second capture provided Bletchley Park with the July key list, which arrived on July 4.

With the daily setting lists, Bletchley Park was now reading naval Enigma traffic as easily as the Germans. The operational consequences were immediate. On June 10 a naval Enigma message revealed that the pocket battleship *Lützow* might be preparing to break out into the Atlantic; two days later the ship transmitted a signal giving her precise location as of 8:30 P.M. as she steamed from Germany toward Norway; by 11:00 P.M. fourteen Coastal Command torpedo bombers were on their way. One torpedo hit home, putting the ship out of action for seven months.

The real test came in August when the lists of daily settings expired. A change in the bigram tables had occurred on June 15, but armed with experience and the June key list, Turing and the Hut 8 staff had the new tables reconstructed by the end of the month. August 1 would thus be the first test of the complete cryptanalytic system that Turing had worked out a year and a half earlier. The bigram tables and Banburismus would reduce the number of wheel orders; a crib run on the bombe would find the stecker, ring setting, and message setting for one message; EINSing would recover settings for additional messages; a short menu run on the bombes would then recover the Grund; and then they would be home free. Miraculously, it all worked. Hut 8 began solving settings from August with an average delay of about three days for the first of each "paired day"; the second day was then usually broken in under twenty-four hours.

The breaking of naval Enigma contributed to a sharp decline in sinkings by U-boats in the Atlantic that began in the summer of 1941. Shipping losses had been steadily rising throughout the spring; in March, April, and May the dozen or so German U-boats on station at any time in the North Atlantic sank 142 ships, totaling 818,000 tons. In June losses reached 310,000 tons. But then the tide turned. Some U-boats were diverted to the Baltic and Arctic for the war against Russia. Convoy escorts were strengthened as the U.S. Navy took over patrolling its half of the Atlantic. Hurricane fighter aircraft were mounted on catapults fitted to ordinary merchant ships; these Catapult Aircraft Merchantmen were a stopgap until escort carriers arrived, and the Hurricane pilots took off with the knowledge they would have to ditch their aircraft at the end of their mission and hope to be fished out of the sea. But they countered the Focke-Wulf 200 Condor long-range bombers that worked with the U-boats to spot merchant vessels. It was impossible to know for sure how much of a difference reading the German U-boat signals made, but on more than one occasion information came through in time to divert convoys around the waiting U-boat wolf packs. By July, U-boat sinkings had dropped to 94,000 tons; by August, to 80,000. The German stranglehold had been broken.

———

The German attack on Russia had been foreshadowed not only by the railway Enigma decrypts in March but also by the first of a series of Japanese diplomatic messages in the Purple cipher, messages that would soon become the Allies' single most reliable source of intelligence on the Nazi inner circle. General Oshima Hiroshi, the staunchly pro-Nazi Japanese ambassador to Berlin, enjoyed the unusual confidence of Ribbentrop, and frequently received briefings from the top echelons of the government, including the Führer himself. Oshima duly reported back to Tokyo, and from the time he arrived for a second tour as ambassador in Berlin in February 1941 until the very end of the war his messages were read almost currently by SIS and OP-20-G. On June 3 Oshima reported that Hitler had told him "in every probability war with Russia cannot be avoided." The United States added its warnings to those Churchill had already sent to Stalin, but the Communist dictator saw only a further conspiracy and refused to alert his armed forces, even as he continued to appease Hitler with shipments of raw materials that fueled the Nazi war machine. In the early morning hours of June 22, 1941, German soldiers watched the last Russian freight trains trundle over the bridges across the Neimen River bearing their tribute. At 3:00 A.M. the Germans attacked, with a force of 3,500,000 men, 3,350 tanks, and 7,200 guns.

Churchill, the lifelong anti-Bolshevist, at once publicly declared Britain's total support for Russia. Privately, one of his first acts was to begin pressing "C" to forward to Moscow ULTRA intelligence on German intentions and plans. On June 24 "C" replied that given the insecurity of Russian field ciphers such an action "would be fatal"; "it would only be a matter of days before the Germans would know of our success, and operations in the future would almost certainly be hidden in an unbreakable way." Churchill retaliated with a barrage of memoranda. An Enigma message in the Red key broken on July 16 reported that the Luftwaffe was planning to bomb railways in the Russian rear at Smolensk to prevent their withdrawal. Churchill scrawled on the bottom: "C. Surely we ought to give them warning of this."

Though "C" still insisted that he must personally approve every message transmitted, Churchill won that round; that same day a message (attributed to "most reliable source") was on its way to the British Military Liaison in Moscow to be passed on to the Soviet General Staff. Throughout September Red key messages indicated that the Germans were shifting forces from the Northern and Southern Army Groups in Russia to concentrate for a push in the center. On October 2 Churchill was again pressing "C": "Are you warning the Russians of the developing concentrations? Show me the last five messages you have sent out to our missions on the subject."

Hitler had been planning this turn against Russia for months—indeed years. On December 18, 1940, he issued Führer Directive No. 21, Operation BARBAROSSA, which ordered the German armed forces to be prepared to "crush Soviet Russia in a quick campaign before the end of the war with England." Three months later he summoned his top commanders and explained something else about the Russian campaign: This would not be a fight in the "knightly fashion." There was no place for such "obsolete ideologies." Soviet "commissars" were to be liquidated. Civilians "suspected of criminal action" would be shot at once. German soldiers who committed offenses against enemy civilians were to be treated leniently; "prosecution is not obligatory" in such cases. The Army was to directly control and administer as little occupied territory as possible. That job would fall to Heinrich Himmler's SS and police forces, who were to follow behind, completely seal off the areas under their control, and carry out "special tasks" that "result from the struggle which has to be carried out between two opposing political systems." As the German forces swept into Russia, the police and SS battalions received more specific instructions. At first, orders to round up and execute Jews were rationalized to the troops as part of an operation against "looters" and "Bolshevik" partisan fighters. By the end of July orders were less circumspect: All Jewish males were to be shot. By mid-August the SS and police were methodically rounding up and killing Jewish women and children in "cleansing" operations.

Hints of these horrors began arriving at B.P. in the first weeks of the fighting. Bletchley Park had begun working on the German Police ciphers in 1939, under John Tiltman's direction. Germany's principal police force, the Order Police, included the municipal and local police, firemen, gendarmerie; it also included some militarized battalions that had been established in the early years of the Weimar Republic to fight political violence and also to serve as an unofficial army reserve to evade the Versailles Treaty limitations. It was thus a hodge-podge of local cops and militia, hardly the kind of sources that could be expected to provide intelligence coups. But GC&CS believed the police traffic might give an occasional snippet of information about morale and dissent within Germany. And working on the police ciphers was useful training in any case; similar hand ciphers were used for medium-grade Army and Air Force communications and were a standby substitute for the Enigma. The traffic might also possibly provide cribs useful in the attack on Enigma keys. In any event, it couldn't hurt to try.

There was a huge amount of police traffic, carried on many different radio frequencies and from many different parts of Germany, and it was difficult to intercept from Britain. Early in the war GC&CS struck a deal with the French to collaborate on the German Police ciphers, and in December 1939 a party from B.P. joined the French headquarters at La Ferté-Sous-Jouarre that was working on the problem. The encryption method involved a hand cipher—double transposition, in which the letters of the message are scrambled and rescrambled according to a pattern determined by a key—and the cryptanalytic challenge was not that great. (Though the British did complain that the French viewed their allies more as rivals than collaborators, and also took a rather leisurely attitude toward the work: "Lunch at their mess was a matter of three hours with little likelihood of much exacting intellectual effort being possible thereafter.")

The fall of France resulted in a significant loss of intercept capability; the invasion of Russia brought a new key and different radio frequencies for the German Police operating in Russia, reducing Bletchley Park's success to about 50 percent; in August 1941 the police introduced two daily keys for Russia that cut the code breakers' success in half again. But the police messages from the eastern front that Tiltman's group began reading in July and August left no doubt that something extraordinary and horrifying was happening. On July 18 a police message came through reporting that 1,153 "Jewish plunderers" had been shot. In August a spate of reports came in, now joined by messages on the principal SS Enigma key ("Orange"):

August 8	SS Cavalry Brigade	3,274 partisans and Jewish Bolshevists shot
August 24	Police Battalion 314	294 Jews shot
August 25	Police Regiment South	1,342 Jews shot
August 25	1st SS Brigade	283 Jews shot
August 27	Special Task Force with Police Battalion 320	4,200 Jews shot
August 31	Police Battalion 320	3,200 Jews shot

On August 7, SS Gruppenführer Erich von dem Bach-Zelewski, who had been placed in charge of the Police Regiment Center, had already sent a dispatch boasting, "The figure of executions in my area now exceeds the 30,000 mark." A B.P. intelligence summary noted that "the tone of this message suggests that the word has gone out that a definite decrease in the total population of Russia would be welcome in high quarters and that the leaders of the three sectors [i.e., north, center, and south] stand somewhat in competition with each other as to their 'scores.'" The messages revealed another bizarre preoccupation with the Nazi occupiers. Decrypts referred to the work of a special SS battalion that swept up behind the panzers and, under Ribbentrop's guidance, was assigned the mission of plundering art works for the Nazi leadership.

There is little trace in the files of Bletchley Park or Whitehall indicating how these reports were circulated to higher authorities in Britain. The only immediate response that Tiltman's section received from its reports on German atrocities—which it sent to MI14, the research section of the War Office's Military Intelligence branch—was a shirty and pedantic memorandum from one Lieutenant Colonel Clarke in MI14: "The author has overlooked the elementary fact that the Gestapo *is* a fundamental part of the Security Police. He also confuses SD and SHD, two entirely different bodies." The colonel did not see anything else in the report worth commenting upon.

A month later GC&CS decided it would not bother including such details in future reports:

> The execution of "Jews" is so recurrent a feature of these reports that the figures have been omitted from the situation reports and brought under one heading. Whether all those executed as "Jews" are indeed such is of course doubtful; but the figures are no less conclusive as evidence of a policy of savage intimidation if not of ultimate extermination.

Churchill was one Whitehall figure who did receive at least summaries of the German Police reports. And while he, too, seems not to have paid particular attention to the mention of Jews, he took to the airwaves on August 24 to issue an emotional warning and denunciation of the German atrocities. "We are in the presence of a crime without a name," Churchill declared. As Hitler's armies advance, "whole districts are being exterminated. Scores of thousands—literally scores of thousands—of executions in cold blood are being perpetrated by the German police-troops upon the Russian patriots who defend their native soil. Since the Mongol invasions of Europe . . . there has never been methodical, merciless butchery on such a scale, or approaching such a scale."

It was a calculated risk, for there was a grave danger that in going public this way Churchill would expose the source of his information, and there is some evidence that that was precisely what happened. Three weeks later, on September 13, the chief of the Order Police sent instructions to the Higher SS and Police Leaders in Russia, instructions that Bletchley Park promptly decoded: "The danger of decipherment by the enemy of wireless messages is great," the order read. Henceforth, information "containing state secrets," including "exact figures of executions," are to be sent by courier.

The order was not immediately obeyed, in part because courier service to SS and Police headquarters in Russia was erratic and spotty, operating at best twice a week. Another security change was made more promptly, however. That same month double transposition was dropped, and the Police networks began using in its place an entirely different hand cipher, double Playfair. This however proved a blessing in disguise for the code breakers; as B.P.'s German Police section noted in a postwar history, "The result was exactly the reverse of what the Germans intended, for whereas the retention of Double Transposition with a still further splitting up of keys would have soon put us out of business, Double Playfair quickly proved to be a most breakable cipher and it became the exception to fail to break a day." The Playfair system had been devised by the British in the mid–nineteenth century and, though ingenious, it was hardly the most secure of systems—especially after a century of use. Instead of a one-for-one substitution of letters, the Playfair system replaces bigrams with bigrams, using a 5×5 key square, such as this:

```
B  L  A  Q  I
M  N  V  C  X
K  Z  H  G  F
D  S  T  O  U
Y  P  R  E  W
```

The plain text is divided into pairs of letters, and each pair is replaced in the cipher text with the pair of letters that make up the opposite diagonals in the key square. Thus the word *bozo* would become QD GS (BO and QD form opposite diagonals, as do ZO and GS). If both plain-text letters appear in the same row they are replaced with the letters that stand immediately to their right; if in the same column, by the letters immediately below (thus *ah* would become VT; *ou*, wrapping around to the start of the row, would become DS).

The German Police double Playfair system used two key squares, with the first letter of each bigram coming from the first square, the second from the second. The Germans added another complication by writing the plain text in a series of rows of twenty-six letters, then encoding the bigrams *vertically*. The key squares changed every day, and with good cipher security practices it would have posed a challenging problem. But the Germans did not follow good cipher security practices, and their insistence on spelling out in full the elaborate SS and Police ranks in the address and signature again and again provided a nearly infallible crib that allowed Tiltman's section to recover the key squares. The most frequent bigrams found in a cipher text usually stood for *ee* or *xx* (X's were used as padding to fill out a line); the next most common ones would occur—and occur right next to one another—whenever repeated letters such as the *pp* in *obergruppenfuehrer* were aligned above repeated letters in a word in the next row of text. Other useful cribs included ranks such as *untersturmbannfuehrer,* place names such as *dniiepropetrowsk,* and words that appeared in daily prisoner-of-war returns, such as *gefangengenommen* ("captured"). Since the first letter of each cipher bigram has to be in the same row or column of the first square as the first letter of each corresponding plain-text bigram, such cribs allowed the code breakers to begin reconstructing the tables. With trial and error the process was fairly straightforward, and it was rare for a day's key not to be broken.

The problem became considerably more difficult starting in November 1942, however, when the German Police began changing key squares every three hours; in September 1944 the problem became almost insoluble for a time when an entirely new cipher system known as "Raster" was introduced. Raster was a stencil or "grille" system, a form of transposition in which the text was written out horizontally into a grid in which a number of the squares on each line were blacked out as in a crossword puzzle; then the columns were transposed in a prescribed numerical sequence and the reshuffled letters copied back out of the grid vertically. Bletchley Park's German Police section grew to a staff of some five hundred to deal with it. At some point someone worked out that it was costing £4 per decode; when the message turned out to be "a request by a police NCO in the Ukraine for an extra issue of underpants" they didn't feel like they were getting their money's worth. But at other times perhaps they, and the other sections at Bletchley Park that handled SS traffic, wished they weren't getting such a return on their investment. Poor security practices by the SS permitted the SS Enigma keys to be broken with relative ease, and from spring 1942 to February 1943, Hut 6 was able to decrypt daily messages reporting concentration camp populations and "departures by any means," which the GC&CS analysts had no difficulty interpreting to mean deaths. The concentration camp reports followed such a terrible predictability that Hut 6 was able to use them as cribs to break the SS Orange Enigma key.

After February 1943 the SS ceased to send these reports by radio. In the spring of 1945, as the Reich was collapsing, a handful of references to concentration camps reappeared in the Police and SS Enigma decrypts. An SS report to Himmler on May 1 bristled at the "most monstrous statements"—

it was a flagrant violation of orders, the report stated, to discuss what went on in the concentration camps—that had been made by SS and SD officers captured by Allied troops in the liberation of Dachau. And several Police messages in April had told of orders to evacuate Jewish prisoners from concentration camps in the German withdrawal so that they could be subjected to "special treatment" before the Allied troops arrived.

In October 1942, the Foreign Office had suggested to GC&CS that a careful file be made of all reports of German atrocities, for use in a future war crimes tribunal. This was done. By the time of the Nuremberg trials, however, captured documents provided far more detailed evidence of the Nazis' methodical work of genocide than did signals intelligence, and the decision was quietly made to keep Bletchley Park's report under wraps. It was released in 1996.

———

Success precipitated a crisis at Bletchley Park in the fall of 1941. The organization had grown chaotically. New arrivals were pouring in each day. An organization used to contemplative and scholarly research was faced with myriad practical problems it had never before had to deal with. Finding food, housing, and transportation for hundreds of people in wartime Britain was by no means the least of it. There was also growing evidence that Alastair Denniston was simply incapable of managing a rapidly expanding establishment that was suddenly in the business of producing a huge volume of a product on a daily basis. "Denniston has shown an amazing lack of imagination and pettiness of outlook," Admiral Godfrey complained. There was a severe shortage of typists and clerks to handle the sorting of traffic, filing of decodes, indexing, and countless other clerical tasks. The Mess Committee struggled to "carry out the wishes of the late Admiral" to provide "a good meal," but the days of the London chef and elegant dinners laid out in the mansion dining room were gone, and the committee was reduced to squabbling over whether one or two helpings of vegetables would be served and whether soup should be put back on the menu. Recreational facilities were nil. There were no more billets available in Bletchley; the nice hotels and inns where the staff had stayed on first arriving were long ago filled up, and new arrivals were being farmed out to private houses. In the Naval Section, Frank Birch complained that it was impossible to recruit girls to work as typists given the "old-fashioned" standard of Bletchley plumbing that prevailed in most private homes: "No baths at all and the W.C. at the bottom of the garden." Meanwhile, only six bombes had been delivered; even so it was proving impossible to find the skilled "high grade men" required to maintain and operate the machines.

Denniston reacted to these and other complaints with mounting irritation and testiness. He shot back a memorandum to Birch complaining of "your somewhat destructive memorandum" and dismissing every one of Birch's suggestions for improving conditions. It was impossible to get more or better food. There was no need to raise the pay offered typists. Only some billets have outdoor plumbing. "What does Birch suggest, that we should move to Harrogate or some such place? We did not ask to come to Bletchley, we must make the best of it. I should say that bad as it is there are worse places in the country where there is not even a cinema. There is certainly one good cinema in Bletchley."

The strains were telling on other members of the old GC&CS staff, bewildered and resentful at the new production-line ethos that the times demanded. Knox, who was seriously ill with cancer by this time but continued working, became petulant and almost selfishly irrational, writing a long and bitter letter to Denniston in November 1941 railing against having to be a mere cog in the machinery. "As a scholar, for of all Bletchley I am by birth breeding education profession + general recognition almost the foremost scholar, to concede your monstrous theory of collecting material for others is impossible . . . had the inventor no right to the development and publication of his discourses, we should still be in the Dark Ages. . . . There are occasions when disobedience is a primary duty." Knox said that he had solved one of the unsteckered Enigma machines only to have the work handed over to Birch, who, he said, is "not even competent" to see that his staff receives traffic when it comes in. Denniston in reply tried to tell him that there was a war on: "If you design a super Rolls-Royce, that is no reason why you should yourself drive the thing up to the house of a possible buyer, more especially if you are not a very good driver." Denniston defended his decision of December 1939 to shift Knox to research and leave the running of Hut 6 and Hut 3 to Welchman and others. But the fact that two years later these tensions were still erupting spoke volumes about what was going wrong at Bletchley Park.

The management situation was even more confused on the naval side, where the code breakers of Hut 8 were battling Birch and the Naval Section in Hut 4, and where Turing was nominally head of Hut 8 but in reality lost in theoretical clouds while Hugh Alexander was attempting to run things on a day-to-day basis but without the actual authority to do so.

And so Edward Travis, Denniston's deputy, and Nigel de Grey, a veteran of Room 40 who was now another senior administrator, began plotting to oust Denniston. It was Welchman who forced the crisis. Despairing of getting action if he went through channels, he conceived the audacious plan of appealing directly to the Prime Minister. Turing, Welchman, and Alexander had been presented to Churchill in July 1941 when they were summoned to the Foreign Office and "C" handed them each a check for £200 for their achievements. Two months later, on September 6, Churchill made a surprise visit to Bletchley Park, driving through the town and up to the gates in a large motorcade with flags flying ("another instance of our wonderful security," grumbled Nobby Clarke). Standing on a tree stump by the lake he had looked over the assembled staff. "You all look very . . . innocent," he began. Churchill expressed his thanks for the work they were doing, and made several gestures during his visit that showed the personal gratitude he felt to the toilers of Bletchley Park and his concern for their welfare. Looking out a window of the mansion, he had seen two girls playing catch with a ball on the lawn and had asked Denniston what the workers did in their spare time. Denniston admitted that there was not much to do. Churchill immediately ordered that a tennis court be built—probably the only tennis court built in Britain during the entire Second World War. Travis took the Prime Minister on a tour of several of the huts, and when the party turned up behind schedule at Welchman's office, Travis leaned over and said in a stage whisper, "Five minutes, Welchman." Welchman had been told to prepare a ten-minute talk, so he launched into his opening remarks—"I would like to make three points"—and had galloped through the first two points when Travis interrupted to say, "That's enough, Welchman." Churchill seized the opportunity for another show of solidarity with the troops; as Welchman recalled, the Prime Minister gave him a "grand schoolboy wink" and innocently asked, "I think there was a third point, Welchman?"

Encouraged by these signals, Welchman drafted his personal appeal to Churchill to do something about the staff shortages and organizational problems. "Some weeks ago you paid us the honour of a visit," his letter began, "and we believe that you regard our work as important. . . . We think, however, that you ought to know that this work is being held up, and in some cases is not being dealt with at all. . . . Our reason for writing to you direct is that for months we have done everything that we possibly can through normal channels, and that we despair of any early improvement without your intervention." The letter went on to itemize the bottlenecks: The breaking of naval Enigma keys was being delayed at least twelve hours every day because of a lack of staff to run the IBM equipment, necessary to supply the data for the Banburismus procedure. Luftwaffe Light Blue traffic from the Middle East was going undeciphered because of a shortage of trained typists. Testing the bombe solutions was putting a huge strain on Hut 8 and Hut 6 staff. Although the letter did not directly criticize Denniston, it did so by implication, praising Travis and underscoring that he was not responsible for their difficulties.

Welchman was a practiced composer of such letters to higher authority—

his "screeds," he called them—but he had never contemplated anything quite so brazen. He talked Milner-Barry, Turing, and Alexander into appending their names, then at the last minute was seized with uncharacteristic doubts and suggested calling the whole thing off. But the other three argued him back, and Milner-Barry, for reasons he could later not recall, was deputed to be the one to carry the letter to Churchill—in person—on October 21. The date was Trafalgar Day, the anniversary of Lord Nelson's defeat of the French and Spanish fleets, and Milner-Barry later described his surreal adventure:

> What I do recall is arriving at Euston Station, hailing a taxi, and with a sense of total incredulity (can this really be happening?) inviting the driver to take me to 10 Downing Street. The taxi-driver never blinked an eyelid. . . . Arriving at the entrance to Downing Street, I was again surprised by the lack of formality: there was just a wooden barrier across the road, and one uniformed policeman who waved my driver on.

Milner-Barry had nothing with him to prove that he even worked for the government, but he gamely marched in and announced that he had come from a secret war station and needed to see the Prime Minister immediately on a matter of national importance. Churchill's principal private secretary, Brigadier Harvie-Walker, explained that that was quite impossible without an appointment. Milner-Barry said that for security reasons he could not deliver his letter to anyone but the Prime Minister in person. It seemed like an impasse, but Milner-Barry was able to refer to Churchill's recent visit to Bletchley Park, which Harvie-Walker knew of, so Churchill's secretary at last promised to see that the letter would be delivered directly to the Prime Minister.

Churchill responded the following day in characteristic style. As First Lord of the Admiralty he had had red labels printed up that read ACTION THIS DAY, which he would plaster freely across memorandums, and it was with those words that he began his order, scrawled at the bottom of Welchman's memorandum: "Make sure they have all they want on extreme priority and report to me that this has been done."

———

Action occurred, if not precisely that day, then at least with a rapidity that was unknown to Bletchley Park. The military services made more men and women available. The Ministry of Labour was ordered to meet with "C" and Denniston and to come up with whatever manpower they needed. Orders for more bombes—at some £5,000 apiece—were placed with the British Tabulating Machine Company.

And by February 1942 Denniston was out. To ease the blow he was allowed to keep his title as Deputy Director of GC&CS ("C" had all along held the title of "Director"), but Denniston henceforth would be "DD(C)," the C standing for civil, meaning diplomatic and commercial traffic. Those sections, with a staff of about seventy, were moved to Berkeley Street in London. The remaining thousand or so workers at Bletchley Park were placed under the direction of Travis, who became "DD(S)"—S for service, as in military services.

Though it was more a matter of coincidence than a direct result of these changes, other things began rapidly falling in place, too. The Army Enigma keys, which had stubbornly resisted cribbing, suddenly broke open. Chaffinch, the main Army key for North Africa, had yielded its first substantial break on September 17, 1941, and decrypts grew at a steady pace from that date on. Easily broken traffic carried on a new Italian machine, the Hagelin C38, began to provide a flood of information on Mediterranean convoys resupplying the German forces in North Africa.

Success bred success. Breaking one Enigma key frequently led to breaks in others when the same text, often orders from higher command or general informational messages, was rebroadcast on several networks. Reading the Enigma traffic revealed that such "reencodements" also occurred between Enigma channels and hand ciphers, which could henceforth provide a source of cribs. One of the most prolific sources of reencodements during the fall of 1941 was the *Werftschlüssel,* or "dockyard cipher," a hand cipher used mainly by small ships in the Baltic; Hut 8 quickly discovered that it routinely carried messages that would be repeated verbatim in naval Enigma traffic sent to larger vessels and U-boats—weather messages and mine warnings in particular.

Direct reencodements were fairly easy to match up with Enigma cipher text, since a long crib that contained no "crashes" was almost a sure bet. On several occasions when the dockyard cipher failed to provide a likely crib, and Hut 8 was under pressure to achieve an early break of the day's naval Enigma setting—for example, when an Arctic convoy was about to sail—the RAF authorized mine-laying operations in areas specifically chosen to generate wordy cribs. Mine laying by air was already known as "gardening"; in support of Bletchley Park's needs these missions were known as Operation Garden.

The proliferation of separate Enigma keys as German operations expanded on fronts poorly served by landlines was both a blessing and a curse: It greatly increased the workload for the bombes and the decoders, but it provided more possibilities for reencodements that could be used to "cross-ruff" from one network to another. The color-coding system was quickly becoming unwieldy and a new system of naming keys was instituted at Bletchley: Luftwaffe keys were given names of insects (Hornet, Wasp, Cockroach); Army keys were given bird names (Chaffinch, used for communications between the African Panzer Army and higher authority; Phoenix, for use between army, corps, and division levels in Africa; Vulture, for the Russian front); the Navy, predictably, got fish (Dolphin, the "home waters" key used in the Atlantic and Baltic; Shark, the separate U-boat key that would be introduced in the Atlantic on October 5, 1941; Porpoise, for the Mediterranean).

Finding "straight cribs" was more challenging than "cross-ruffing" with reencodements, but the continuous reading of traffic and the identification of the foibles of individual operators soon filled a growing larder of likely cribs. In general even the best cribs only lasted a few days at a time, which was why continuity was so important. But there were notable exceptions. In North Africa, several Luftwaffe and Army operators could be relied upon for weeks or even months at a stretch to begin a daily message with exactly the same phrase, sometimes sent at exactly the same time each day. Standbys included LAGE UNVERAENDERT ("position unchanged"), DAS IST EIN ABSTIMMSPRUQ ("this is a coordination message"—in the German wireless procedure *q* could be used in place of *ch*), BEFEHL FUER DIE KAMPFFUEHRUN ("instruction for the conduct of operations"), AM TAGE UND IN DER NAQT KEINE FEINDTAETIGKEIT ("Through the day and in the night no enemy activity"), and TAGESMELDUNG ("daily report"). Gordon Welchman forever regretted the capture of a German unit that reported in every morning with the words VERLAUF RUHIG—situation unchanged—and expressed a wish that he would be consulted before the British Army took any more prisoners.

Through the course of 1941 Bletchley Park had gone from being stymied by maddening cryptanalytic obstacles to being overwhelmed by its own success; by the end of the year it was at last hitting its stride.

CHAPTER 7

The Machines

T HE U.S. Army was nothing if not punctual, and at precisely nine o'clock on Saturday morning, August 16, 1941, a meeting in Room 3341 of the Munitions Building began. Seated at the table were all the leading lights of the Signal Intelligence Service, assembled to welcome Alastair Denniston to Washington. Denniston had been dispatched to try to smooth some of the frictions that had arisen in the U.S.–British relationship. Cooperation in the Far East on the Japanese naval codes was working well. The Enigma was another matter. The British view was simple: As Denniston laid out the situation to "C," the Enigma traffic was "almost life-blood" to the British war effort, whereas for the American cryptanalysts it was at most a "new and very interesting problem." GC&CS had been "aghast" to receive a request from the Americans for a bombe. Britain had only six in operation and could not possibly spare one. The very fact that the Americans had put this request in writing was a breach of security precautions. It was out of the question to permit an American firm to manufacture bombes because of the security risk. In any event, GC&CS had no intention of letting SIS or OP-20-G handle the actual breaking of Enigma traffic and distribution of decrypts. If the American cryptanalysts wanted to study the Enigma from a theoretical point of view and possibly come up with new ideas on how to tackle it, that was fine. But handling operational traffic was a British prerogative, and would remain so.

Meanwhile resentment was growing in Washington. The British had received the Purple machine; they kept talking about full cooperation; but now they seemed to be throwing obstacles in the way. A modest request for Enigma traffic and recovered key settings had been sent from Washington on July 15; a month later Bletchley Park explained that the material "was still being copied."

Denniston decided that a trip to Washington "to clear up the position on E traffic" was called for. Denniston hoped to placate the Americans by explaining the shortage of bombes but also by proposing that the U.S. Army and Navy send some "young mathematicians" to Britain to work at Bletchley Park directly on the Enigma. "C" vetoed both ideas the same day. "I am a little uneasy about the proposal for young mathematicians," he wrote back, and "I should feel inclined not to mention" the bombe situation, lest that be used as an argument to have them built in America. Denniston's mission, in other words, was to continue making polite noises whenever the subject of the Enigma came up.

There was one piece of unfinished business that Denniston *was* authorized to do something about, however. That was Herbert O. Yardley. After his sensational departure from the Black Chamber a decade before, Yardley had tried writing a few adventure novels, one of which, *The Blonde Countess,* was made into a movie in barely recognizable form by MGM. He had then failed in a brief attempt to become a real estate speculator. But then he was approached by Chiang Kai-shek, who offered him $10,000 a year to enter his services as a code breaker. Yardley took up the offer and spent months being bombed by the Japanese in Chungking while tackling a few low-level Japanese Army field codes. In May 1940 Yardley's contract was up and he went to see the American military attaché, Major William Mayer, to ask a favor: Would Mayer back up the story he had told his Chinese employer—that the War Department was eager for him to return to Washington? It wasn't clear whether Yardley was hoping to get a better deal on his contract or was trying to cook up a tale that would persuade the Chinese simply to let him leave the country, but in any case Mayer told him to go jump in a lake and sent a blistering report back to Washington. Yardley's discretion was as unreliable as ever, Mayer wrote. Yardley went under the name of Osborn and was "ostensibly a dealer in hides," but in fact "his real name and occupation are an open secret in Chungking, probably due to his own indiscretions." He "drinks a great deal" and is obsessed with two topics: sex, and a violent hatred for William Friedman. "I recommend we have as few official dealings with him as possible."

Yardley nonetheless managed to get out of China without Mayer's help, arrived back in Washington, and convinced his old colleague General Mauborgne, who was soon to retire as Chief Signal Officer, to give him a contract to write up a series of reports on his recent work. Friedman, desperate for any scraps of information about the Japanese military codes, reluctantly went along with the idea, but would have nothing personally to do with a man he considered a blowhard and a betrayer of trust. Frank Rowlett was deputized to handle negotiations with Yardley and to oversee his work.

Six months and $4,000 later, Yardley turned in six reports that Rowlett deemed worthless, and that was that. But Yardley was to have one final hurrah to his long and odd cryptanalytic career. In the spring of 1941 two mathematicians from the Canadian National Research Council showed up in Washington and called on Mauborgne and Abraham Sinkov at SIS. Canada wanted to set up its own code breaking bureau; could the United States help? Mauborgne was not very encouraging, but then had a brainstorm. Why don't they hire Yardley? A hasty interview was arranged at the Canadian Embassy, and two weeks later Yardley was on his way to Ottawa to set up Canada's new "Examination Unit."

In their innocence, the Canadians took Mauborgne's endorsement as a sign that Yardley enjoyed the confidence of the American secret services and that all had been forgiven. In fact, no one in SIS or OP-20 wanted to touch Yardley with a ten-foot pole, and when GC&CS discovered in the summer of 1941 that Yardley was the head of the Canadians' new bureau, the British promptly stopped sending Ottawa any intercepted traffic. The Admiralty had been pushing Washington to establish a liaison with Ottawa but had found the American response "frigid," and suspected they now knew why. So when Denniston arrived in Washington in August, one of the first things he did was to reassure Friedman that British cooperation with the Canadians "would be wholly dependent upon the elimination of Mr. Yardley from the latter organization." Yardley was both a security risk and an embarrassment as far as GC&CS was concerned, and the British told the Canadians in no uncertain terms to get rid of him, softening the blow by offering to supply a British expert to take his place.

Still under the illusion that Yardley enjoyed the support of the U.S. government, the Canadians hastened back to Washington in a feeble attempt to play off the United States and Britain and were crestfallen to learn what the Americans really thought of Yardley. Admiral Leigh Noyes, the director of OP-20, said that Yardley ought to be locked up. Friedman hinted that packing Yardley off to Canada had been a convenient way to get rid of an embarrassing presence in Washington. By January 1942 Yardley was out of a job again. Yardley never took life too seriously but his remaining years could not have been easy for him. He bought a Washington restaurant, which stayed in business for nine months before closing, then took a job in the Office of Price Administration. After the war he bought a small electrical appliances business, which folded up a year later, started a vacuum cleaner business, which failed, tried his hand at construction, and finally went back to writing, completing a classic poker instruction book, *Education of a Poker Player,* that appeared in November 1957, a month after he suffered the stroke that would lead to his death the following year.

One source of marvel to Denniston during his August 1941 visit, and to other British cryptologic visitors who followed, was the extensive and, to them, extravagant use of IBM equipment by both SIS and OP-20-G in analyzing enemy codes. "They make far greater use of these machines to avoid use of personal effort," Denniston reported, "but I am not convinced that these mechanical devices lead to success. Close personal effort makes one intimate with the problems which, when served up mechanically, fails to appeal." Despite the proven success by late 1941 of the bombes, and of the mathematical ingenuity behind them, the old British disdain for science and technology had not completely vanished. It was true that the Americans probably overdid it, but that was in part a reaction to the wilderness years when they had to do without. Back in the early 1930s Friedman had recognized the enormous advantage of IBM machines for compiling codes. With an IBM employee, he had worked out a detailed scheme for using punch cards to carry out the task. Rowlett, Sinkov, and Kullback had spent months placing sixty thousand index cards in alphabetical order by hand to prepare the new War Department code; that was precisely the sort of task that the IBM sorters could do automatically. But Friedman also saw how valuable punch-card equipment might be for cryptanalysis. Finally, on October 30, 1934, he wrote to Major Akin pleading for a few of the machines:

> In many years service here I have never once "set my heart on" getting something I felt desirable. But in this case I have set my heart on the matter because of the tremendous load it would lift off all our backs. . . . Please do your utmost to put this across for me. If you do, we can *really* begin to do worthwhile *cryptanalytic* work.

Friedman figured that to compile all the other codes he had been assigned would take all six of his employees two and a half years of full-time work. With IBM machines he could cut it to six months. Renting two punches, a card sorter, and a printer for six months would cost $2,250. If that was beyond the War Department's means, then Friedman somewhat desperately suggested that two months' rental, $750, would at least allow them to get through their biggest task, compiling two editions of the War Department Staff Code. Akin passed Friedman's request up the chain of command. A month and a half later the answer came back from the Adjutant General: No funds were available, period.

To make things worse, Friedman learned "by a devious route" that the Navy's OP-20-G had been able to acquire several IBM machines. His "chagrin was almost unbearable," he later recalled. But then by an equally devious route Friedman discovered that a new officer had just been assigned to take charge of a division in the Office of the Quartermaster General in the Munitions Building that had been using IBM equipment to manage the accounts of the Civilian Conservation Corps. The new officer had no time for the newfangled ideas of his predecessor and wanted to be rid of the machines at once. Friedman went to him and suggested that he just let the lease continue until the expiration of the existing contract, which ran through June 30, 1935; in the meanwhile Friedman would take the machines off his hands. The officer happily agreed, and SIS was in business.

Although by the standards of the postwar computer era IBM punch-card equipment was primitive in the extreme, it was surprisingly powerful and in many ways perfectly adapted to the work of code makers and code breakers. Punch cards had first been used over a century earlier in the Jacquard loom, which wove complex patterns by following a sequence of operations specified in a stack of as many as thousands of cards. In the late nineteenth century the American engineer and inventor Hermann Hollerith had the idea of combining the punch card with an automatic sorter and tabulator that would allow census data to be compiled automatically. For the 1890 census Hollerith provided the Census Bureau fifty machines at $1,000 apiece for a year's rental, plus 100 million cards at 31.7 cents per thousand. Information such as age, sex, home ownership, and native language was entered on the cards, one per person, by punching a hole in the appropriate column. The cards were then placed one at a time, by hand, in a "press"; a top plate was swung down and wherever a hole had been punched a spring-loaded metal pin would pass through the card to make contact with a little cup of mercury below. That would complete an electrical circuit causing a counter to advance one click.

The system automated the task of tallying endless columns of figures, but that was not its real power. The fact that one card stood for one person was the real brilliance of Hollerith's concept, for it made it possible to reshuffle the raw data at will according to any criterion that a statistician wanted to analyze. In addition to the "tabulator" that did the counting, Hollerith invented a "sorter" that could, in a simple sense, be programmed to divvy up cards into categories. A card would be placed in the press, and depending on which columns were punched, a relay would trip the spring-loaded cover on one of several different boxes. The operator would then place the card in the opened box. Once sorted, the cards could be run back through the tabulator to tally, say, how many non-English-speaking males under age thirty live in Massachusetts.

Besides the Census, another early user of the Hollerith machines was the War Department. The Surgeon General's office rented one of the first tabulators in 1888 to compile monthly health statistics on individual soldiers. Over the subsequent decades Hollerith provided his customers with increasing flexibility by adding more and more relays to the tabulators. Relays could be wired in a custom arrangement to trip a counter only when several simultaneous conditions were met. At first this involved resoldering wires, but in 1905 Hollerith borrowed an idea from telephone switchboards and equipped the tabulators with plugged cables for quick reprogramming. Automatic feeders for the sorters and tabulators were also introduced. By 1907 the company began to attract sizable commercial customers who were putting the system to use to keep track of accounts or freight waybills or payrolls. (At one point Hollerith toyed with the idea of supplying the machines for free but charging only for the cards—a large commercial customer such as the Marshall Field department store was using ten thousand cards a day, and at 85 cents to one dollar per thousand cards each such user represented a pure profit to Hollerith's Tabulating Machine Company of several thousand dollars a year.)

In 1911 the company merged with two other firms to become the entity that would later change its name to International Business Machines. That company was narrowly saved from extinction ten years later by the timely introduction of a printing tabulator. Another innovation was a coding system for the cards that allowed either numbers or letters to be represented. Eventually the standard IBM card contained eighty columns; in each vertical column a series of holes would be punched in a unique pattern that stood for one particular letter or number. A typewriterlike keypunch machine did the actual punching. But other than these improvements, the basic concept of the punch-card system had by the 1930s changed little from Hollerith's original idea. Neither had the company's business strategy altered: It refused ever to sell its equipment, and its rental schedule was steep. A printing tabulator went for $250 a month, $1,500 a year. Sorters and punches rented for $25 to $60 a month.

When the code breakers at last were able to get their hands on some of this equipment, Rowlett and Kullback at SIS and, separately, Thomas Dyer at OP-20-G (which had managed to get the Bureau of Ships to provide $5,000 in 1931 to rent IBM machines) worked out several cryptanalytic procedures using the machines. One of the earliest applications was what were called "language studies": a large body of plain text would be punched onto cards, one word per card. The sorters and tabulators would then be used to print a list of words in alphabetical order showing the frequency of each; the cards could also be resorted and placed in order of descending frequency. Likewise, frequencies of individual letters and digraphs, trigraphs, and so on could be compiled.

The other basic use of IBM equipment was to produce catalogues of message groups that appeared in enciphered codes. This was a huge leap forward in the search for "double hits," the first step in building up a depth, stripping off the additives, and figuring out the meanings of the underlying code values. The usual procedure was to punch a card, one for each message, containing the first half-dozen cipher groups of the text. This was the part of a message most likely to contain stereotyped words such as addresses or message numbers, and thus to contain repetitions from one message to the next. The cards would next be sorted in numerical order of the first cipher group and a list printed; then resorted by the second cipher group; and so on. A portion of a printout sorted by the second group, for example, might look like this:

91245	89610	73950	65210	19234	message 723
62390	89612	43563	01836	72950	message 12967
30017	89612	51234	08324	81563	message 9720
73956	89612	61027	01836	91739	message 18378
29762	89613	98623	25649	35612	message 5290

All messages containing the same cipher group in the same position would thus appear together in the printout; the tabulator could be set up to print an extra blank line between entries with different cipher groups to delineate them into groups. Scanning the lists by eye would then reveal double hits, such as between message 12967 and 18378 in the example above, where 89612 appears as the second cipher group and 01836 as the fourth.

Another frequent early use of IBM equipment was to produce indexes of recovered code groups that listed the code groups that preceded and followed each group in every message in which it appeared. This was of enormous help to the book breakers whose job it was to figure out the actual linguistic meanings of recovered code groups; whenever a code group appeared in a message they could immediately look it up in the index to find other instances where it had occurred and in what context. This was the forerunner, by several decades, of the standard technique that would be later known as the Key Word In Context or KWIC index that would be used in many fields beyond cryptology.

And, finally, for tackling machine ciphers or other polyalphabetic ciphers, there was the indexing procedure that Samuel Snyder and Lawrence Clark of SIS had worked out in 1937. The aim was to find any strings of repeated cipher text within a large body of messages. (This same problem was what Turing's Banbury sheets performed in a far more labor-intensive fashion.) The procedure was to punch each message on cards in fifty-character increments; the other thirty columns on each card would be used for identification numbers that would keep track of which message each string had come from. The next step was to produce a whole series of additional card decks that contained every sequence of fifty characters found at every possible point in every message. Snyder worked out a way to do this automatically using a machine called a reproducing punch. This machine would read the first two message cards, which contained characters 1–50 and 51–100 of the first message, and reproduce the data on fifty new cards by shifting one column over at a time: the first new card would contain characters 2–51, the next 3–52, then 4–53, and so on. The punch would then repeat the procedure for the second and third cards of the original message deck, and so on until the end. The result was tens of thousands, or even hundreds of thousands of cards that contained every sequence of fifty cipher characters that appeared anywhere in the entire body of traffic. Those cards were then sorted and printed out in an alphabetical catalogue, which would make repeats of any length immediately obvious.

All of this was virgin territory. Some of the things that SIS and OP-20-G wanted the machines to do nobody had ever tried before. It was a condition of the lease contract with IBM that the user would not touch the inner wiring of the tabulators. But Rowlett and the others eventually couldn't resist pulling the cover plate off and plunging in. One of them would always first scout out the machine room to see if the IBM service representative was on the premises; if the coast was clear the rest of them would quickly descend on the machine with screwdrivers and soldering irons to make the latest adjustment they had cooked up.

The SIS, still strapped for funds, had managed at last to get a budget to rent its own machines, though by the beginning of 1940 it still had only six machines of its own. But it was continually running out of IBM cards, and when that happened Snyder would be dispatched to the adjacent Navy Building to beg or borrow more. The Army's purchasing authorities refused to order more than 100,000 at a time, so whenever a new order arrived it was usually just enough to pay back the Navy, who were then promptly hit up for a new loan. Finally in March 1940 Friedman managed to convince the powers that be to order 500,000 cards, and they at last caught up on their debt.

Denniston's mild disdain for mechanical methods was shared by some of the American cryptologic pioneers. Mrs. Driscoll would have nothing to do

with the IBM machines at first. The view was not entirely that of curmud-
geons or Luddites. There were some essential steps in breaking a code, especially an enciphered code-book system such as the Japanese naval codes,
that depended more on feel and insight than anything mechanical. The essential break in JN-25 came from exactly such an inspiration. The initial attack on JN-25 had begun at OP-20-G not long after the new code first
appeared on June 1, 1939, with extensive IBM runs to search for double hits.
But after a year of effort, and lavish expenditure of machine time, only a
small number of messages had been placed in depth, allowing only a few
dozen of the thirty thousand groups of the additive book to be recovered.

The hits that linked these messages together, and thus to shared stretches
of the additive book, had all occurred in the first few cipher groups of each
message, which raised the possibility that the underlying code groups in
each case stood for numbers: "Reference your message no. 374" was a typical, stereotyped way to begin a dispatch. Someone at OP-20-G suddenly recalled that in an old four-digit code that had been among the booty stolen
from the Japanese consulate in New York in one of the Navy's black bag
jobs, numerals were represented in highly patterned fashion. The JN-25 team
went to the files, pulled out the old Japanese "S" code, and sure enough, zero
was assigned the code group 0000, one 0102, two 0204, three 0306, and so on.
Placing messages in depth and extracting additives and code groups gave
only relative values for the groups, and one of the cherished goals of the
cryptanalysts was to somehow anchor these recovered values to the true
five-digit numbers that appeared in the actual Japanese code and additive
books. One reason this was so important was that code books usually included a "garble check" feature, some pattern or formula in the way code
groups were assigned that served to ensure that a message had been copied
and decoded correctly by its recipient. Discovering the garble check would
be a breakthrough of the first order, for it would immediately provide a
check on whether a recovered code group value was valid, and so eliminate
many false leads. Several of the most frequent code groups that appeared in
the JN-25 traffic were 13343, 13445, 13547, and 13649. The JN-25 team immediately realized that these had a common difference of 00102, which perfectly fit the "S" code's pattern for numerals. If 13343 stood for the numeral
one, with a true value of 00102, then the other four high-frequency groups
would be 00204, 00306, and 00408. The whole story came together in a single
day in the fall of 1940. In a flash, the code groups for all numbers from 0 to
999 were known; these alone made up 3 percent of the entire book. The garble check feature had also been busted open: In all of these valid code
groups, the sum of their digits was divisible by three.

But once the initial break occurred the job was more a matter of rote than inspiration, and in the end it was only a lack of manpower—and machine power—that prevented the Navy from reading JN-25 in the critical months before Pearl Harbor. Recovering 50,000 additives (a new additive book issued October 1, 1940, had grown to 500 pages from the previous 300) and 30,000 code values was a job perfectly suited to the brute-force abilities of the IBM machines. Message upon message needed to be differenced and searched against tables of already identified code groups. The JN-25 team grew to about ten people; Station Cast, the Navy's cryptanalytic unit at Cavite Naval Base in the Philippines, was thrown into the assault; and hopes were high they would begin reading current traffic by the end of the year. But on December 1, 1940, the Japanese Navy threw out its old code book (Able, in the American code name) and introduced an entirely new one (Baker). Luckily the Japanese also made an incomprehensible error: For two more months they kept using the same additive book (number 5) that had been brought into service in October. The additives recovered throughout the fall from the Able-5 version of JN-25 could thus be used to strip the new Baker-5 traffic to begin recovering the Baker code groups.

But even with the help of Cavite, and now the British, the clerical workload was simply overwhelming. Adding to the difficulty was that the new Baker code (JN-25B, in the later U.S. nomenclature) was a two-part code, whereas Able (JN-25A) was a much simpler one-part code. In Baker, that is, the numerical order of the code groups no longer followed the alphabetical order of the words they stood for. On February 10, 1941, while the Sinkov mission was still at Bletchley Park, Admiral Godfrey radioed to the Far East to have the British intercept and decryption unit at Singapore exchange liaison officers, and results on JN-25, with Station Cast at once. The British and Americans found they had each made about equal progress, though on different parts of the additive book, so by pooling their results they nearly doubled the number of recoveries. By August 1, 1941, they had recovered 4,800 additives in the current book. Then it was back to square one yet again as Baker-6 was superseded by Baker-7. A few days before the Pearl Harbor attack only 2,500 additives and 3,800 code groups had been recovered. The subsequent crash effort in Washington and Honolulu that consumed millions of IBM cards and cracked Baker-8 in time to begin current decryption in mid-March, and so turn the Battle of Midway, amply vindicated the American obsession with the machinery. The only American fault, indeed, had been not being even more dependent on IBM methods earlier. From December 1941 to June 1942 Station Hypo recovered 25,000 additives, using IBM machines to automatically calculate and search for common differences between cipher groups in aligned messages. OP-20-G in Washington came up with another 16,000. Only by the extensive use of machines were such scores racked up.

———

The other major area of U.S.–British cooperation that produced quick results also would have got nowhere were it not for the American enthusiasm for IBM equipment. This was the attack on the German diplomatic code that GC&CS called "Floradora" and SIS called "Keyword," and it was the needle-in-a-haystack search to end all such searches. Floradora was a doubly enciphered code; two separate starting points in a 10,000-line additive book would be chosen, the lines of additive drawn from each place would first be added together and then to the code text. That made for 50 million possible unique key sequences, which the Germans were confident gave the system a degree of security approaching that of a one-time pad. When British forces occupied Iceland in May 1940 they rescued a complete copy of the basic code book from a pile of burning papers at the German consulate. When the book was replaced with a second edition, it was the United States' turn to contribute. In July 1940 the Germans dispatched a copy of the new book to South America via a courier who lacked diplomatic immunity—a businessman named Dr. Emil Wolff who worked for the I. G. Farben chemical company. According to one account, the FBI was tailing him; as his boat, the Japanese steamer *Yasukini Maru*, passed through the Panama Canal, Wolff became nervous and threw the briefcase with the code book overboard. The FBI recovered it intact. (The haul also included several thousand pages of one-time-pad key for the other main diplomatic code, known as "GEE.") But even with the basic book, breaking the double additive seemed impossible, even after a French agent managed to obtain a worksheet that contained twenty-five lines of additive. So hopeless did it seem that the British at one point stopped intercepting traffic and discarded their files.

SIS had begun intercepting the German diplomatic signals in earnest in fall of 1939, and by 1941 had punched all of the messages onto IBM cards. With the additive lines supplied by GC&CS, a team under Kullback then went to work on a huge machine search. Because of the way the additive book was constructed, the twenty-five lines of additive immediately gave twenty-five more lines. Kullback's group, which included Frank Lewis and Delia Taylor, combined those fifty lines in every possible combination to produce 1,250 different possible lines of double additive. That was 1,250 out of the 50 million total possible strings of double additive, which meant that perhaps 1 in 40,000 messages might have been enciphered by one of these strings. But the odds were skewed by the habits of the code clerk at the German Embassy in Brazil, "a lazy son of a bitch," in Kullback's words, who reused the same lines over and over, including a number of the lines that GC&CS had happened to obtain. Kullback's team added each of the twelve most commonly used code groups to every possible position in the 1,250 lines of double additive to produce 15,000 lines of "hypothetical cipher." The machines then begin the Sisyphean task of combing through every intercepted message to see if any of these hypothetical cipher groups appeared. About a quarter of a million hits were found. These were printed out in an index and double hits were searched for by eye; there were a total of twelve. But that was enough to start: It allowed fifty-five messages to be placed in depth, and the laborious task of recovering additional additives could begin.

The job took over a year. Starting in August 1941 the SIS team began working night shifts, mainly so they could use the IBM machines in the Adjutant General's offices in the basement of the Munitions Building to supplement their own small installation. William Lutwiniak, who reported to SIS on February 1, 1941 (he was one of the winners of American Cryptogram Association contests whom Friedman spotted and recruited), remembered running the equipment so hard that they would put five or six of the IBM machines in the Adjutant General's office out of commission in a night: "But we'd get the job done. We'd go back upstairs and I don't know what they thought of us when they had to clean up the next morning." The German diplomatic section of GC&CS, now headed by a new recruit, William Filby, the former science librarian at Cambridge University, was inspired to renew the charge. Ernst Fetterlein, the czarist cryptanalyst who worked at GC&CS between the wars, came back from retirement to help. Breaking the indicator system would of course speed the process immensely, for that would reveal precisely which lines of additive were being used on each message. And again a remarkable bit of luck came to their assistance. One day Denniston received a package from the British consul in Portuguese Mozambique. "Dear Alastair," the cover letter began, "this was dropped in my box by a sailor, thinking it was the German Consulate. Are the contents of any use to you?" They were. The misdirected sailor had been charged with delivering the daily indicator keys for the next three months.

With the keys in hand, the task then was to build up chains of messages that shared a line of additive. If a message contained one line of additive that had been recovered, it was possible, by guessing at stereotyped beginnings and endings of messages, such as BERLIN STOP or GLEICHLAUTEND AN ———— ("copied to," followed by the name of other embassies), to recover the other line of double additive. Hunting for another message that contained that second line as one of its two double additives would split apart the encipherment of that message to yield yet another additive line. The additive recoveries snowballed. Kullback spent several months in Britain during 1942 working with Filby, and Filby came to Washington in December 1942 for several weeks. Finally on February 15, 1943, Kullback sent a message to the American liaison officer at GC&CS: "Captain JOHNSON. Floradora. We DOOD it." They had gotten out the last line of additive. Within two months the traffic was being read currently. When Filby brought the first current message in to show Denniston, his chief looked blankly at it for a while until Filby pointed to the date of transmission. Denniston almost jumped. He at once dialed the scrambler phone on his desk. "Denniston here, may I speak to 'C'?" A pause, and then: "'C,' I have some good news for you, may I come over?" Denniston put on his hat, folded the message neatly and tucked it into his inside coat pocket, and practically ran from the building.

Later in the war a system was worked out to perform the decipherment of the Floradora messages completely automatically. Often the keys were known because of the German habit—such was their faith in the system's security—of transmitting future sets of keys in the messages themselves. In such cases the intercepted cipher text was punched on cards; decks of prepunched cards for the two indicated additive lines were pulled from the file and run through the tabulator along with the message text; the tabulator would automatically perform the subtraction and print the resulting code groups. When keys were not known, the cryptanalysts would guess the first two or last two code groups of the message, a task made easy by the stereotyped phrasing that was almost always used, calculate the resulting additive that had been applied, and an IBM search would be made through all possible double additives to find one combination that matched. The IBM machines could perform that search in about two hours. Brute force carried the day.

––––––

The Japanese attack on Pearl Harbor on December 7, 1941, had caught America unprepared on countless fronts. The millions of dollars that Roosevelt was pouring into munitions industries to build tanks and planes and ships were barely beginning to show results. The draft bill that had been approved by Congress in 1940 provided for only a single year of military service; in the summer of 1941 Congress began debating whether the Army, come October, would simply pack up and go home. There was strong sentiment to do just that. A *Life* magazine reporter visited an Army camp in Alabama to see what the "selectees"—no one called them soldiers—thought about it. The answer was not much. "We came here to learn how to fight a blitzkrieg," grumbled one private, "instead we get close order drill and kitchen police." Maneuvers were carried out with wooden rifles and cardboard boxes marked "tank." Everywhere—on the walls of latrines, on trucks, on artillery pieces—the word "OHIO" was chalked. OHIO stood for "Over the Hill In October," and "over the hill" was Army slang for going absent without leave. On August 12 the momentous vote came in the House of Representatives. The galleries were packed as Speaker Sam Rayburn struggled to keep the Democrats in line, but they were deserting their President in droves and voting against extending the draft. After the initial roll call it looked like the bill had failed to muster a majority. House rules allowed members to change their vote by approaching the rostrum one at a time and seeking the Speaker's recognition. This went on for a while until the total finally edged up to 203 in favor, 202 opposed. Rayburn banged his gavel, ignoring the cries of outrage from Republicans, and declared the bill had passed.

The narrowness of the vote disconcerted many in Britain who saw it as another sign of America's lack of resolve and weakness; even if America entered the war, would she fight? But Churchill saw the essential truth about America. "No American will think it wrong of me," he said in explaining his reaction to the Japanese attack on Pearl Harbor four months later, "if I proclaim that to have the United States at our side was to me the greatest joy." It meant, simply, that the war was won. "Silly people—and there were many, not only in enemy countries—might discount the force of the United States," believing that Americans were soft, paralyzed by fractional politics, unable to stomach bloodshed. But Churchill had studied the Civil War and how it had been fought to "the last desperate inch"; he recalled what the British politician Edward Grey had told him thirty years earlier, that "the United States is like a gigantic boiler. Once the fire is lighted under it there is no limit to the power it can generate."

Churchill hastened to confer directly with his new ally, setting sail across the Atlantic in the battleship *Duke of York*. Anchoring at Hampton Roads on December 22, Churchill was seized with impatience to see the President at once; abandoning arrangements that had been made to cruise up the Potomac the next day, he ordered an airplane to fly his party to Washington that very night. As the plane approached the city they looked down and, for the first time in years, beheld a city blazing with light. "Washington represented something immensely precious," recalled Churchill's aide-de-camp, Commander C. R. "Tommy" Thompson. "Freedom, hope, strength. My heart filled."

Like the might of the country itself, Washington's strength was more potential than reality in December 1941. The city still slumbered with what new arrivals jocularly called its northern charm and southern efficiency. The American capital was woefully ill-prepared for the role it was about to assume; it was more like a nice middle American town than the nerve center and command post of the free world. By the end of 1941 it was a nice middle American town bursting at the seams. Construction of the Pentagon, the world's largest office building, had begun in September on a muddy flat across the Potomac only after the Army had managed to appease outraged members of Congress, who demanded to know what the country would do with all that space once the emergency was over. (The suggestion was advanced that government records could be lodged in its seventeen miles of corridors.) For now the offices of the Army and Navy remained scattered around the city. The General Headquarters of the new Army was miles away from the Munitions Building, at the Army War College building. The city was only just getting a real airport at last. The old airport, Hoover Field, which had been dislodged by the Pentagon construction, consisted of a grassy strip wedged between an amusement park on one side and a warehouse area and dump on the other. For years the Arlington County authorities had refused pleas by the airport's manager to close off a county road that cut directly across the field. The danger of a collision reached such a point that the manager at last took it upon himself to install a flashing red light that would warn motorists when planes were taxiing. He was haled into court and fined for illegally obstructing traffic. The light came down.

The city was finding itself totally incapable of coping with the flood of new clerks and typists that had already begun arriving in 1941 as the peacetime military began expanding. Washington's population had almost doubled since 1940, to over one million. The city's streetcars were able to carry only a third of the government workers trying to get to their jobs each day. In desperation the streetcar operators brought out of retirement some wooden, nineteenth-century trolleys with celestory windows around the roofs. Housing was, simply, a nightmare. An agency called the Defense Housing Registry tried to help; new arrivals would walk up to the desk and explain they were looking for a nice one-bedroom apartment with a private bath within walking distance of the Munitions Building and would quickly learn, as a *Washington Post* headline reported, NEWCOMERS DISCOVER PRIVATE BATHS WENT OUT WITH HITLER. A joke made the rounds about the man who was crossing the Fourteenth Street bridge when he looked down and saw another man struggling in the waters of the Potomac below. "What's your name and address?" he shouted down, then ran back to the drowning man's landlord—only to find that the now-vacant room was already taken. "But I just left him drowning in the Potomac," the man protested. "I know," the landlord replied, "but the guy who pushed him in got here first." Zoning laws that prohibited renting out single rooms in residential neighborhoods were suspended after the war began, and some new arrivals were able to find spare bedrooms for thirty dollars a month, but others doubled or tripled or quadrupled up in single-room apartments. D.C. housing officials reported "Dickensian" conditions in some tenements. In one townhouse with peeling paint and sagging staircases they found nineteen tenants sharing twelve rooms and two tiny bathrooms; two lived in the basement right next to the coal furnace.

The other joke about Washington real estate had to do with the Navy's insatiable appetite for it. General Marshall had invited the Navy to share the Pentagon, and the Navy chewed that over for a while. But the war presented an opportunity that was unlikely to be soon repeated, and the Navy had its eyes on some of the choicest spots in town. Armed with the power to take over property required for the national defense, it did so with admirable energy. People were saying that if the U.S. armed forces could seize enemy territory as quickly as they seized land in Washington the war would be over in a week. Managers of the new Statler Hotel going up on Sixteenth Street, not far from the White House, knew that the Navy was casing the place and planned to claim it as soon as construction was completed; the hotel avoided announcing an opening date and just quietly began doing business, and by the time the Navy caught on, the rooms were full of businessmen and lobbyists and Mrs. Roosevelt was coming in for lunch regularly. But few other private property owners slipped through the Navy's grasp. Commander Wenger, who lived in northwest Washington, had had his eyes on what he thought was the perfect new home for the rapidly expanding OP-20-G for some time, for he passed it every day as he drove to work. Mount Vernon Seminary was a finishing school for girls from wealthy families—or, as its founder had put it, "a place where young girls should continually be inspired and aided to grow towards noble, helpful, gracious, Christian womanhood." Mount Vernon did not exactly stress education, but it had a lovely campus of red brick buildings and lawns, built in 1917 in one of the nicest areas of the city. There was a gym and tennis courts; a rustic "tea room"; a chapel; and, a few hundred yards away, a stately brick residence where the headmistress lived.

In the summer of 1942 the Navy went through the motions of examining other sites for OP-20-G, including the Pentagon, and even considered moving in with the Army's SIS at a new, shared facility; officials made a show of gravely weighing security considerations and the "fire and possible bombing hazard under present conditions," and even consulted a naval camouflage

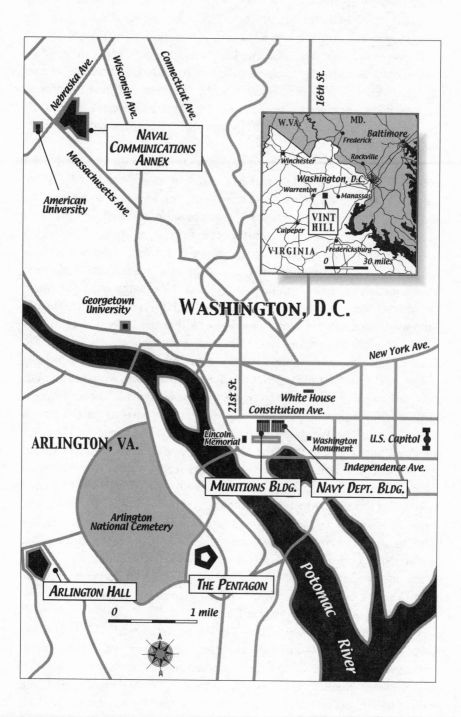

Nebraska Ave.

Wisconsin Ave.

Connecticut Ave.

16th St.

NAVAL COMMUNICATIONS ANNEX

W.VA.   MD.

Baltimore

Frederick

Winchester   Rockville

Washington, D.C.

Warrenton   Manassas

VINT HILL

Culpeper

VIRGINIA   Fredericksburg

0   30 miles

Massachusetts Ave.

American University

Georgetown University

WASHINGTON, D.C.

New York Ave.

21st St.

White House

Constitution Ave.

ARLINGTON, VA.

Lincoln Memorial

Washington Monument

U.S. Capitol

Independence Ave.

MUNITIONS BLDG.   NAVY DEPT. BLDG.

Arlington National Cemetery

ARLINGTON HALL   THE PENTAGON

Potomac River

0   1 mile

expert about how visible various sites were from the air. But it was clear this was all window dressing. The Navy wanted Mount Vernon Seminary, and the Navy got it. (The camouflage expert duly concluded that Mount Vernon was by far the best place for the code breakers to hide from any Japanese Zeros that came looking for them.) On November 20, 1942, acting Navy Secretary James Forrestal informed the school that the Navy Department was acquiring the site by condemnation "in the interest of the war effort." The school was offered $800,000, a fraction of what the buildings and grounds were worth. The school had Christmas vacation to find a new home; when they returned in January, the 162 students were boarded in private homes and classroom space was rented on the second floor of the nearby suburban branch of Garfinkel's department store. The city's planning commission mumbled that this was a historic neighborhood and asked the Navy to explain its plans; the Navy replied that it was exempt from local ordinances and that the site was "vital to the war effort and its nature should not be publicized." On February 7, 1943, OP-20-G took up residence at its plush new quarters, now named the "Naval Communications Annex." A double line of barbed wire fence went up around the perimeter and construction crews descended on the school's courtyard and began ripping out colonnaded walkways to clear the ground for new buildings that would accommodate the rapidly expanding staff and its equipment. The headmistress's house was converted to the Navy Exchange store, its polished hardwood floors and decorated walls soon scuffed and soiled by hundreds of sailors tramping in and out each day to buy cigarettes and Cokes and sundries.

The Army had actually beat the Navy to the punch this time, for it had taken over its own posh girls' school in the meanwhile. A group of SIS officers had discovered Arlington Hall Junior College one day in April 1942 while driving back from Warrenton, Virginia, where they had been inspecting the site of the Army's new intercept station at Vint Hill Farms, thirty-five miles west of Washington. Arlington Hall offered a hundred acres, a gymnasium, a stately residence hall, tennis courts, indoor and outdoor horseback riding arenas, as well as convenient access to the Pentagon and Vint Hill and a parklike setting of manicured grounds that ensured ample security. The school's trustees willingly offered to turn over the campus to the Army for the duration of the war, but the Army was not going to let its chances slip by, either, and promptly invoked the Second War Powers Act to condemn the site for $650,000. The sum barely covered the school's outstanding mortgage obligations. On June 10 a guard detail consisting of one armed second lieutenant and fourteen troops carrying sawed-off broom handles took possession. It had all happened so swiftly that several of the school's young women students were still in the dormitory when the Army showed up to take charge.

The staff of the SIS moved in over the next two months. Movers wrestled a dozen enormous safes, each weighing a ton, into the building, damaging the driveway when two of the safes fell off trucks while being unloaded. The school building was almost completely unsuited to the work of a government agency, much less one engaged in highly secret activities. Each pair of dorm rooms had a shared bath between them; the water was turned off and the tubs pressed into service as filing cabinets. The room where the Purple machine was installed contained the only working bathroom on the second floor, and every hour the small team of men and women who worked to decode Japanese diplomatic traffic would cover the equipment and their work while the rest of the staff on the floor was allowed in to use the facilities.

Construction work began at once at Arlington Hall on two more practical, if charmless, office buildings in the traditional government style of long blocks with a series of wings coming off at right angles. These would become the operations buildings, each surrounded by its own barbed wire fence and security checkpoints. The buildings were dreary, every bit as stifling in summer as the old Munitions Building, and the legions of rodents that swiftly invaded became legendary among Arlington Hall workers. But at least they provided ample space, a luxury that SIS had never enjoyed in the past.

Although Arlington Hall Station acquired the usual trappings of a military post, with guardhouses, enlisted barracks, and drill fields, it was impossible to make a large cadre of bright and nonconformist civilians behave in regimental fashion, and for the most part the Army had the good sense not to try. Friedman, following his nervous breakdown in early 1941, had returned to work that spring in a much-reduced capacity as a technical adviser; the actual day-to-day management of SIS subsequently passed through the hands of four different military commanders from 1941 to 1943, until Colonel Preston Corderman, who actually knew something about cryptanalysis, having served prewar tours of duty at SIS as a student and instructor, took charge on February 1, 1943. The commanders who tried to "run" things soon discovered that the staff had its own ideas and just got on with its job in its customary unmilitary way, for the most part ignoring the military hierarchy. Anyway, the civilians outnumbered the military personnel by better than 3 to 1; by April 1943 the civilian work force at Arlington Hall totaled 2,300, versus 766 military personnel. And even the military personnel were almost all "civilians in uniform," draftees assigned to Arlington Hall because of linguistic or mathematical skill, or prewar civilian employees of SIS who had received commissions or draft notices. The three principal cryptanalysts, Rowlett, Sinkov, and Kullback, were all in the civilian-in-uniform category; Kullback was about the least military of the three, looking most of the time "like an unmade bed," as one coworker recalled. Certainly none of them knew how to shoot a weapon. One naval officer who knew all three said he would have had no qualms about standing in front of Sinkov and daring him to hit him with a .45. The Army had a standard procedure whenever one of its cryptanalysts at Arlington Hall received a draft notice or decided to enlist. When Bill Lutwiniak was about to be drafted he went to see a Signal Corps captain at SIS, who "rigged the whole thing." Lutwiniak was told to appear at the Army Recruiting Office at the Greyhound bus station in Washington at a specified time. He did so, and was swiftly sworn in as a private and handed his orders: Report to SIS immediately for active duty. Arriving back at the Munitions Building, Lutwiniak reported to the captain, who immediately promoted him to sergeant and told him not to worry about basic training.

The atmosphere was very much one of a "meritocracy," said Cecil Phillips, who had started at Arlington Hall in July 1943 as a bottom-rung clerk doing such chores as stapling papers together and within a year was directing the cryptanalysis of Soviet diplomatic codes. If you could do the job you were given increasing responsibilities, and for the most part that was all that anyone cared about. Cryptanalytic talent was what determined status, both up and down the ranks, far more than whether one was a private or a lieutenant colonel. After the move to Arlington Hall Lutwiniak was suddenly put in command of an enlisted barracks, a purely "military" duty in addition to his professional duties as a cryptanalyst and IBM machine expert. It all worked out fine when the troops covered for his deficiencies as a drill master. "The barracks I inherited was full of these privates who were all Japanese [language] students, all out of places like Yale and Harvard and NYU. They'd never seen more than two stripes . . . I had that going for me," Lutwiniak recalled. "Then they found out I had a reputation as a cryptanalyst"—and that was what really smoothed the way.

The occasional regular Army type assigned to Arlington Hall was generally made mincemeat of by the staff, civilian and "military" alike. Frank Lewis was once assigned as his military adjutant a major who, his first day on the job, issued "Executive Order No. 1" requiring all personnel to respond with a hearty "Good morning!" when greeted in the corridor—rather than looking at their shoes or staring off into space, as the major had disapprovingly observed. The major then decided to "up productivity" by requiring that everyone in the section sign in and out on a sheet whenever they left the wing. This of course was exactly the sort of thing that bright college boys knew how to deal with; everyone in the wing immediately began going over to the log sheet and signing out, and then immediately back in, whenever they had to get up from their desks for any reason at all. The major was quickly reassigned to new duties. Another favorite activity was tormenting the security-obsessed officer who would sweep the rooms at night looking for infractions of the regulations. The cryptanalysts would leave notes for him under their desk blotters (and, on one memorable occasion, as a faint impression on a note pad) bearing messages such as, "If you're so smart, why are you reading this?"

The Navy was more conventionally hierarchical and wanted everyone who worked in a Navy office to be in uniform to reinforce military discipline and the chain of command. It was particularly antithetical to Navy tradition for an officer to have to report to a civilian. At the time of America's entry into the war OP-20-G had 65 officers, 124 enlisted personnel, and 36 civilians; two years later the numbers were almost 800 officers, 3,000 enlisted, and 88 civilians. The Navy cryptanalytic units at Hawaii and the Philippines were made up almost entirely of service personnel.

But the Navy did not particularly stand on ceremony when it came to handing out commissions, and OP-20-G, too, quickly became an organization of civilians in uniform. The Navy had established a cadre of faculty members at Harvard, Yale, University of Chicago, and Temple University who had received some initial training in cryptanalysis by correspondence course, and when war came they were brought into the Naval Reserve and ordered to Washington. Ranks were doled out strictly according to age. If you were twenty-eight or under you were an ensign; twenty-nine to thirty-five, a lieutenant; over thirty-five, a lieutenant commander. These recruits in turn recruited other mathematicians, physicists, and linguists, largely by word of mouth. Additional Japanese linguists were rounded up almost literally right off the boat; Navy officials met ships bringing repatriated Americans back from Japan after the declaration of war, and these included a number of children of missionaries who had grown up in Japan and spoke the language with a native fluency.

The difference between the Navy and the Army was that while OP-20-G had its share of civilians in uniform, and some very unconventional civilians at that, it remained firmly under the control of regular military types at the top. The power play by the Redman brothers and their push for central authority in 1942 had only reinforced that. The regular-Navy outlook affected the day-to-day work in subtle but odd ways. For one thing it made dealing with the Army far more difficult than it should have been. At one point the Navy wanted the Army to forbid civilians to touch the SIGABA cipher machine, in conformity with the Navy's policy; the Army protested that such a rule would be rather difficult to implement since the Army employed over five thousand civilians in communications and cryptanalysis, and since it would even mean that the man who oversaw the SIGABA's development, William Friedman, would have to be transferred to another line of work. The regular Navy's clannishness and bigotry was breaking down under the rush and necessities of wartime, but not entirely; when Commander Sandwith, the Royal Navy's intercept specialist, visited OP-20-G in spring 1942 he observed that "the dislike of Jews prevalent in the U.S. Navy is a factor to be considered" in the difficult relations between the two code breaking bureaus, "as nearly all the leading Army cryptographers are Jews."

Relations between the military and civilian sides was strangest of all at Bletchley Park, where the chain of command was so loose that it bordered on anarchy. Technically GC&CS was part of the Foreign Office yet it was reporting on a day-to-day basis to the military services while at the same time issuing orders to the intercept stations that were technically under the command of the Army, RAF, and Navy. Some key personnel such as Denniston and Travis had long been members of the naval reserve and others, such as Tiltman, were regular Army officers who for years had been detached for assignment to GC&CS. But none would be expected to have the outlook of a normal uniformed officer. Tiltman, outwardly at least, was very much a military man; he had been awarded the Military Cross for his service in France in the First World War as an officer in the King's Own Scottish Borderers, and had been severely wounded at the Somme. It was really only a fluke that he had been assigned to signals intelligence work in 1920. But while friends and colleagues said he always remained "every inch a soldier," the extraordinary cryptanalytic talent that had so unexpectedly emerged reflected a mind that simply cut through the conventional, and certainly the conventional military, ways of looking at the world. (Tiltman's flexibility of mind was evident in his being one of the very few "old-fashioned" cryptanalysts who managed to bridge the gulf between the familiar world of hand ciphers and the new world of machines. He led the GC&CS teams that broke Soviet codes in the 1930s and the German Police hand ciphers; he also made crucial and brilliant discoveries that helped to break the railway Enigma and the German teletype ciphers.)

After Welchman's letter to Churchill, the military services began assigning increasing numbers of officers and men to Bletchley Park. The enlisted men were lodged in barracks built nearby and run with the usual military discipline, but when they walked through the gates of B.P. all conventional notions of discipline went out the window. Most of the officers did not wear uniforms at all except on ceremonial occasions. William Filby enlisted as a private in the Army in June 1940 and in September was ordered to report to Lieutenant Colonel Tiltman at Hut 5 at Bletchley Park. On the appointed day he presented himself, saluted in smart regimental fashion, and stamped the wooden floor with his army boots in a thundering crash. Tiltman looked up and then down at Filby's feet. "I say, old boy," he said after a pause, "must you wear those damned boots?" It wasn't usual for privates to wear battle dress and white running shoes, but then it wasn't usual for colonels to call privates "old boy," either, and both became the norm in Tiltman's section, much to the disgust of the regular Army major who served as military adjutant at Bletchley Park. In the pubs of Bletchley everyone mingled without regard to rank, and in many of the huts, somewhat unusually for Britain as a whole in those days, everyone was on a first-name basis.

Occasional attempts were made to keep the Park on a military footing; these generally ended in farce. Air raid wardens were organized to keep watch and blow a whistle whenever a German plane was in the vicinity; this was supposed to trigger a procedure whereby everyone dumped their secret papers into large green bags and headed for the shelters. The air raid watch was very soon abandoned when it proved a nuisance to sort out the work when the alarm had passed. Drills were sometimes organized pitting the civilians against the military personnel; on one occasion, the civilians were supposed to try to infiltrate the Park while the military guarded the perimeter, protected with a high fence topped with barbed wire. "But the civilians decided to end this nonsense very quickly," Filby recalled, "and they tunneled under the fence during the day, so they captured the Park five minutes after the opening of hostilities."

The most farcical episode in military–civilian relations at Bletchley Park involved the requirement that everyone except the heads of sections enlist in the local Home Guard unit. The Home Guard was Churchill's inspired name for what had been called Local Defence Volunteers—civilians, mainly older men, put in uniform and placed under the command of retired soldiers. The idea was to prepare for the German invasion, and it is not impossible that the force could have done something if called upon; at least they had rifles and a few First World War cannons. (Though a memorable Noel Coward song of the period told of one Colonel Montmorency writing politely to the Minister of Supply: "Could you please oblige us with a Bren gun / or failing that a hand grenade will do . . . with the vicar's stirrup pump, a pitchfork and a spade / it's rather hard to guard an aerodrome / so if you can't oblige us with a Bren gun / the Home Guard might as well go home.")

Alan Turing, as nominal head of Hut 8, was exempt from Home Guard duty but decided he would rather like to learn how to shoot a rifle and signed up. One of the questions on the form read, "Do you understand that by enrolling in the Home Guard you place yourself liable to military law?" Turing answered "No." Of course no one noticed. He accordingly went through the drill and learned to become a first-class shot with a rifle, at which point he stopped showing up for parade.

He was then summoned before the colonel in charge who furiously demanded why he had been absent. Turing explained that he had signed up only to learn to shoot.

"But it is not up to you whether you attend parades or not!" the colonel sputtered. "When you are called on parade it is your duty as a soldier to attend."

"But I am not a soldier," Turing replied.

"What do you mean you are not a soldier! You are under military law!"

Turing calmly pulled out his ace. "You know," he began, "I rather thought this sort of situation could arise," and explained that if they would consult his form they would discover that he was *not* under military law. The colonel and his aides dug out the form, stared at it a bit, and concluded that all they could do was declare that Turing had been improperly enrolled and indeed was not a member of the Home Guard.

Arlington Hall was a hierarchical organization that tolerated a certain amount of insubordination against rigid rules as long as the work got done. At Bletchley Park, by comparison, it was not even clear who was in charge in some sections. Turing was officially head of Hut 8 but Alexander increasingly ran the show, and there was a story that the change happened officially only when Turing was away at one point and a form came around asking who was in charge and Alexander answered that he guessed he was. In some ways, though, this casual attitude reflected an enormous strength of the institution. People pitched in and got the work done; they cared about the job and not in the least about conventional rules, and the atmosphere was one that still managed, at least at moments, to resemble a senior common room at Cambridge or Oxford. Stuart Milner-Barry said that "no one ever really regarded anyone as their boss" and that as head of Hut 6 he never once had to discipline anyone or tell them to do their work. Alexander and Milner-Barry were not only friends but shared a billet at the Shoulder of Mutton, a thatched-roof inn in Old Bletchley, and they informally worked out issues that in a more formal and hierarchical organization would have required endless meetings and would likely have erupted into open bureaucratic warfare. When there was a question of how to allocate limited bombe time between naval and military or air problems, Milner-Barry and Alexander would amicably talk it over and agree with little fuss. A more formal organization would never have tolerated even the mild eccentricities of a Turing, who bicycled to work in his gas mask during pollen season to fend off his allergies, nor would it have had the resilience to adapt to the often swiftly changing circumstances of the war. Some of this began to rub off on the services themselves, and it was a measure of how swiftly social attitudes had changed from the between-the-wars norm that the Royal Navy's own intel-

ligence organization, the Operational Intelligence Centre, had early in 1941
placed in charge of its Submarine Tracking Room at the Admiralty a most
non-naval thirty-seven-year-old barrister. Rodger Winn suffered from a terrible limp and a twisted back, the result of childhood polio, and would even
under wartime standards have been rejected for military service, but the
Navy gave him a direct commission as a commander of the Royal Navy Volunteer Reserves. His fierce intellect and almost uncanny ability to anticipate
the moves of U-boat captains earned him considerable respect in an organization that had just a few years earlier had little respect for or patience
with "outsiders." Needless to say, in Nazi Germany insubordination of the
sort that Turing casually displayed was generally met with a firing squad,
while antisocial misfits and cripples were sent off to concentration camps.

But uncertainty over the larger chain of command at Bletchley Park did
cause repeated problems. This came to a head in Hut 3, the section responsible for translating and interpreting air and military decrypts and drafting signals to commanders in the field. As ULTRA became increasingly valuable, the
War Office began to realize it had been a bit hasty in its offhand decision to
allow Hut 3 to communicate directly to commanders in Cairo and elsewhere.
The War Office thus began lobbying to wrest the entire operation away from
"C," starting with control of the intercept stations, which as a practical matter were being told what to do directly by GC&CS via an informal liaison
that Welchman had worked hard to establish. Whitehall also began to insist
that in Hut 3, Army officers should have the final say on the wording of signals sent to Army commanders. This led to an increasingly tense tug of war
within Hut 3 throughout the winter of 1941–42. A complete reorganization
of the hut in the spring finally worked out an acceptable compromise in
which a Duty Officer, who was either an RAF or Army officer and answerable to the War Office or the Air Ministry, gave every outgoing signal a final
review. Welchman had meanwhile argued with considerable conviction and
merit that the decision of what frequences to monitor was an hour-by-hour
matter that had to be left to the technical experts; it depended on weighing
together highly technical questions of radio propagation, German habits of
shifting frequences and call signs, the current cryptanalytic needs of Hut 6 for
breaking various keys, and the intelligence requirements of Hut 3 as to which
keys were the most important sources of information at the moment.

The flip side was that this meant the services were at last taking signals
intelligence seriously—it was a prize now worth fighting over. The shock of
Pearl Harbor had certainly swept aside any doubts on that score within the
United States Army. Immediately after the attack Secretary of War Stimson
asked Assistant Secretary John J. McCloy to find someone—preferably a
lawyer experienced in handling complex matters—who could thoroughly

review Army signals intelligence procedures with the aim of making sure that nothing fell through the cracks again. McCloy's former law partner Alfred McCormack had meanwhile come to Washington to ask McCloy for the "toughest assignment he had." McCloy obliged, and McCormack proved to be an inspired choice for the job. Armed with a commission as a colonel and an extremely tough legal mind, McCormack showed not the least hesitation in throwing out of his office several regular Army officers who were assigned to help him but who did not meet his standards of intellectual rigor. By March 1942 he had drawn up a blueprint for a sweeping expansion in SIS's mission. Not just military, but economic, political, and psychological intelligence would be collected, and about enemy, neutral, and friendly nations alike. To make sure this information reached top officials throughout the government, a new organization, "Special Branch," would sift, analyze, and digest all SIS decrypts and issue a daily report, the Magic Diplomatic Summary. McCormack's plan gained swift approval, as did his choice for the man to run Special Branch: Colonel Carter W. Clarke, a regular officer with twenty-five years of service who had spent much of that time, prevailing attitudes in the Army notwithstanding, in intelligence. McCormack at once became Clarke's deputy. Special Branch marked a huge step forward in the professionalization of intelligence.

The U.S. Navy took the simplest course of all in setting up an intelligence system for handling the output of its code breakers: OP-20-G just took over the job itself, and given the clout of the Redman brothers, it stuck. The Office of Naval Intelligence, which might have seemed to have an obvious claim for the job, was in no position to argue, for one thing. For months before Pearl Harbor, ONI had been paralyzed by a bitter bureaucratic struggle that ended with the Navy's war plans chief, the aggressive and short-fused Admiral Richmond K. Turner, wresting away effective control of operational intelligence. Unfortunately Turner had no experience in intelligence and the sum total of his knowledge of Japan was derived from the few days he spent there in 1939 when he headed a delegation returning the ashes of Japan's ambassador, who had died in Washington. But Turner was an expert intellectual bully—he had once taken a crash course in music so he could bawl out the band on a ship he commanded. In July 1941, Turner, without even consulting ONI, issued his own warning to the fleet about Japanese intentions, advising all commands "to take appropriate precautionary measures against possible eventualities." (Commander Arthur McCollum, the genuine Far East expert on the intelligence staff, retorted to the absurd vagueness of it, "What are you going to do? Be ready to shoot in all directions at once?") The split authority certainly contributed heavily to the Pearl Harbor debacle, but it was ONI that took the blame, and it never recovered for the rest of the war. In June 1942 the Vice Chief of Naval Operations, Admiral Frederic J. Horne, ruled that all signals intelligence would henceforth be the exclusive domain of OP-20-G, and so it remained throughout the war.

―――――

America's entry into the war was swiftly followed by two disasters on the Atlantic front. The hold on naval Enigma that had been gained by Turing's breakthroughs in the summer of 1941 was tenuous in the extreme. One crisis had already occurred on November 29, when the bigram tables changed. All twelve bombes were set to work running a single crib, and the old tedious process of EINSing and building up the tables begun again. The providential capture of the new bigram tables from the armed trawlers *Donner* and *Geier* (VP 5904 and VP 5102) at the end of December cut what would have been a much longer story short.

But the sword of Damocles that hung over Hut 8 throughout the fall and winter of 1941 took another form. A captured document dated January 1941 had revealed that a new naval Enigma machine was being issued to U-boats. It was called M4, and it was to be an adaptation of the three-rotor machine; the standard reflector would be replaced by a new thin reflector paired with an additional fourth rotor, designated by the Greek letter beta. The addition of the fourth rotor would increase by a factor of twenty-six the number of possible settings of the Enigma rotors. It was in theory possible to tackle signals sent by M4 on the standard three-rotor bombes, but one run would now be required for each of the twenty-six possible positions of beta. An all-wheel-order run would thus require $336 \times 26 = 8{,}736$ bombe runs in the worst case—4,000 hours, or more than 150 days of bombe time. Even running all twelve bombes at once could require two weeks of flat-out effort to break each day's key. And that would leave no time at all for running the Army and Luftwaffe keys that also needed to be broken each day.

What was obviously needed was a high-speed bombe equipped with four-rotor Enigmas, and what was obvious, too, was that this posed an almost insurmountable technical challenge. For such a "four-wheel" bombe to work through all $26 \times 26 \times 26 \times 26 = 456{,}976$ rotor positions in a reasonable amount of time (fifteen minutes or so) would require that the fast rotor spin at speeds approaching 2,000 rpm. That in turn would mean that each rotor position would make electrical contact for less than a thousandth of a second. Normal relays were simply not fast enough to sense such a short pulse of electricity. Some exploratory work had begun, and C. E. Wynn-Williams, a physicist, was leading a project at the Post Office's Telecommunications Research Establishment in Malvern to build a new electronic sensing unit that would use vacuum tubes in place of relays. It would be plugged into the existing bombes along with another attachment that would contain the high-rpm fourth rotors. A huge cable with nearly two thousand wires would carry the electrical connections between the third and fourth rotors of the bombe's thirty-six Enigmas. The thick black cable gave the unit its name: the Cobra.

But when M4 came into service on the Atlantic U-boat Shark key on February 1, 1942, Wynn-Williams's machine was not even in the prototype stage. The captured documents had revealed that the M4 Enigma was designed with a feature to make it compatible with the M3 machines still used on other naval networks with which the U-boats would need to communicate: When beta was set to A, the combined effect of beta and the new thin reflector was identical to that of the old M3 reflector. Hut 8 exploited this fact brilliantly when it picked up a number of "duds," messages for which the daily key had been broken but which nonetheless proved unreadable. Duds were sometimes the result of errors made by British intercept stations in transcribing the message indicator; more often, they were caused by a German operator who had sent the indicator incorrectly or had enciphered the message with his machine set incorrectly. But some of these U-boat duds were clearly "four-wheel duds": The messages had been sent in three-wheel keys but the operator had accidentally set beta at a position other than A. In late 1941 and early 1942 these duds were used by Hut 8 to recover the wiring of the thin reflector and beta. But the machinery to put this knowledge into practice was still lacking. The change to M4 completely blacked out Bletchley's reading of the U-boat signals.

The blow that this struck to the safety of the Atlantic convoys would have fallen much harder but for the easy target that American merchant shipping along the East Coast of the United States presented: That was the other disaster of the winter and spring of 1942. The Admiralty's Submarine Tracking Room had lost its most valuable source of intelligence for diverting convoys around U-boat patrol lines in the Atlantic; but just at the moment the U-boats were not particularly interested in sitting out in the middle of the ocean waiting patiently to intercept convoys when there were such rich pickings elsewhere. In two weeks in January, the German submarines dispatched thirty-five ships, more than two hundred thousand tons' worth, between Halifax and Cape Hatteras. Miami and other resort cities resisted the blackout, fearing it would be bad for tourism, and night after night the U-boats surfaced to find their targets silhouetted against a backdrop of bright city lights. On the Carolina sea islands, residents could hear the diesel engines of the German submarines churning through the waters at night. In February, Admiral Karl Dönitz, Commander in Chief of U-boats, ordered the offensive extended into the Caribbean, and tankers began to erupt in

gaudy fireballs night after night. Dönitz pleaded with Hitler to send more
U-boats to the American coast; the Führer, by now convinced of the infalli-
bility of his intuition, which had so often been right when all his military
commanders were wrong, congratulated Dönitz on his success—and ordered more U-boats to Norway. That, he said, was where the Allies were
planning to move next. Still, the toll on shipping mounted month by month;
in June more than six hundred thousand tons of Allied shipping was sunk,
close to Dönitz's goal of seven hundred thousand tons, which, he calculated,
would be enough to starve Britain out.

Against this slaughter, ULTRA would have been little help in any case. The
U-boats communicated by radio only when they needed one another's help
in findings convoys in the midst of the vast ocean and when converging for
an attack. On the American coast they operated singly. The captain of *U-
124*, Johann Mohr, was however unable to resist breaking onto the airwaves
to issue a report to headquarters, in doggerel:

> *The new moon—night is as black as ink*
> *Off Hatteras the tankers sink.*
> *While sadly Roosevelt counts the score,*
> *Some fifty thousand tons.*
> *—by Mohr*

The United States Navy lacked sufficient numbers of escort vessels; it
also stubbornly insisted that to institute coastal convoys without adequate
escorts would only make things worse (incorrect) and that it was impossible
to forecast U-boat movements and thus evasive routing was not even worth
trying (patently incorrect). So the ships continued to go down. The amateur
yachtsman Roosevelt showed a Churchillian streak in bombarding his naval
commanders with his own ideas for improvising defenses. In February he
ordered the Navy to establish an armada of civilian volunteers in yachts and
fishing boats to patrol the shores. The Coast Guard called it the "Corsair
Fleet." The press dubbed it the "Hooligan Navy." Ernest Hemingway was
one of its more enthusiastic members, cruising about with a machine gun
and dreams of dropping hand grenades down the open hatch of a conning
tower. The move may have boosted civilian morale, but the only results it
produced were hundreds of false sighting reports.

Churchill had no luck in urging Roosevelt that a British admiral be appointed to command all Allied operations in the Battle of the Atlantic, but
Rodger Winn ultimately had better success in pleading the case for the U.S.
Navy to establish its own Submarine Tracking Room, which would work in
close cooperation with the British. He had a rough time of it for a while.

Winn arrived in Washington on April 19, 1942, and after several days of battering against skepticism and hostility finally got in to see Admiral Richard S. Edwards, the deputy chief of staff. Edwards began by repeating the company line that trying to track U-boats and reroute convoys on an operational basis was a waste of effort. As the discussion became hotter Edwards grew increasingly provocative and belligerent, telling Winn at one point that if America wanted to lose ships that was America's business. Winn, who had spent time in America as a student at Harvard and Yale, figured that the American bent for straight talking was called for on his part, and bluntly told Edwards that that was all well and good, but a lot of the ships were British—adding with well-calculated fury, "We're not prepared to sacrifice men and ships to your bloody incompetence and obstinacy." But Winn offered a carrot, too, hinting to Edwards that if the United States did as he suggested then Britain might have "better information to impart." Winn's tough line turned the tide, and after an "alcoholic luncheon," Winn and Edwards "parted the best of friends" and things began to happen. By the time Winn left Washington in May, the American U-boat tracking room was, he reported, "a going concern." (As the flow of U-boat signals increased, a special "Secret Room," which could be entered only though the main Submarine Tracking Room, was established on December 27, 1942. Its job was to receive, analyze, and plot intelligence from the Enigma decrypts. Entry to the room, which was officially known only as F-211, was limited to Admirals King and Edwards, three other top staff officers, and the five junior officers and yeomen who staffed the operation.)

The only trouble was that the British did *not* have "better information to impart" so long as the blackout of the Shark four-rotor key continued. Some of the more suspicious minds in the U.S. Navy began to wonder if the British were deliberately holding back information they did have. But everyone at OP-20-G who was in the know about Enigma felt that, whatever the case, the time had simply come for the British to take them on as full and equal partners. The failure to break the Shark key, and the fact that America was now a full ally in a shooting war, undercut the principal British claims for keeping the Enigma to themselves. In April, Tiltman was sent to Washington to assess the situation; he wasted no time in urging that the Americans be given at least some of what they were demanding. "In view of the fact that they are now at war and have a vital interest in submarine traffic," Tiltman cabled back to Travis, "they are entitled to results or a detailed statement as to why traffic cannot be read at present and what are prospects for future. Unless a rapid and satisfactory solution is found . . . the high command will insist on their naval cryptanalysts attempting to duplicate our work on E." Countering the British excuse of concerns over poor American security as a reason to keep the Enigma to themselves, OP-20-G professed itself worried over what would happen if the Germans invaded England, or if the Luftwaffe bombed Bletchley Park, and urged that a "skeleton party should be formed here with some machinery . . . to act as an insurance."

Travis fought a rearguard action for a few more weeks. "Hardly think necessary to form skeleton party as if real danger arose of present facilities being lost we would certainly send experts other side," he cabled back to Tiltman. But he knew it was a losing battle, and on May 13, 1942, Travis informed OP-20-G that "higher authority has agreed future policy regarding E solution . . . we will continue exploiting but will send you a machine for solution in August or September and lend you a mechanic to instruct in working. We will also give full instructions and try to spare some one to explain our method." Travis also agreed with Tiltman's suggestion that several Navy experts who had been working on ideas for speeding up the bombes be sent over to Bletchley. Lieutenants Robert Ely and Joseph Eachus arrived on July 1 and, after a few more delays and excuses, were finally given what the United States had been seeking since February 1941: complete wiring diagrams and blueprints of the actual bombes.

But even with these concessions, Travis still hoped to keep control over the "product." The key phrase in his message was "we will continue exploiting," and "exploiting" meant decrypting operational traffic and distributing the resulting intelligence to operational commands. As the British saw it, the Navy could build a few bombes and learn how to operate them, and perhaps even use them to break traffic that could be intercepted from the East Coast of the United States, but GC&CS would remain the senior partner.

The more the British tried to put the Americans off with such provisos and qualifications, the more suspicious the Americans became. By August, Wenger, now head of OP-20-G, was convinced—wrongly as it happened—that the British were concealing success they had achieved against Shark. But it was clear in any case that the British were getting nowhere on the four-wheel bombes, and OP-20-G had had quite enough. It was also clear the British were unable or unwilling to make good on Travis's promise to send over a bombe. A few weeks later GC&CS was suddenly confronted with a fait accompli by the Americans. On September 3, 1942, Wenger proposed to his superiors that the Navy take matters into its own hands and spend $2 million—"it must be understood that it is a gamble," he stressed—to build 360 of its own four-wheel bombes. The British liaison to OP-20-G caught wind and protested; his rather Jesuitical argument to Wenger was that Tiltman had promised only to provide results or a detailed explanation of why traffic could not be read, and by doing the latter the British had fulfilled their obligations. But the fact was that GC&CS only had about thirty bombes built by this point, and there was no doubt that they desperately needed the help. In July and August the demands of trying to break the Shark keys had overloaded the available machines, and work on some Luftwaffe keys had fallen dangerously behind. (Even a half-year later the situation remained close to desperate. On January 5, 1943, Welchman warned Travis: "An analysis of probable and possible requirements for bombes during 1943 is most alarming." At the new year there were 49 machines in operation; Welchman calculated that as many as 120 three-wheel bombes would be needed for "urgent work" on Luftwaffe and Army traffic, while breaking the U-boat and other naval keys might take as many as 134 high-speed, four-wheel bombes.)

Faced with the inevitable, Travis and Frank Birch, head of GC&CS's naval section, traveled to Washington in September and acceded to a negotiated surrender. GC&CS and OP-20-G would establish "full collaboration" on attacking the German naval Enigma—exchanging traffic, recovered keys, and cribs to feed the bombes. The British monopoly was over.

––––––

A lot had happened in bombe design since machine No. 1 had arrived at Bletchley Park, and Travis agreed to give the Navy complete blueprints and samples of mechanical components. One of the greatest improvements was the result of an astonishing discovery made by Gordon Welchman in the spring of 1940. The original bombe had required that the menus derived from cribs produce closed loops in order to generate enough feedback to detect incorrect positions. Even so, they produced many "false stops," rotor positions that could not immediately be rejected as a contradiction without hand testing. In fact, the very poor showing of the first bombe revealed this weakness; it was barely practical to use it at all.

Welchman's discovery was that it was possible to exploit the reciprocal nature of the stecker pluggings to greatly speed the bombe's operation, reduce the number of false stops, and most important reduce the need for closed loops in the menu. The idea was this: If for example the output cable from one Enigma in the bombe represented the steckered letter E, then if the wire K in that output cable were energized, that stood for the assumption that E was steckered to K. But if that were so, it meant that K had to be steckered to E as well. So if another Enigma in the bombe fed into the cable representing K, you would be completely within your rights to apply a current to the E wire of that K cable in this instance. Welchman quickly designed a matrix of wires which he called the "diagonal board" that would automatically make these reciprocal interconnections. The effect was to vastly increase the feedback within the plugged loops, and so reduce the number of false stops. A recent calculation had shown that a menu of eleven letters would produce 676 stops without the diagonal board, but only 4 with this improvement.

Welchman excitedly went to Turing with his discovery. Turing, astonished, at first thought it could not be possible. It was one of those too-simple, cutting the Gordian knot sort of insights that seem too good to be true. But he quickly saw Welchman's point, and beginning with Bombe No. 2 all were fitted with the diagonal board. Bombe No. 1 was ultimately refitted with one as well.

The bombes were still crude affairs; when a stop was reached the method at first was to reach behind the unit and feel for which relay was tripped. The danger of receiving an electrical shock in the process was not insignificant. Later models indicated steckers with a panel of lights, or even printed out the information. Another later refinement was a series of relay devices called the "machine gun" (named for the rattling noise it made as the relays clattered through their tests) that automatically hunted for contradictory steckers in the menu when a stop was reached. If a contradiction was found, the bombe would automatically start up and resume its search. A series of bombes called "Jumbos" were equipped with the device. Another simple shortcut was provided by the discovery that the Luftwaffe had a strict rule that a letter could never be steckered to the immediately following letter; that is, A could not be connected to B, or B to C, and so on. It was another ostensible security rule that backfired. So when running Luftwaffe menus, the bombe operators could simply flick a "Consecutive Stecker Knock Out" switch that filtered out any solutions that had consecutive steckers. All points on the diagonal board that implied a consecutive steckering (the B wire on the A cable; the C wire on the B cable; and so on) were connected together, so if current appeared on any one of these points, it would instantly flow to all twenty-six letters and the bombe would not stop.

The mechanical troubles GC&CS was having with its four-wheel bombes at first inspired OP-20-G to try to build an all-electronic version of the machine. Besides the Cobra, which was intended only as a stopgap, GC&CS had set the British Tabulating Machine Company to work designing a complete four-wheel bombe, dubbed the High-Speed Mammoth. It was not as far along, but in theory it was superior to the Cobra. Although it was to use high-speed relays instead of vacuum tubes to detect a stop, it would be able to test each "stop" without stopping at all, while the machine continued running, and punch out the results on IBM cards. But tests in September and October made clear that the problem of getting good electrical contacts on the rapidly spinning fourth rotor was plaguing both designs. The Cobra suffered from "brush bounce" as the wire brushes swept over the contacts at high speed, and the first tests were a total failure. Various fixes were tried, but each only created new problems. When the engineers tried to overcome the resistance of the poor contacts by increasing the voltage, problems with the electronic sensors cropped up. They added a bank of resistors to cut the voltage to the electronic unit, but this failed to work as intended. The only other option was to slow down the whole works. That meant slowing down the attached three-wheel bombe, which was driven by a DC motor, and the only way to do that was to place a large resistor in series with the motor; that in turn reduced the power of the motor, causing it to stall when a multiple turnover of the rotors occurred. Meanwhile the electronic sensing unit was acting up because its power supply was not steady enough. The engineers attempted to smooth out the ripple by adding huge electrolytic capacitors to the power supply. That did work—except when the capacitors exploded, which they did with alarming regularity. Meanwhile open warfare was breaking out between the rival groups at BTM and the Post Office; Wynn-Williams's group sneered that the relays of the High-Speed Mammoth could not possibly be up to the job of rapid sensing and would wear out in no time.

The U.S. Navy's all-electronic design would avoid all of these mechanical and electrical problems by using twenty thousand vacuum tubes in place of the rotating wheels and relays. But no one had ever tried anything like that before, and it was not even clear that so many tubes could be purchased or that a power supply could be built to handle such a huge load, so the Navy quickly shelved that plan and decided to blend the best of the two British designs, combining a complete four-wheel bombe with an electronic sensing device. The Navy signed a contract with National Cash Register in Dayton, Ohio, to build the machines, and work began at once.

Getting the high-speed fourth rotor to behave was the crux of the problem. The other rotors had a "carry" mechanism just as the actual Enigma machine did; after each rotor made a complete turn it would advance the adjacent rotor with a pawl. But given the geometry of the rotors in the bombe, this required a rather complex system of levers, ratchets, and solenoids, and at the speed the fast rotor was to turn it would have to activate the carry mechanism more than thirty times a second, and no piece of hardware was up to such an ordeal. So instead, the third and fourth rotors were both to be driven continuously by gears off of the main shaft. But that in turn introduced new problems. Ideally, the third rotor would stay at position A until the fourth rotor had fully rotated once, then advance to B and repeat the cycle. But the independent gearing of the fourth rotor meant that the third rotor would gradually creep from A to B while the fourth rotor was turning through one revolution. Thus it would be moving through the gap between the A and B contacts during part of the cycle, and some settings would be missed. To overcome this problem, the fourth rotor was geared to turn at a 37 to 1 ratio so it would be sure to cover all twenty-six settings while the third rotor was still making contact.

Just stopping the machine was a problem, too, given the inertia of rotors spinning at over 1,000 rpm. The Navy design used an electronic memory to record a stop; a brake and clutch would then be triggered by a relay, the machine could come to a jolting halt, and a reversing motor would slowly back up to the indicated stop. The steckers would then be tested and if it was a valid stop, a record printed out. The machine would then automatically start back up.

The final design was a machine that weighed five thousand pounds and consumed two and a half kilowatts of electricity. It contained sixteen four-wheel Enigmas and could do a complete four-wheel search in twenty minutes, or a three-wheel run in fifty seconds. In place of the cumbersome cables that the British bombe operators had to plug by hand to set up a menu, the Navy bombes had uniselector switches that could be rotated to the desired letter of the alphabet. The cost of the project quickly doubled to $4 million. To provide the manpower to help assemble and solder the machines, three hundred enlisted Navy women—WAVES—were sent to Dayton and put up at a summer camp that NCR maintained for its employees. The first prototypes were to be ready by spring.

The British astonishment with American technological overkill was reinforced when Turing came to Washington in late 1942 to consult with the Navy on its bombe design. OP-20-G had decided it would simply build one bombe for each of the 336 possible wheel orders. Turing said he "smiled inwardly" when he heard this, for it showed that the Americans really had not grasped the role of Banburismus or what bombe room procedure was like. Not to have the flexibility to run several machines simultaneously trying multiple menus on the same wheel orders was a terribly rigid limitation. As a result of Turing's advice and further discussions, the Navy settled in March 1943 on ninety-six machines.

What the Americans lacked in finesse they did make up for in sheer technical ability to get things built. National caricatures often contain a core of truth, and American know-how and industrial might were very real indeed. Within months after America's entry into the war Liberty Ships were coming out of American shipyards like cars off an assembly line; in September 1942, Henry Kaiser, a charismatic and aggressive industrialist who had built dams and bridges but never a ship until half a year before, set a world record by completing the ten-thousand-ton Liberty Ship *John Fitch* in twenty-four days, start to finish. The next month *Joseph N. Teal* slipped out of the ways in ten days. A month later it was the turn of *Robert E. Peary* to claim the record: four days and fifteen hours. Stalin's toast at the Teheran summit meeting in late 1943 was no exaggeration: "To American production, without which this war would have been lost."

The technically minded civilians-in-uniform who swelled the ranks of Arlington Hall and the Naval Communications Annex set off a boom in cryptanalytic technology. Friedman made a blanket arrangement with IBM: If any of their engineers received a draft notice, they would simply contact Friedman and he would arrange for them to receive a commission and a posting to Arlington Hall. The Navy also cut a deal with IBM in February 1942: The company would supply new equipment on an experimental basis at no charge and would detail engineers to assist in developing new methods. Not that this stopped IBM from doing very well indeed off of their new clients. By the following year the Army was renting more than 350 machines, at a monthly cost of over $40,000, while OP-20-G was expanding its machine room to about 200 machines. At the peak of the war, IBM was taking in more than three quarters of a million dollars a year from the two code breaking bureaus.

IBM cards, along with the procedures devised to stack and sort them and run them through the tabulators, were the closest thing to a general purpose programmable computer that existed in the early 1940s. There was a lot the machines could do, but there were obvious limitations as well, and one of the most serious was that the machines themselves had no memories beyond a few counters, nor even any real calculating hardware. For a task like searching through thousands or even millions of characters of cipher text to find repeats, the best the standard IBM methods could do was print catalogues, which would then have to be searched by eye. So there began to emerge from the workshops of the Army and Navy research groups, Arlington Hall's Section F and the Navy's OP-20-GM, a whole zoo of odd-looking contrivances designed to go where no computing machines had gone before. One family of gizmos consisted of huge banks of relays with cables that snaked out to plug into the standard IBM equipment. The Slide Run Machine, for example, contained relays wired to automatically recognize the 250 most common code groups used in an enciphered code. The cipher text from several thousand messages would be punched on cards and the slide run unit would then automatically subtract five given groups of additive from each message, and whenever two or more of the high-frequency code groups emerged, they would be printed out.

Add-on units were also devised for the bombes; these "grenades" were principally used to handle "duds" and similar problems in which the stecker, wheel order, and ring setting were known but the initial rotor setting was not. The "drag grenade" was particularly effective in taking over the task that had been performed by the "EINSing" method: Instead of just the word "EINS," any crib of four letters or less could be set up and automatically dragged through every possible position in sixteen letters of cipher text.

The use of relay banks or electronic circuits (such as the vacuum tube sensors in the high-speed bombes) as simple memory storage devices anticipated by several years one of the crucial components of the modern digital computer, and it gave the machines a speed and flexibility not available through ordinary card-by-card comparisons. The key in a device like the Slide Run unit was that rather than having to first strip the additive and punch thousands of new cards, then compare every one of those thousands of cards against a stack of 250 cards representing the high-frequency code groups, the subtraction was done electrically and the search for the code groups was carried out simultaneously by the relay bank.

The only other technologies available at this time to carry out rapid and simultaneous searches seem exceedingly quaint by modern standards but were actually powerful ways to store and compare large amounts of data. The Navy took the lead in building a series of machines that encoded message texts either as a series of punched holes in long strips of paper tape or photographically as dots on film. The genesis of these devices went back to the mid-1930s, when Captain Stanford Hooper, the technologically minded Director of Naval Communications, began meeting with Vannevar Bush of MIT to explore ways that the Navy could benefit from the latest in scientific research. (Bush subsequently became the leading force in mobilizing scientists in support of the war effort; after the war he was instrumental in advancing the government–academic partnership formalized by the establishment of the National Science Foundation that would pump billions of federal dollars into basic scientific research.) At that time, Bush was trying to develop a pioneering general-purpose computer that used high-speed paper tape readers to enter data and fully electronic counting circuits to perform the calculations. Bush suggested that a device that incorporated some of these same principles could have applications to cryptanalysis, and somewhat grandly proposed that the Navy pay him a consulting fee of $10,000 for the privilege of picking his brains further. OP-20-G was interested. The Navy Department bureaucrats were appalled. They said they wanted nothing to do with a "college professor" who demanded outrageous fees and was not even proposing to deliver a piece of equipment for the price. In January 1937 a scaled-down contract was finally approved and Bush promised he would try to build an actual machine and give it to the Navy at no cost, except for shipping. The device eventually arrived in a huge crate at the main

Navy Building in December 1941 and sat around for three months gathering dust until Wenger finally assigned a bright young lieutenant to get the thing working.

Bush's "Comparator," and the later generation machines it spawned, looked like a movie projector that had gone slightly wild sprouting extra sprockets and pulleys. The basic idea was that it would hunt for repeated strings of cipher text in multiple messages by feeding two loops of tape simultaneously through a pair of scanning heads. The tape or film recorded each cipher letter or cipher group of a message in a row of holes or dots that ran across the tape. Photocells in each scanning head would detect which holes had been punched at each position. An electronic circuit would then compare the two outputs; if they matched, the machine would either record the position, or add it to a running tally, or stop the tapes so an operator could manually note the result. With a whole bank of photocells in the scanning heads, several rows could be read simultaneously, allowing for searches to be made for repeats several letters or groups long. After the tapes had completed one loop the first tape would be advanced one position to try a new offset and the process begun again. Other variants of the machine could search a message tape for a specific, fixed sequence that would be punched on an IBM card that controlled the pattern of lights illuminated in the scanning head.

Several of the machines used optical principles to perform calculations directly, a concept that would shortly be left by the wayside in the digital electronics revolution but that would be revived a half-century later. The I.C. Machine was designed to place two polyalphabetic cipher texts in depth using Kullback's principles of searching for the alignment that yielded the highest index of coincidence. The two messages were recorded on a pair of glass plates; a single, clear dot in each row was placed in one of twenty-six positions that stood for the letters A to Z. The two plates were then placed one atop the other over a light source. Wherever the same letter appeared at the same position, the dots in the two plates would line up, allowing light through. The total amount of light coming through the entire plate was thus directly proportional to the number of coincidences across the entire length of the message. By shifting the two plates relative to one another, every possible overlap could be examined; a single photocell measured the amount of light emerging, and when it exceeded a set threshold, a signal lamp would illuminate. This was an analog rather than a digital device, but it eliminated the need for complex electronic circuits to carry out a row-by-row comparison and permitted a huge gulp of data to be digested in a single swallow.

Several other machines (the Navy "Copperhead" was one) were designed to carry out brute force searches for double hits in enciphered codes, and they employed a clever "complementary" encoding system that similarly took advantage of simple optics to do the work of complex electronics. The digits one through nine for each code group were each represented by a specific pattern of dots across one row. On the first message tape the dots would be encoded as clear spots on opaque film; on the second tape they would be opaque spots on transparent film. The two films would then be superimposed and run through a scanning head. In any row where an exact match occurred, the light would be completely blacked out. That meant only one photocell per row was required; if the cell detected any light at all emerging, there was no match. As many as one hundred cipher groups could be scanned simultaneously with a hundred photocells as the tapes passed over the head.

One of the odder machines—and one that underscored the inadequacies of IBM equipment or other technologies of the day for recording multiple channels of data simultaneously—was the Navy's "Mike." This read two punched tapes simultaneously and recorded a complete tally of how often each of the 676 possible alphabetical bigraphs occurred. It did this with a twenty-six by twenty-six bank of mechanical dials. When a run was completed, the most efficient method of recording the data was to photograph the dial board so the machine could be promptly reset and readied for another run. The photographs would then be developed and the data manually transcribed.

To handle the conversion between all of these various media, the Navy had IBM build a special electric typewriter called the Letterwriter that could transfer data between paper tape and other formats such as IBM cards. Automatic cameras were built to prepare the film for comparators directly from these data tapes. Eastman Kodak, National Cash Register, and Gray Manufacturing received contracts from the Army and Navy to manufacture the actual Rapid Analytical Machinery ("RAM" became the standard shorthand for the whole field of machine cryptanalysis, excluding IBM methods), and late in the war the machines played a vital role in attacking Japanese naval attaché machine ciphers and Japanese Army codes and naval weather codes, as well as in "dudbusting" Enigma traffic. Some of the machines were capable of making literally thousands of comparisons per second as the tape or film fairly flew through the air; for some applications such as "brute forcing," RAM could in principle outpace IBM methods by a factor of ten or a hundred. And for some applications RAM was the only way to carry out the sort of extensive brute force or slide runs required.

The machines were expensive, $50,000 or more apiece, which could buy a fighter airplane in 1942, and temperamental; they pushed to the very limits electro-optical technologies that within a few years were to be rendered completely obsolete by the digital computer, and it is unlikely that without

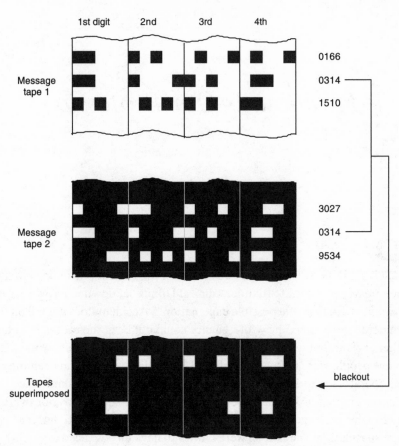

Copperhead and other optical machines employed a complementary coding system: opaque spots on a transparent tape for one message, the reverse for the second message. When the tapes were superimposed, a line would be completely blacked out if the same cipher group appeared in the same position in both.

the emergency of the war these technological avenues—or perhaps they were more accurately dead ends—would ever have been explored. But the high-speed electronic circuits the machines incorporated were quite another matter, and so were the general concepts of data processing that the cryptanalysts pioneered in close collaboration with IBM, Eastman Kodak, National Cash Register, and other firms that would play the central role in the impending information revolution. These were forerunners of technologies that would soon shake the world.

Meanwhile, however, there were codes to break, and the situation in 1942 was not altogether encouraging. Success had produced the spectacular victories at Matapan and Midway. It had also done exactly what the code breakers had feared, in arousing Axis suspicions.

CHAPTER 8

Paranoia Is Our Profession

PUZZLING over the repeated failure of his U-boat patrol lines to find their quarry in the mid-Atlantic, Admiral Dönitz set down in his war log on November 19, 1941, the possible explanations. "Accident does not fall on the same side every time," he wrote; chance could not be at the root of it. Yet no other explanation seemed adequate, either. Treason was extremely unlikely; so few people were in the know about U-boat dispositions, and even the radio operators who handled the messages were now in the dark since the Navy, beginning in June, had begun disguising grid square positions within the Enigma messages. Some new British radar was a possibility, but that was pure speculation. The British were known to have an advanced capability to fix the location of signals through direction finding, but the German Navy had taken extraordinary steps to foil that. U-boats were forbidden to send long signals except in certain cases of absolute necessity. Special short signal books were issued; U-boats could send an Enigma message containing a half-dozen or a dozen letters to transmit routine weather reports or sighting reports of enemy ships. Their radio transmitters would be on the air for perhaps ten seconds at a time. When more involved communications were required, a shore station would sometimes pose a question and list several possible answers, allowing the ship to respond with only a "yes" or "no." But even if these precautions were insufficient, Dönitz noted, direction finding still did not seem quite able to account for the British success. "In many cases it has been shown that the enemy has not drawn the conclusions which might have been expected from [direction finding] data which were certainly known to him." The one thing the German admiral was certain of was that the British had not broken the Enigma: "This possibility is continuously checked by the Naval War Staff and regarded as out of the question."

248

Suspicions about the security of the Enigma had been aroused several times in the course of 1941. Following the sinking of *Bismarck*'s supply ships, the War Staff had launched a full security investigation, noting with particular interest that *Esso Hamburg* and *Egerland* had been ambushed by the British at a planned rendezvous point whose latitude and longitude had been transmitted in an Enigma message. Even more alarming was the unsuccessful attack by a British submarine on three U-boats at a rendezvous point off the Cape Verde Islands in late September. It seemed incredible that a British submarine just happened to show up at such a remote rendezvous purely by coincidence.

But there were two watertight arguments for the security of the Enigma, or so it seemed to the German high command. First was the cryptographic security of the Enigma itself. The methods used to disguise the machine settings for each message meant that even if the enemy had a machine he also needed daily key lists; even with the daily lists, a special procedure called the *Stichwort* would foil any attempt to put those key lists to use. When the *Stichwort,* or cue word, was broadcast, the Enigma operators aboard the U-boats would open a sealed envelope. Inside was a slip of paper on which was written a key word, and the operators were to follow a complex and highly secret procedure for adding various letters of the key word to the wheel order, ring setting, and steckers specified on the daily key lists. The comfort that the high command derived from this procedure completely ignored the possibility that anyone could obtain the wheel order, ring setting, and stecker directly, by cryptanalytic means, from the cipher text itself—a possibility that seemed too fantastic to entertain.

The other proof of security was that the German Navy's own code breaking bureau, the Beobachtungsdienst (the "observation service," or B-Dienst for short), was reading the British naval codes with almost ridiculous ease. The Admiralty used two main encryption systems, both enciphered codes: One, for officers, was called Naval Cypher and the other, for enlisted personnel, the Naval Code. Before the war the Naval Code had been used both unenciphered for long administrative messages and enciphered with an additive book for confidential signals, and that sloppy practice had allowed the Germans to recover both the basic code book and the additives with little trouble. For a long time the Royal Navy did not bother to encipher the indicators transmitted with each message that specified the starting point in the additive book, which made life still easier for the B-Dienst. By early autumn of 1940 the Germans were reading 50 percent of British naval signals with little delay.

Some belated changes in the code book and enciphering procedures introduced in late 1940 and early 1941 had set back the German effort, but in October 1941 the British, American, and Canadian navies began to make heavy use of a special new code, Naval Cypher No. 3 (also called the Anglo-U.S. Cypher or Convoy Cypher), for coordinating escorts in the Atlantic. B-Dienst recognized soon that this new code worked in a way very similar to the Admiralty codes it had broken earlier, and by March 1942 the German Navy was receiving up-to-the-minute reports of what the Allied convoys were up to. As much as 80 percent of the traffic in Cypher No. 3 was read during some periods, and the German Naval Staff was quick to draw two lessons from this. One was that if the British *were* breaking the German codes, there was precious little to show for it in the signals they were sending out to their ships. "Decyphered British Admiralty appreciations of the dispositions of our own U-boats give proof that these . . . are not enriched by any results of deciphering," the War Staff concluded. But there were reasons for this absence of signals intelligence in British reports that the War Staff did not realize. One was British luck, the other British cleverness. As fate would have it, B-Dienst's breaking of Cypher No. 3 coincided exactly with Bletchley Park's loss of Shark. Cypher No. 3 contained no hints of British success in decrypting German signals for the simple reason that at that time the British were *not* decrypting German signals. But even when they had been, Bletchley Park and the Admiralty were extremely careful to disguise the fact in the intelligence appreciations that were sent out to general recipients. ULTRA in less disguised form went out only on one-time pad systems, which B-Dienst was not reading.

The other lesson the Naval War Staff drew from the ease with which they cracked the British codes was that anyone so inept as to make the security mistakes the British did could not possibly possess the cryptologic sophistication to break the Enigma. The "British reputation for stupidity" that had served Room 40 so well in the First World War—or perhaps it was more accurately German arrogance—came to the rescue again. It was telling that success made the British and American code breakers wary; it made the German code breakers smug.

———

Using signals intelligence tactically was always a calculated risk, but at least there were potential benefits that could be weighed against the risk of arousing enemy suspicions. The real nightmare for the code beakers was that simple carelessness, or gossip, or bragging might put all of their work in jeopardy for no gain at all. Or, worst of all, that someone in the know might fall into enemy hands and, under interrogation or torture, reveal the state of Allied success against enemy codes. Bletchley Park solved that problem by simply refusing to let anyone leave once they showed up. The U.S. Army and Navy were not quite that strict, but they generally would not let anyone who had worked on high-level enemy codes enter a combat zone. That was a sound principle, but it couldn't do anything about the situation in the Philippines, where the combat zone had come to the code breakers. Just before the Japanese attack, Station Cast had moved from Cavite Naval Base to one of the tunnels blasted out of the solid rock of Corregidor, an island fortress that stood squarely in the mouth of Manila Bay. The original War Department plan to defend the indefensible archipelago of islands that made up the Philippines was for the entire, small U.S. Army contingent to withdraw to Corregidor and the adjacent Bataan Peninsula and wait for the Navy to come to the rescue. MacArthur, who had served as adviser to the Philippine Army since 1935, had rejected that plan as defeatist; he decided he would whip his Filipino troops into shape and, backed by American air power, meet any invaders on the beaches. That was inspiring, no doubt, but MacArthur's plan lasted for all of two weeks after Pearl Harbor—until the Japanese actually landed on the beaches. MacArthur then retreated to Bataan and Corregidor. A hundred thousand troops and civilians made it there. Their ammunition and food didn't; MacArthur had refused to stockpile supplies there when he planned to fight on the front.

Station Cast's move to the tunnels of Corregidor had nonetheless been planned for some time, given the critical and sensitive nature of its work; its underground bunkers were equipped with diesel generators and air-conditioning, and when its ten officers and fifty-one men arrived with their radio gear, IBM machines, and kana typewriters, they were able to resume their work without delay. But almost immediately the island's food supplies were overwhelmed by the influx of retreating American and Filipino troops. Japanese bombers began pounding Corregidor on December 8 and continued without letup. Station Cast's houses and a handball and tennis court were almost immediately wiped out, and the staff took to sleeping at "the office," wolfing down meals right at the tunnel entrance, where the station cook had set up his range and where they could quickly duck for cover if bombers appeared. By late December rations were cut to two meals a day, timed around the twice-daily Japanese bombing runs; breakfast was right after the morning raid, around nine o'clock, and lunch/dinner around three o'clock, either just before or just after the afternoon raid. Breakfast was spaghetti, with or without tomato sauce, bread, and coffee. Lunch was rice, with or without tomato sauce, bread, and coffee. For variety the cook would sometimes serve rice for breakfast and spaghetti for lunch.

The station managed to keep up its work, even requesting a shipment of four hundred thousand IBM cards, which were placed on ships on

December 31. But the situation was quickly deteriorating. Supplies were not getting through, nor were they going to get through as long as the Japanese command of sea and air remained so one-sided. On January 15 Japanese troops launched a frontal assault against the line the Allies were holding across the middle of the Bataan Peninsula. At the end of a week of savage jungle fighting the Americans held their positions, then fell back, exhausted. Rations were cut to thirty ounces of food a day.

By the end of the month it was clear the situation was hopeless, and Rudolph Fabian, the commander of Station Cast, had a long talk with his deputy, Lieutenant John Leitwiler, about what to do. They finally decided to ask permission to be evacuated so as "not to lose the whole unit in one crack." On the night of February 4 the submarine *Seadragon* slipped into Manila Bay and docked at Corregidor. Fabian, four other officers, and thirteen enlisted men came on board (after managing to wrestle down the hatch six receivers and nine kana typewriters), and under cover of dark left "the Rock," bound for Java. The idea was to establish a joint station with the Dutch, and upon arriving at Surabaya on the thirteenth the party traveled by train across the entire length of the island to Bandung. There, at the Dutch War Department building, they managed to set up a few antennas before coming under Japanese air attack again; on the twentieth they were ordered to evacuate on four hours' notice. After driving through the night they boarded a second submarine, bound for Australia this time. Their orders were to disembark at a place called Exmouth Gulf and proceed by car or truck. On arriving there Fabian discovered that Exmouth Gulf had a population of one human, several million voracious desert flies, and no cars or trucks. They finally made it to Perth on March 3, almost a month after leaving Corregidor.

MacArthur himself was ordered out in late February. As he boarded a PT boat on March 12 with his wife and son, the man most responsible for the disaster he was escaping exercised his customary talent for self-dramatics by declaring, "I shall return." (Couldn't you at least say, "*We* shall return," the War Department cabled, but MacArthur declined.) With the situation rapidly deteriorating, Admiral King in Washington sent orders on March 5 that the remaining Cast personnel get out, too, any way they could:

EVACUATE PERSONNEL RADIO INTELLIGENCE UNIT SOON AS
POSSIBLE X IF SPACE CANNOT BE MADE AVAILABLE IN
SUBMARINE USE ANY MEANS OF TRANSPORTATION TO GET
THEM AT LEAST TO SOUTHERN PHILIPPINES OR TAKE ALL
STEPS POSSIBLE TO PREVENT LOSS OF PERSONNEL OF RADIO
INTELLIGENCE UNIT

The second group to leave, four officers and thirty-two enlisted men, had the more hair-raising journey, as well as a very personal introduction to the dilemma of protecting their source. The submarine *Permit* had embarked them on March 16. As a result of confusion at the dock, two evacuation parties were boarded and the submarine was carrying 111 men, twice its capacity. Chief Radioman Duane Whitlock immediately faced a predicament. The submarine's captain set a course through the Philippine Islands for the Sulu Sea. Whitlock knew, from traffic analysis he had just completed at Corregidor, that they were heading straight for a concentration of Japanese destroyers. But he had no idea whether the captain was cleared to receive radio intelligence, and after discussing it with the others from Station Cast they decided that in fact they did not have the authority to pass on the information. The next night the captain surfaced right in the middle of the Japanese force. *Permit* was under depth charge attack for the next twenty-eight hours while she lay on the bottom with her cooling systems turned off.

Several days later there was another close call when the submarine's captain, "a charger," Whitlock said, engaged in a futile chase of a Japanese merchant ship all one afternoon and finally, just before dusk, fired two torpedoes at extreme range. A few seconds later the intercom blared with a frantic order: "Crash dive! Crash dive!" One of the torpedoes, exhibiting an alarming defect that plagued the American weapons early in the war, had circled back and was heading straight for the submarine. Whitlock could hear its screws turning as it passed overhead a moment later. The nerve-racking journey ended on April 7, though not before *Permit* was attacked once more, this time by an Allied patrol bomber—"much to the indignation of all on board."

The final party, commanded by Leitwiler, was taken off by *Seadragon* April 8, the day Allied forces on Bataan surrendered. They received special permission to take the wheels of their SIGABA machine with them, but other than that carried nothing but the clothes on their backs. The station's IBM equipment was left behind with instructions to the remaining troops on Corregidor to dump it into the ocean after mixing up parts from different machines into the same packing case.

The Rock was bombarded for another month; it held out until a Japanese assault force came ashore on the night of May 5. The following day General Jonathan Wainwright, MacArthur's successor, opened surrender negotiations. He and the rest of the American and Filipino troops—the ones that survived the infamous Bataan Death March, that is—would spend the rest of the war in prison camps. The personnel of Station Cast all got out in time. Their secret went safely with them, and they were able to carry on their tasks for the rest of the war, operating from the newly established Fleet Radio Unit, Melbourne (FRUMEL), a joint U.S.–Australian operation.

There were closer calls than that in safeguarding the Allies' code breaking secrets. At several points during 1941 and 1942 there had been minor scares when the universities that were training cryptanalysts for the U.S. Army and Navy couldn't resist blowing their own horns. In June 1941 the University of Illinois Alumni News ran a photograph of seven of its ROTC graduates who were "now on duty in the Signal Intelligence Section, Office of the Chief Signal Officer, Washington," having completed a special course in cryptography. In late 1941 the Navy approached several of the leading women's colleges to run courses in cryptanalysis, but almost immediately had to "blacklist" Brown University's Pembroke College for publicizing its contribution to the war effort. Radcliffe, Bryn Mawr, Vassar, Smith, Mt. Holyoke, Wellesley, and Goucher managed to get their courses under way without similar publicity.

The constant danger of leaks from the field as more and more people were put in the know led to a formalization of the rules for handling "special intelligence" in mid-1942. On June 3, Travis proposed regulations to be issued to all commands in the field:

1. Utmost secrecy to be used in dealing with special intelligence. If from any captured document, intercepted message, prisoner of war examination, or ill-considered action based on special intelligence, enemy suspect existence of source, source would instantaneously cease probably for ever.

2. This would vitally affect operations on all fronts not only that on which you are commanding.

3. Avoid giving information as information to lower commands, use only for passing operational order to lower commands. If passing information unavoidable make no reference secret source use only prefix ULTRA. Such messages only to be handed to designated officer only to be deciphered by designated officer using only highest grade cipher: no record of messages to be kept at forward commands, messages to be destroyed by fire when read; same for ships at sea.

4. In briefing pilots only use information essential to success of operation and only such as possibly obtained from other sources.

5. If any action is based on source local commander to ensure action cannot be traced to source alone. Momentary tactical advantage not worth risk of compromising source.

6. No reference to this information to be made in summaries however limited the circulation—no discussion of it.

"Special intelligence" was a catch-all term for internal use only; it referred to all high-grade and machine ciphers. ULTRA was to be used for intelligence signals originating at Bletchley Park; ZYMOTIC for Washington and other field stations. Eventually ULTRA came to be the classification code word for the entire area of high-grade enemy codes and ciphers, and documents discussing both cryptanalytic theory and the results of breaking enemy codes were stamped in huge letters TOP SECRET ULTRA or TOP SECRET (U).

Just four days after Travis's memorandum, there occurred what was probably the single most serious leak of the war, and it amply proved what Travis was worried about. Commander Arthur McCollum, who headed the Far East section of the Office of Naval Intelligence, arrived at his office at the Navy Department Building on the morning of Sunday June 7 to find the place "shaking." The cause was a front-page story that had appeared that day in the *Chicago Tribune,* the *Washington Times-Herald,* and the New York *Daily News.* Bearing a Washington dateline and running alongside other stories about the Battle of Midway, it began:

NAVY HAD WORD
OF JAP PLAN TO
STRIKE AT SEA

The strength of the Japanese forces with which the American Navy
is battling somewhere west of Midway Island in what is believed to
be the greatest naval battle of the war, was well known in American
naval circles, several days before the battle began, reliable sources
in the naval intelligence disclosed here tonight.

The story went on to list specific Japanese ships involved in the action and stated that the United States was even aware that a Japanese move against another "American base" was a feint.

McCollum was unnerved to find that he was the prime suspect in the leak. He rushed to his office to compare the newspaper story with his own "bootlegged" copy of the intelligence appreciation that had been sent by Admiral Nimitz to his commanders at Midway, and saw at once that they bore a striking similarity. Both itemized the "striking force," "support force," and "occupation force" of the Japanese fleet, both listed the ships in identical order, both even misspelled their Japanese names identically in places. Even though McCollum realized it might get him into hotter water if he revealed he was in possession of an unauthorized copy of the message, he felt he needed to bring his discovery to his boss, Admiral T. S. Wilkinson, director of naval intelligence. McCollum recalled what happened next:

[Wilkinson] grabbed the secret dispatch out of my hand and the newspaper clipping out of the other and he went charging down the hall toward Admiral King's office with me behind him hollering, "wait, wait, wait, don't take that down." He paid no attention—he was a little deaf anyway. He went charging through the outer office, into King's office, and [there was Captain] Carl Holden, who was King's communication officer. I got into the office just in time to hear Carl Holden say, "Well, they can't point the finger at me. There are only five copies of that dispatch in existence, and I've got all five of them." Well here was Wilkinson going in there with Number Six.

The Navy was apoplectic over the leak, and immediately compounded the disaster by initiating legal proceedings against the *Chicago Tribune* for violation of the Espionage Act. It was a bad move on several grounds, not least that the fanatically anti-Roosevelt paper would be sure to fan the flames with more unwanted publicity, which it did. The *Tribune,* which reminded its readers each day on its front page that it was "The World's Greatest Newspaper," was run by the acerbic and eccentric Colonel Robert McCormick; it was part scandal sheet and part mouthpiece for the Colonel's crackpot political views, and nearly every day it carried front-page attacks on Roosevelt, his Cabinet, and his policies, domestic and foreign. (The paper's other claim to fame was the Colonel's quixotic crusade for "simplified" spelling: "tho," "thru," "staf," "jaz," "agast.") The paper had on several earlier occasions published secret government documents, and the government had more than once seriously considered bringing charges against Mc-Cormick. This time, the Navy apparently believed that the Japanese could hardly have missed either the story or its implications, and was determined to make an example of the paper once and for all.

The *Tribune* had a field day. Denouncing the government's grand jury investigation as Roosevelt's "Get the *Tribune* Offensive," the paper kept the story alive week after week. Editorials indignantly declared the paper's innocence, explaining that its correspondent, Stanley Johnston, was a great devotee of *Jane's Fighting Ships* and had derived all of the information for his story by perusing that personal "Bible" of his. The situation was all made rather more complicated by the fact that Secretary of the Navy Frank Knox owned Chicago's rival newspaper, the *Chicago Daily News.* A front-page story in the *Tribune* on August 9 charged that Knox was earning $60,000 a year from his newspaper and $15,000 as Secretary of the Navy, adding: "How much time he puts in on each is probably a military secret."

The Navy's own investigation did not take long to piece together the truth of the matter. Johnston had been on the carrier *Lexington* during the

Battle of the Coral Sea; along with a number of survivors from *Lexington,* he was returning to San Diego aboard the transport *Barnett* at the time of Midway. The senior officer among the Coral Sea veterans asked *Barnett*'s communication officer if he could decode broadcasts to U.S. forces engaged in the current fighting so they could know what was happening. He did so. He was able to do so because a cardinal violation of security rules had occurred: Nimitz's intelligence summary had been repeated in a lower-level code for the benefit of subordinate commanders. Johnston, a mustachioed Australian adventurer (in the 1930s he had staked several gold claims in the jungles of New Guinea, among other pursuits), quickly befriended the ship's officers, who proceeded to show him the signals, too, another obvious security breach. A series of small but fatal gaffes then completed the fiasco. Johnston submitted his story for censorship approval on the ship; it was cleared. He filed his dispatch from San Diego and advised his editor to again have the story cleared, this time with Navy headquarters in Washington; the editor, seeing that the story referred only to enemy ships and not to the movements of U.S. forces, decided no censorship approval was required and ran the piece.

Although the original *Tribune* story didn't actually say that the source of the American intelligence was decoded Japanese signals, Washington gossip was rife with such stories, and Walter Winchell's column of July 7 explicitly referred to the claim "that the U.S. Navy decoded the Japs' secret messages." After several months of this, when it became clear that the Japanese had *not* changed their JN-25 code in response to the leak—and with the Justice Department's prosecutor insisting that if he was to proceed with the case, the grand jury would have to hear testimony from naval intelligence officials on the Navy's code breaking operation and on the potential damage that the *Tribune*'s disclosure had done to it—the Navy did what it should have done from the start, and dropped the case.

———

Although the Japanese never did appear to take notice of the *Tribune* leak, a month later they did react to another U.S. move that pointed up the unique and perverse dangers associated with signals intelligence. On August 7 the Marines landed on Guadalcanal in the Solomon Islands. Only a token Japanese force was there at first, and the landings were largely unopposed. The Marines almost immediately came across a trove of abandoned Japanese code books. OP-20-G and Station Hypo (now renamed FRUPAC, Fleet Radio Unit Pacific) had been working feverishly since the Battle of Midway to recover the newest version of JN-25, which had gone into effect on May 27, just before the battle. The new version, designated C-9, had both

an entirely new code book and new additives. But many messages were erroneously enciphered using the previous additive tables, which allowed the
code breakers to make rapid progress, and by early July they were again
sending out decrypts. In the normal course of events the new code book
would stay in effect for perhaps six months. The Marines' capture of the
books was a disaster. "They sent that thing to me," Rochefort recalled, "and
all I could do was curse the whole Marine Corps. You see, that's the last
thing we wanted. We didn't want to see this lousy book. We were reading the
stuff. But when the Japanese realized it was lost they made a big change."
On August 15, after just two and a half months of use, JN-25C was replaced
with a new code book and new additive tables, as well as new operating procedures, and OP-20-G was again playing catch-up.

   Other such zealous action in the field caused repeated panics at OP-20-G
and Arlington Hall, especially when several agencies with little code breaking experience tried to get into the espionage and counterespionage act. The
FBI was eager to expand its overseas role and had already staked a claim to
fight Axis espionage throughout the Western Hemisphere. Colonel William
"Wild Bill" Donovan and his Office of Strategic Services—the OSS,
wartime predecessor of the CIA—was another loose cannon, demanding
access to ULTRA intelligence and planning various undercover operations in
neutral and enemy countries that made the cryptanalysts extremely nervous. Even the Federal Communications Commission, which had a responsibility under the law to locate and shut down illegally operating radio
transmitters, tried to get into the business of monitoring enemy radio activity and hunting down spies.

   Since 1931 the Coast Guard had been intercepting and breaking illicit
short-wave communications used by smugglers, and it, too, was jockeying to
play the lead role in cryptanalysis of enemy agent systems in the Western
Hemisphere. The Coast Guard routinely circulated its decrypts to other interested agencies, including the State Department and the FBI, but in February 1942, under orders from OP-20-G—in wartime the Coast Guard
became a part of the Navy—it decided to cut off everyone but the Navy's
Office of Naval Intelligence and GC&CS, with which it had already forged
a close working relationship. As it happened it was already too late. Several
hundred messages transmitted by German agents in Brazil had been intercepted by a U.S. Army monitor in Rio de Janeiro, forwarded to the Coast
Guard for decryption, and then passed on to the FBI. The FBI in turn furnished them to the State Department, which turned them over to the Brazil-
ian government, which, on March 18, 1942, and with great fanfare,
conducted a nationwide roundup of German agents operating clandestine
radio stations.

Using signals intelligence to arrest and prosecute spies was a doubly clumsy move: Not only was there the usual risk of alerting the enemy that its codes were being read, but also, given how inept and harmless most of the German agents were, it was far better policy just to leave them in place and under surveillance. The Abwehr, the German espionage service, relied chiefly on pro-Hitler German-Americans for its agent network in the United States, and most of their intercepted messages consisted of a pathetic litany of pleas for more money interspersed with occasional scraps of stale news. "Your messages 187 to 193 unfortunately worthless because they are from old newspaper reports," read one signal from Germany to an agent in New York, and that was par for the course. At a meeting on April 2, 1942, the FBI agreed not to move against any clandestine transmitters in the United States without the approval of the Navy and War departments; after a few more months of wrangling the three departments agreed, on June 30, that the Navy and FBI would share responsibility for scrutinizing clandestine traffic in the Western Hemisphere. (At the same time, the Navy agreed to shift full responsibility for Japanese diplomatic traffic to the Army, putting an end at last to the even–odd day arrangement; the thirty-eight OP-20-G personnel who had been handling the Purple watch were shifted to Japanese naval problems. The Navy correctly observed that the Army had little to do other than diplomatic work at this point, while OP-20-G was overwhelmed trying to keep up with the Japanese naval codes.) The three agencies also drafted for FDR's approval an order shutting down the small code breaking bureaus in the offices of the Director of Censorship, the FCC, and the OSS, which the President signed on July 8.

But the wrangling continued, and so did the bungling. A committee that was supposed to sort out the exact division of labor between the Navy and the FBI had a few acrimonious sessions and disbanded, and the Navy basically continued its policy of refusing to provide verbatim decrypts to the FBI. At one point J. Edgar Hoover threatened to begin seizing foreign agents in South America and shutting down clandestine stations if the Navy continued its policy of providing only paraphrases of agents' messages.

The OSS meanwhile was shut out entirely from receiving ULTRA, which sent Donovan into a tirade; he demanded to know if the "loyalty, discretion, or intelligence" of his agents was being questioned. The Joint Chiefs of Staff replied that since the OSS staff included Army and Navy representatives, any necessary intelligence could be passed through that channel. The truer answer was that while the OSS's loyalty was probably not in doubt, its discretion was. One of its worst gaffes was its infiltration of the Japanese consulate in Lisbon the following year, during which an agent scrounged a message sent by the Japanese military attaché. On May 5, 1943, OSS proudly circulated to the Army and Navy intelligence divisions in Washington a report on "the type of intelligence gathered by the Japanese Military Attaché in Lisbon," and proceeded to quote a Japanese message, whose text had been broken and read by Arlington Hall just a few days before, reporting on the disposition of Allied military forces in Tunisia. Worse news followed on June 29, when Arlington Hall intercepted a Rome–Lisbon message sent in the Purple code. The Japanese ambassador in Italy was warning his colleague in Portugal that there was a story circulating in Rome that "an American espionage agency in Lisbon not only knows to the minutest detail all the activities of the Japanese Ministry in Lisbon, but is also getting Japanese code books, etc." Luckily, the investigation that followed focused only on physical security, with the ambassador in Lisbon insisting that his code books were locked up and the building well guarded, and at no time did the Foreign Ministry raise questions about the cryptographic security of the Japanese military attaché code—which the British and Americans had in fact just broken after a lengthy effort. But as a result of the incident the Joint Chiefs proposed to the Anglo-U.S. Combined Staff on July 27 that both OSS and the British Secret Intelligence Service be forbidden "to penetrate enemy embassies, legations, or attaché offices in neutral countries. The discovery of such activity by the enemy will result almost certainly in special efforts on his part to improve his cryptographic security."

A final amateur turn was offered by the FCC. The Army and Navy were both unhappy that the FCC's "Radio Intelligence Division" was continuing to serve as the FBI's ears and lobbied to have its functions turned over to the military services, but they were unhappier still with some of the distinctly wet behind the ears attempts by the FCC's radio monitors to play intelligence officer. In Hawaii the FCC issued several breathless reports on its efforts to locate the source of transmitters sending "a new type of Japanese code." The location of the mysterious Japanese transmitters shifted from day to day and caused a considerable stir until the Navy discovered that the FCC was in fact tracking the Navy's own weathersonde balloons. In Alaska and on the West Coast, the FCC monitors issued so many warnings of approaching Japanese landing parties that the Army and Navy learned simply to ignore the reports. But then one warning went from the FCC field station on the West Coast to Washington, D.C., where it was passed to the Canadian authorities, who relayed it to the Canadian West Coast, where it was picked up by the U.S. forces who were coordinating with the Canadian military on hemispheric defense, and a military alert was ordered. Knox and Stimson pressed these and other examples as a good argument to get the FCC out of the game once and for all, but the President said the FCC monitors were doing valuable work for various civilian agencies, particularly the FBI, and refused to endorse their proposal. The only consolation was that all of this bureaucratic jockeying did not result in far worse leaks, as it so easily could have.

———

The Navy moved swiftly after the Midway leak to improve security, and to its credit it did so with a scalpel rather than a bludgeon. Its major step was to adopt almost word for word Travis's proposed regulations on the use and dissemination of ULTRA, and an order went out from Admiral King on June 20, 1942, to the commanders of the Atlantic and Pacific fleets and the Southwest Pacific Force that began, "The extreme importance of radio intelligence as a reliable source of enemy information has been repeatedly demonstrated. From no other form of intelligence can the enemy's intentions be so positively determined." King then underscored the need to strike the essential balance: "Any disclosures in the past with regard to the source of radio intelligence have invariably resulted in an immediate change in the enemy's communications and the consequent loss of weeks or months of painstaking effort on the part of our radio intelligence personnel. It is recognized, of course, that radio intelligence is of no value unless proper operational use can be made of it. However, momentary tactical advantage is seldom worth the risk of compromising the source." He then went on to prescribe Travis's procedural rules for how ULTRA was to be handled and distributed.

The British that summer also moved to plug leaks with precision, resisting the urge simply to encase the whole ship in concrete. A special danger became apparent in late August when Hut 3 read a message in which General Erwin Rommel announced he was unwell and asked to be relieved as commander of the Afrika Korps while he took "fairly long" leave. A week later rumors of Rommel's illness began appearing in newspapers, and Churchill was alarmed enough to order a formal inquiry into the leak. The fact was that such tidbits of inside information were precisely the sort of thing that no one could resist talking about, and a memorandum went out on September 12 establishing a new procedure whenever "gossipy" items were encountered in intercepts: Hut 3 was not to teleprint them to the Whitehall ministries, but to type up four copies, sending three by bag to "C" and one to Travis, marked "personal."

The fact was that there were astonishingly few leaks from the code breaking establishments themselves, despite an uneven and at times almost comical approach to safeguarding and securing their facilities. The obvious measures of fences and guards and badges were instituted, and a vivid memory for many at Arlington Hall was the end-of-the-day drill of emptying the trash cans into large canvas hoppers and poking through the candy wrappers and used Kleenexes with a wooden ruler in search of any classified documents that might have slipped through. But there were constant gaffes and security lapses. Two WACs dispatched in civilian clothes to test security at Arlington Hall were able to obtain visitor's badges at the guardhouse, walk through the compound, swipe two staff badges, and then wander unchallenged through the operations buildings, picking up a load of top-secret documents. At both Arlington Hall and the Naval Communications Annex, the code breakers were pressed into performing guard and escort duty until the obvious dangers of having people running around with firearms, but with no firearms training, became manifest. More than once, the quiet of the night was interrupted by gunfire at Arlington Hall when one of these guards-for-a-night dropped a pistol or confused the engaged and disengaged positions on a submachine gun. The Navy for the most part, though, predictably ran a tighter ship, and at the Communications Annex a Marine detachment patrolled the double line of barbed wire fence. There was no doubt they at least took their job seriously. The only infiltrators who ever tried to get over the fence were fraternity boys from American University, whose campus was just across the street. This initiation rite was abandoned after the commander explained to the university's president that his Marines carried submachine guns and were going to use them. The commander's other headache relating to the physical security of the grounds came later in the war when the Assistant Secretary of the Navy sent word that his daughter, who had attended Mount Vernon Seminary, was about to be married and would like to have her wedding in her old school chapel. Her old school chapel was now the Navy Chapel, and it also now happened to be located within the security fence—and no more than thirty yards from Building 4, where the Navy's ultrasecret bombes were housed. The commander said he was sorry, but given the security problems it was impossible. He was overruled: The wedding would go on. Guests arrived on the appointed day to find the chapel ringed with a line of Marines armed with submachine guns. After having their names checked off a list, the guests were permitted to enter, one by one, through a hole that had been cut in the perimeter fence. The Marines stood guard throughout the ceremony. When it was over the fence was sealed back up.

There were also background checks of a sort to weed out obvious German sympathizers and Communists. (Though Bletchley Park, at least, applied a generous rule of common sense even here, and while there was a general rule against hiring anyone even whose parents had been born in Germany, some notable exceptions were made. Walter Eytan and his brother Ernest were both born in Germany themselves but the powers that be had the perception to realize that as Jews they were unlikely to be pro-Hitler, and both became translators in Hut 4. Patrick Wilkinson, who worked on Italian naval codes, recalled once being summoned into Denniston's office and asked about a certain person. "We know he's a communist," Denniston said, "but is he the sort of communist who would betray his country?" The FBI was never so subtle in its screening of people assigned to OP-20-G.) Late in the war Bletchley Park and Arlington Hall would each be infiltrated by at least one Soviet spy apiece.

But Travis and his American counterparts were well aware that, in reality, a much greater threat to secrecy than spies sneaking in over the fence was human nature. It was especially difficult for young, able-bodied men to resist the urge to boast to family and friends about the importance of their work when the inevitable questions, spoken or not, arose about why they were not overseas fighting. The real enemy was rumors and public knowledge. New recruits at Bletchley Park were required to sign the Official Secrets Act and were told that they must not speak of their work to anyone, including their immediate family, to their dying day. WAVES arriving at the Naval Communications Annex were ushered into the Navy Chapel and given an indoctrination lecture the high point of which was the threat—which many remembered vividly a half-century later—that they would be shot if they ever breathed a word of what went on inside the fence. For the most part that was the last mention that was made of security, and it was enough. Partly it was the times, partly it was the still untarnished trust in government, but thousands of workers of all educational and social backgrounds kept their secret, not only through the war but for decades after. Walter Eytan said his wife found it difficult to marry a man who would not tell her what he did in the war; when in the late 1970s information was at last released about Bletchley Park's work, many husbands were astonished to learn that their wives had had a role in the Enigma secrets that were coming to light.

To deflect the inevitable questions, workers were drilled in how to respond when family or friends asked what they did. If you were stationed at OP-20-G you could say you worked in "naval communications." A circular sent around at Bletchley Park recommended the following script:

Question: What are you doing now?
Answer: Working for the Foreign Office (or other Ministry as appropriate).
Question: But what do you *do*?
Answer: Oh—work.

"A gay reticence in that vein will win you far more real respect from anyone worth respecting, than idle indiscretion or self-important airs of mystery," the memorandum concluded. The same notice also warned, DO NOT TALK AT MEALS, DO NOT TALK IN THE TRANSPORT, DO NOT TALK TRAVELLING, DO NOT TALK IN YOUR BILLET, BE CAREFUL EVEN IN YOUR HUT ("Cleaners and maintenance staff have ears, and are human") and DO NOT TALK BY YOUR OWN FIRESIDE ("If you are indiscreet and tell your own folks, they may see no reason why they should not do likewise. Moreover, if one day invasion came, as it perfectly well may, Nazi brutality might stop at nothing to wring from those you care for, secrets that you would give anything, then, to have saved them from knowing. Their only safety will lie in utter ignorance of your work.")

A security officer at Bletchley who was given the arduous assignment of making a complete tour of the local pubs, hotels, and clubs found a certain amount of curiosity about what went on inside the Park and an occasional complaint about all of these civilians who were dodging military service, but no knowledge about the work itself. Although there were minor infractions of people overheard talking on the platform at Bletchley station, only two serious and deliberate security violations were ever recorded. Both occurred in early 1942. In one case a woman told her family what she was doing and one of them repeated it at a cocktail party. In the other case "one of the most vital tasks in which the organisation is engaged" was disclosed by a Bletchley Park worker in after-dinner conversation at his old senior common room at Oxford. (Luckily, as Walter Eytan observed, "there were enough sensible people around to keep the secret to themselves" on that occasion.) Travis sent a terrifying memo to everyone at Bletchley Park outlining the two cases and reporting that they had been investigated and turned over to the Director of Public Prosecutions, who had concluded that ample evidence existed for conviction under the Official Secrets Act; only the intervention of "C," "who wished to prevent the organisation from coming under public discredit," spared them from prosecution, conviction, and imprisonment. No such leniency would be shown again, Travis warned.

The most basic principle of security at Bletchley Park, Arlington Hall, and the Naval Communication Annex, and the most effective, was compartmentalization. The cardinal rule was "need to know," and if you didn't need you didn't know. The bombe rooms in the Navy's Building 4 were off limits even to people in OP-20-G who had the highest security clearances, but who worked on different problems. Many did not learn until years later even of the bombes' existence. OP-20-G went so far as to make sure that the staff working on additive recoveries in JN-25 did not know anything about the content of the messages themselves; they were even cautioned against studying Japanese or buying dictionaries. That degree of compartmentalization was dropped after the first year when it was recognized as probably self-defeating—and unnecessary, since the real secret was not what the messages said but rather the very fact that the Navy was successfully tackling JN-25 at all. But many of the most sensitive projects were completely hidden from the view of workers in other sections, and the principle of not asking one's coworkers what they were doing was second nature. At Bletchley Park there was similarly a small list of people who were allowed into Huts 8, 6, 4, and 3. The main problem in enforcing compartmentalization was not within the Park itself, but with outside intelligence officials who tried to crash the gates. Travis and other top officials repeatedly had to bar high-level visitors from access to the sensitive areas, and on one notable occasion B.P. officials informed "C" that a certain Captain Hugh Trevor-Roper, who was working in counterespionage, was henceforth banned from the site after he had barged his way into Huts 3 and 4. ("He is not a suitable person to whom I should give, so far as I can see for no reason, the run of Huts 3 and 4," the memorandum to "C" concluded.) Later a special joint U.S.–British war room was set up in London specifically dealing with counterespionage, and Trevor-Roper was among a small number of officers permitted to come a few times a week to look over relevant ULTRA material kept there.

———

The other source of security was ULTRA itself. It provided a window not only on the enemy's military plans and intentions but also on his fears and suspicions, and as such it was the first line of defense against leaks about the Allied code breaking operation. Shortly after the alarm over the rumors of Rommel's illness, an Enigma decrypt of September 11, 1942, revealed that the Germans were extremely suspicious upon learning from British POWs that the British Eighth Army had known in advance the date and general plan of the Axis attack at Alam el Halfa in the desert west of Cairo that began on August 31. General Bernard Montgomery, taking command of the demoralized British forces on August 13, had at once sized up the tactical situation and began a vigorous tour of confidence building. He had good reason to be confident, for just four days into the job he received from GC&CS a decrypt of the Panzer Army's complete plans. By the summer of 1942 Bletchley Park was reading Rommel's main Army Enigma key (Chaffinch), often with a delay of only one day; it had also broken Scorpion, a special Luftwaffe key used for air–army cooperation, which was being intercepted and now decrypted by a special GC&CS unit sent to Cairo— the only time during the entire war GC&CS allowed Enigma traffic to be decrypted in the field. (Breaking Scorpion was made considerably easier by the astonishing discovery that the daily settings for each month were copied lock, stock, and barrel from the key lists of another Luftwaffe key, Primrose, used in the previous month. Once the first day's Scorpion was broken each month, no further bombe runs were needed to read each day's traffic.) Montgomery, "believing that the confidence of his men was the prerequisite of victory, told them with remarkable assurance how the enemy was going to be defeated," recalled his intelligence officer, Brigadier Edgar Williams. "The morale emerging from the promise so positively fulfilled formed a psychological background conditioning the victory which was to follow [at El Alamein]." Montgomery attributed the information he passed on to the troops to "an Italian POW." When Rommel attacked, Montgomery's men were well prepared, and the German tanks were subjected to heavy RAF attacks as they passed through a British minefield, then brought to a halt by massed artillery and tanks that Montgomery had concentrated on the Alam el Halfa ridge. It was a spectacular victory for ULTRA; it injected an enormous boost of confidence into the Eighth Army; it made Montgomery look like a genius; and among the small circle in the know in the top echelons of the desert army it caused a revolution in attitudes toward signals intelligence.

Passing the information to lower levels posed the same problems as always, however. Only Montgomery, his intelligence officer, and his chief of staff were allowed to see the ULTRA signals "naked"; intelligence officers at the corps level were not cleared, and hanging everything coming out of Army HQ on POW reports ran the danger of either being an obvious pretense or discrediting the quality of the intelligence in the eyes of the recipients. The solution was a series of winks and nods that let the corps intelligence officers know that we had "something up our sleeve," as Williams put it, while maintaining the "polite convention that the Intelligence Staff at Army had a habit of making the correct appreciation by virtue of some remarkable element in their glandular make-up."

The POW story was good enough for the troops. It just wasn't good enough for the Germans when they got wind of it from their own POW interrogations of captured British soldiers, and the Germans launched a security investigation at once. It was in situations like this that reading the enemy's messages became both a blessing and a curse, for by its very nature the intelligence game engendered a mindset where paranoia was never far away. GC&CS began to imagine all sorts of double- and triple-crosses and sent a warning to the Middle East commands that the Germans might lay a trap by sending a deliberately false Enigma message:

> We have impeccable evidence that Germans much perturbed at our knowledge of general Axis situation especially supply situation on your front. They are conducting special inquiry into security QT

source. Matter hangs in balance. All recipients of QT intelligence should be specially warned and rigid security discipline enforced. Curtailment of matter given to press and promulgated in Intelligence Summaries, whatever the camouflage adopted, essential. Possibility enemy will lay trap to verify his suspicions not improbable. Trap might well take form of passing false information of location specially important target or other fact causing us to make tactical movement. . . . Vital at present moment avoid jeopardizing source for sake of minor gains.

The Germans in fact never did resort to such subtlety. But the paranoid mindset could be exercised on allies as well as on enemies, and because reading enemy codes was a way to read what the enemy was able to read of one's allies' codes, things frequently got complicated. American suspicions of the British were inflamed by one complex episode in the Middle East in 1942 when the British learned via ULTRA that the Germans had broken the American military attaché code and were reading dispatches sent from Cairo to Washington by Colonel Frank Bonner Fellers, the American attaché. The problem was delicate. On February 25, 1942, Churchill had written Roosevelt a remarkable confession: "Some time ago . . . our experts claimed to have discovered the system and constructed some tables used by your Diplomatic Corps. From the moment when we became allies, I gave instructions that this work should cease." Churchill was calling the matter to the President's attention to warn that the State Department's codes ought to be tightened up, and the Americans for the most part took it in that spirit. But suspicions lingered, and when in June 1942 the British informed Washington that the Cairo code was compromised, the native paranoia of the cryptanalysts was stimulated into action. What made matters worse was that the Enigma messages GC&CS had intercepted related some especially embarrassing indiscretions on Fellers's part. Several of the messages—which consisted of intelligence reports to German theater commanders bearing the heading "Good Source Reports"—reported information about the British order of battle and other inside details. On May 29, 1942, another message, now citing "a particularly reliable source," came through, and it quoted from an American report of April 16 that severely criticized the RAF for failing to maintain U.S. airplanes properly. Then came an even more embarrassing intercept, quoting Fellers disparaging British military performance in Egypt and predicting that Rommel could easily reach the Nile Delta if he went on the offensive.

While SIS to its credit acted swiftly to tighten up security, dispatching a SIGABA machine to Cairo, Colonel McCormack and others smelled a rat, and wondered if the British were in truth reading not German accounts of Fellers's dispatches but Fellers's dispatches themselves. There was even some suspicion that it was an elaborate scheme to find out more about the SIGABA, which the British wanted the Americans to share but which the Americans were being extremely tight-lipped and touchy about. McCormack finally concluded that there was no such ruse taking place, but meanwhile considerable ill feeling had been aroused by the charges that had been leveled against Bletchley Park by the Americans.

Dealing with the Russians, for whom paranoia came even more naturally, was even trickier. The Russians had failed to reciprocate the confidence Churchill had shown in sending to Moscow intelligence derived from Enigma decrypts; when his forces were facing a crisis or setback, Stalin would retreat into silence and provide no information at all about the fighting. Requests from the British for technical information on captured German weapons, the identity of certain German units, or Russian military action were ignored, even as the Soviet government bombarded the British Military Mission in Moscow with requests for information about the strength and intentions of the German forces. By far the most reliable and complete information that the British obtained about the Soviet order of battle came from decrypting German intelligence reports. Ambassador Oshima toured the Russian front for ten days in July and August 1942 and filed a detailed report that Arlington Hall decrypted, and that too provided far more details on the fighting than Stalin was giving his allies.

The Russians were supplying one critical bit of assistance, however, and that was copies of German Police and Army intercepts, signals that could not be picked up in England. In return GC&CS provided through a British representative in Moscow instructions on how to solve the Police hand ciphers and some general information about German radio procedures. But the Russians kept throwing obstacles in the way of further cooperation, and on October 16, 1942, the British representative in Moscow reported that the Russians "have relapsed into old bad ways and I have had no meeting for a fortnight although they know I have immediate information for them." In December the Russians said cooperation on wireless intelligence should be discontinued; the British representative told London that no explanation was given but it was probably that the Russians "cannot continue indefinitely without showing their hand, and it contains so little." A few weeks later they halted the flow of German Police intercepts, which severely limited Bletchley Park's ability to break traffic on some of the police circuits. In February 1943 the Russians abruptly shut down a tiny intercept and communications station that the Royal Navy had been permitted to establish at

Polyarnoe near Murmansk to support the Arctic convoys that Stalin had so vociferously demanded.

The British were still willing to try to keep the ball rolling, but the major problem was not even so much Russian obstinacy as the continuing evidence from ULTRA that the Germans were reading the Russian codes. It was difficult to know what to do about it, since the Russians would take no steps to tighten up procedures but at the same time demanded more details on the British evidence—which the British were loath to provide, since they were not about to tell Moscow about their success in reading Enigma traffic. One idea was to send the Russians "a theoretical" treatise on how their weather cipher could be broken, but the obvious pitfall was that the Russians might take that as proof that the British were the ones breaking their codes. (Acting on Churchill's orders, the British "Y Board," which set policy for the military services' monitoring stations, had in fact ordered a halt to interception of Russian military traffic upon the outbreak of hostilities between Germany and Russia.) Eventually GC&CS sent along additional proof of Russian cipher insecurity, claiming that it had come from reading low-level German codes whose solution had already been shared with Moscow and from a disgruntled Austrian POW. Discussions sputtered along for a few more months, but the British Military Mission now reported that whereas before the Russians had been secretive in defeat now they were "swollen headed with success" and clearly felt they no longer needed help. It was but a foretaste of the mutual suspicions that would grow unchecked at war's end.

CHAPTER 9

The Shadow War

ROMMEL was brought to a halt at Alam el Halfa in August 1942 by Montgomery's forewarned forces, but by an even more formidable foe, too: his supply line. Rommel's panzers were running out of gas. An acute fuel shortage had begun to pinch the Panzer Army since July. Ship after ship that crossed the Mediterranean bearing fuel for the Italians and Germans was being sunk by RAF bombers and British submarines. By mid-July the Luftwaffe's air transport system in North Africa was at a standstill. During the Battle of Alam el Halfa the Germans took to flying fuel in by air, so desperate was the shipping situation. For month after month the British attacks on shipping continued with uncanny precision, often even seeming to single out ships carrying fuel, ammunition, and tanks, while leaving those laden with less vital cargoes of food or other supplies unmolested. A German staff officer in Italy wrote in frustration that British spies in Italian ports seemed to know exactly what was being carried on each ship: "The convoy sets sail from Naples quietly and peacefully, escorted by three Italian torpedo boats; then just before it gets to Tripoli the British come flying over and of course it is only the German steamer that gets hit, and it carries a cargo for which Rommel is eagerly awaiting. Nothing happens to the Italian ships. The German security service should get to work on this problem."

From July to October 1942, the British sank forty-seven supply ships in the Mediterranean, totaling 169,000 tons; in forty-four of those cases the ship's time of sailing, route, or location in port was relayed by Bletchley Park to the British Navy and RAF commands in Cairo in time to effect the interception. Most of the intelligence came from Italian messages encrypted with the relatively insecure, and badly used, C38 Hagelin machine, but the near-current reading of the Chaffinch Enigma key was now adding an entirely new dimension to the precision with which the British forces were able to pick off Rommel's supplies. Since July, GC&CS was reading almost daily the Panzer Army Operational Staff report, which included a summary of the supply situation. Beginning in October, a far more detailed pro-forma report issued every ten days by the Chief Quartermaster of the Panzer Army was unraveled by GC&CS. The British intelligence staffs in Cairo formed a special office that began to assign target priorities with statistical precision: They would analyze Rommel's supply reports, calculate his daily consumption of diesel fuel, gasoline, 76.2-mm antitank ammunition, 75-mm tank shells, tires, and food and pinpoint the precise supplies the loss of which threatened to hinder his operations the most. They would then compare those calculations with decrypts on cargo ladings to decide which ships to hit.

This was scientific warfare of a new kind, and its effect was devastating. In October, 44 percent of the Axis shipping tonnage that left Italy for Libya was sunk; more important, in the last days of the month three tankers, carrying eight thousand tons of fuel in all, were singled out for destruction, a blow that essentially finished Rommel's ability to continue resisting the British advance that began at the Battle of El Alamein on October 23. A series of Chaffinch decrypts chronicled day by day the deteriorating situation. October 28: Rommel's army reports its fuel situation is "grave in the extreme," with stocks down to about five hundred tons, enough to move only another hundred kilometers. November 3: Rommel informs the German high command, "PANZERARMEE IST ERSCHOEPFT"—the Panzer Army is exhausted. November 4: Afrika Korps strength is down to twenty-four serviceable tanks. November 10: *Eleven* tanks.

Yet with victory in his grasp, Montgomery's instinct for caution suddenly overwhelmed him, and he ordered an abrupt slow-down in the pursuit of the fleeing Germans across Libya. Hut 3 was swept with "fierce indignation and dismay," Ralph Bennett later wrote. Montgomery, who had made such brilliant tactical use of ULTRA, seemed to be unaccountably ignoring the ultimate gift that Bletchley had handed him on a silver platter. It seemed, said Bennett, "to cast doubt on the whole point of our work." The truth, as always in war, was more complicated, and ULTRA itself bore part of the blame. Enigma decrypts throughout November and early December confirmed that Rommel was not being resupplied, and would not be resupplied. Montgomery, facing supply problems of his own, concluded that there was no danger in pausing to assemble an overwhelming force that would finish the Germans off in a final frontal assault against their defensive position at Agheila.

What Montgomery did not consider was the possibility that Rommel would simply cut and run when the British attack began, and that is precisely what happened. When Montgomery decided to repeat the performance and began another methodical buildup against the new defensive line Rommel had established 250 miles to the west, it was not just Hut 3 that was in a rage; the Prime Minister was barely able to control his impatience with what even an official British history of the campaign called Montgomery's "ponderous" preparations. On December 27 Churchill sent a personal signal to Montgomery urging that the general move immediately and not wait another two weeks as he planned:

> Boniface shows the enemy in great anxiety and disarray at Buerat, and under lively fear of being cut off there by an enveloping movement from the south which he expected might become effective as early as December 26. Reading Boniface after discounting the enemy's natural tendency to exaggerate his difficulties in order to procure better supplies, I cannot help hoping that you may find it possible to strike earlier than the date mentioned.

Montgomery was unmoved; attacking at Buerat on January 15 he lost his last chance to cut off Rommel's fleeing forces. The Desert Fox made good his escape into Tunisia as Montgomery's Eighth Army nipped at his heels. Four more months of brutal fighting would be needed to push the Germans out of Tunisia, and so out of Africa once and for all.

———

Under Montgomery's predecessor General Archibald Wavell, the British Army in the Middle East began in the fall of 1940 plotting ways to make deception a basic element of its military strategy. Wavell had served under General Edmund Allenby in Palestine in October 1917 when Allenby, showing an imagination unique among First World War commanders, decided that trickery was preferable to the heroic mass slaughter of his own troops. He arranged for a briefcase containing false plans to be "lost" in no-man's land; then, instead of attacking the heavily fortified Turkish position at Gaza on the coast with the expected head-on assault, he led his troops in a sweeping flanking motion through Beersheba thirty miles inland. This feat left a deep impression on Wavell, and in November 1940, as Commander in Chief Middle East, he brought in Lieutenant Colonel Dudley Clarke to blaze a trail that would lead in time to the massive and successful deception operations in support of the Allied landing at Normandy on D-Day. Clarke was traversing mostly virgin territory, but by the spring of 1942 his unit, known as "A Force," had succeeded in creating so many "notional" British

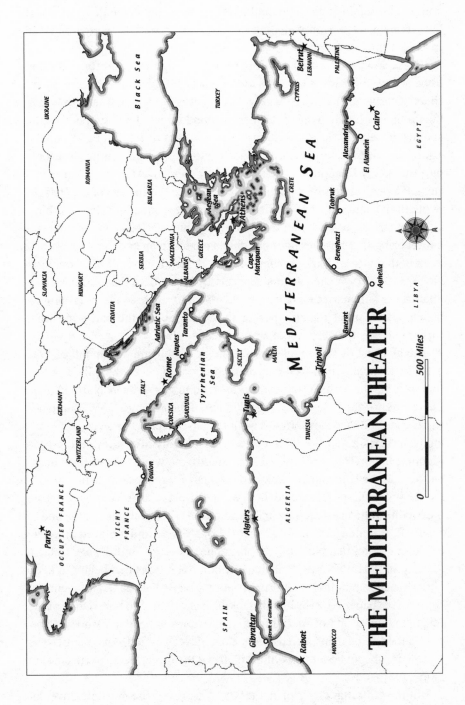

THE MEDITERRANEAN THEATER

UKRAINE

Black Sea

ROMANIA

TURKEY

BULGARIA

Aegean Sea

CRETE

Athens

GREECE

MACEDONIA

ALBANIA

Cape Matapan

SERBIA

SLOVAKIA

HUNGARY

CROATIA

Adriatic Sea

Taranto

Naples

ITALY

Rome

Tyrrhenian Sea

SICILY

MALTA

GERMANY

SWITZERLAND

CORSICA

SARDINIA

Tunis

Toulon

VICHY FRANCE

Paris

OCCUPIED FRANCE

SPAIN

Gibraltar
Strait of Gibraltar

Rabat

MOROCCO

Algiers

ALGERIA

TUNISIA

MEDITERRANEAN SEA

Tripoli

Buerat

Agheila

Benghazi

Tobruk

LIBYA

El Alamein

Alexandria

Cairo

EGYPT

CYPRUS

Beirut
LEBANON

PALESTINE

500 Miles

0

273

forces in the Middle East—including a nonexistent armored division, seven nonexistent infantry divisions, and a nonexistent corps headquarters—that the Germans were overestimating total British strength in the region by 40 to 45 percent.

Clarke soon established two core principles. One was that for deception to be successful it was necessary to know what the enemy was already inclined to believe. The easiest way to mislead was not with a total fabrication but instead by building on a preexisting "foundation of fear" supplied by the enemy himself. Clarke's second principle was that deception had to be designed not merely to plant a false idea but to produce action: "The only purpose of any Deception is to make one's opponent ACT in a manner calculated to assist one's own plans and to prejudice the success of his," Clarke wrote in the summer of 1942. "In other words, to make him *do* something. Too often in the past we had set out to make him THINK something, without realizing that this was no more than a means to an end."

In both matters, Clarke recognized, the British had a tool of incomparable value in ULTRA. The Enigma decrypts, by laying bare the Germans' own intelligence summaries and plans, could reveal to a degree that was unimaginable by any other means whether they were taking the bait—and what sort of bait was the most appetizing in the first place.

The first test of these ideas on a strategic scale came in the fall of 1942 with the launching of a well-coordinated effort to cover the Allied operation TORCH, the British and American landings in French North Africa. There was no way to hide the large buildup of naval forces at Gibraltar, nor was it expected that the Germans would fail to detect the large convoys sailing from Britain and the United States—especially after they entered the Mediterranean. But it was possible to muddy the waters about their destination, and the British plan exploited German apprehensions to the utmost. On the broadest strategic scale there was an effort to distract attention from the Mediterranean altogether; a plan called SOLO ONE, initiated at the end of July 1942, threatened a fictitious invasion of Norway, and it played precisely to Hitler's belief that Norway was "the zone of destiny in this war." Indeed, as early as 1941 elaborate British deception operations had begun to keep Hitler's finger pointing at Norway; reports were fed via tame agents of British newspaper advertisements seeking fishermen with knowledge of the coast of Norway and of special training of troops in Scotland who were reported to have been inspected by the King of Norway. A second plan, OVER-THROW, was begun at the same time; it placed the target as the north coast of France.

But the Gibraltar cover plan, KENNECOTT, proved the most effective, for it was closest to the truth, and would hold its value even *after* the troop convoys had passed Gibraltar. The story was that the British naval buildup was in preparation for running a resupply convoy to Malta, and the garrison there was being pictured as being in desperate straits. It was here that the synergy of signals intelligence and deception paid off so richly. In late August and September intercepted German messages showed their anxiety about the Gibraltar activity and suspicions of a planned Malta convoy, or even an attack on Marseilles. Further reports came in showing German apprehensions about a possible Allied attack on Sicily. Analysis of German intelligence reports had earlier revealed that the Germans had come to associate an upsurge of certain British radio activity with preparations for a Malta convoy, so the British obliged by staging just such a show of signaling at Gibraltar. At midday on November 7, GC&CS decrypted a German naval Enigma message that provided a great measure of reassurance: It was an intelligence appreciation that considered the likeliest destination of the Allied convoys in the Mediterranean to be either Malta or the Eastern Mediterranean—the Tripoli–Benghazi area, or Sardinia or Sicily. The landings at Morocco and Algiers the next day achieved total strategic surprise.

The source of the November 7 message was a recent addition to GC&CS's bag. The Atlantic naval Enigma keys remained unbroken. But as early as August 1941 the Naval Section of Hut 4 had identified a separate naval Enigma key, Porpoise, that was being used in the Mediterranean and the Black Sea. The cryptanalysts of Hut 8 glanced at it, saw that not enough messages were carried each day to enable the use of Banburismus, and put it on the shelf. A year later, in July 1942, Hut 4 again called the code breakers' attention to Porpoise; it was now carrying two hundred messages a day, and wasn't it worth another look? It was, and years later Hut 8 was still kicking itself for what it had overlooked. Porpoise was "the worst episode in the history of the section," a postwar report admitted. The indicators carried on Porpoise messages consisted of a pair of four-letter groups, just as did all other naval Enigma messages that used the complex naval system of bigram tables to encipher indicators. (See Appendix B.) But when, in July 1942, Hut 8 sent the messages to the IBM section to look for overlaps that could be used to build up Banburismuses, a curious phenomenon was noted among overlapping messages. Most had the same second letters as well as the same sixth letters in their indicators; some also had the same third letters and the same seventh letters. It still did not dawn on anyone what was going on until a day or two later when one of the Hut 8 cryptanalysts described this curious phenomenon to a member of Hut 6. He immediately recognized it as nothing but the much simpler pre–September 1938 indicating system that had been used on Luftwaffe and Army keys—the twice-enciphered indicator—padded out with two extra letters. The reason for the repeated letters in the indicators was simple: If two messages overlapped, their initial rotor settings by definition had to be fairly close. They usually had the same left-rotor setting, and often the same middle-rotor setting as well, say S<u>K</u>A and S<u>K</u>L. When enciphered twice in a row on the Enigma at the ground setting, and padded with the two extra key letters (in this example DA and XN), the resulting indicators would be of the form D<u>WG</u>H A<u>UC</u>K and X<u>WG</u>A N<u>UC</u>Z. It was a shattering discovery. It meant that Rejewski's tried and true, and long ago outmoded, method of "females" could be employed directly to break Porpoise.

The Mediterranean U-boats were still using the three-rotor machine, though with all eight naval rotors to choose from, so the Zygalski sheet method would have been extremely time consuming (as well as requiring a huge effort in creating 336 series of punched sheets, instead of the previous 60). But the bombes could be used to run a specialized menu built up from the chains of letters that appeared in the indicators. Beginning in August 1942 a "fairly extravagant" bombe procedure was used in order to break Porpoise as quickly as possible when its daily key changed at midday. All twelve bombes would be pressed into service to begin running through all wheel orders. At the same time, Hut 8 would be working to eliminate certain wheel-order possibilities by using a modified Banburismus procedure on the incoming messages. As these were identified they would be phoned over to the bombe huts to narrow the search.

Why the German Navy adopted this insecure indicating procedure for Porpoise can only be guessed at; it continued to use the twice-enciphered indicators even after Porpoise went over to four wheels on February 1, 1944. The double encipherment was only dropped at last on June 1, 1944. One possibility is that the Navy had to communicate with non-naval personnel in the Black Sea and was loath to give away the secret of its special indicating system to the other German services. In any event it was, in the words of historian Ralph Erskine, "an astonishing blunder."

———

It was a Porpoise decrypt read in the early morning hours of November 11, 1942—three days after the TORCH landings began—which disclosed the urgent news that the Germans were about to occupy Vichy France. Since September, Gustave Bertrand had begun to pick up signs of increasing danger to the clandestine intercept station that he had set up in Vichy. A German official in Paris had revealed to him the German plans for occupying Vichy in case of an Allied invasion of French North Africa. Two motorized divisions in Dijon were to rush to Marseilles; Bertrand's Station Cadix at Uzès lay directly in their intended path. Bertrand consulted with London about evacuation plans. The British suggested the Poles be sent to North Africa, while Bertrand himself would be airlifted directly out if necessary.

Later that month the noose began tightening. Bertrand learned from local French officials that a special German radio detection unit had arrived in the area. The Germans were spotted several times driving around the region in Chevrolet pickup trucks and six Citroen sedans, one an unmistakable purple color. Ten times in October the electrical power at Cadix suddenly cut off—a standard technique used by radio detection units to pinpoint a clandestine transmitter.

At dawn on November 6, in the middle of a transmission to London, Langer looked out the window of the chateau and saw, half a mile away, a dark blue Chevrolet pickup with a circular antenna glistening in the dawn. A black Citroen was in the lead. Both were heading straight for the chateau. Bertrand and Langer immediately hid all of the radio gear and waited. The procession stopped at the gates of the chateau; three policemen with rubber truncheons got out—and headed for two nearby farms, where they roughed up the farmers, and left.

Bertrand decided it was time to get out. On November 9 the station was evacuated. Three days later German troops occupied the town. Bertrand melted into France. The Poles split up into groups of twos and threes with instructions to make contact with the underground network in Marseilles, which would try to get them aboard fishing boats for passage to Gibraltar, or stow them away on passenger ships for North Africa. Rejewski and Zygalski were passed along a sort of underground railroad of safe houses and trustworthy hotel keepers—to Nice, Cannes, Antibes, back to Nice and Marseilles, then to Toulouse, and at last to the Spanish frontier. Late in the evening of January 29, 1943, Rejewski and Zygalski set out across the Pyrenees. After a tense altercation with their guide, who, after hours of trekking through the dark, suddenly pulled out a revolver and demanded all of their money as his price for taking them the last few miles, the two Poles reached Spain—where they were promptly arrested by Spanish border police and thrown into a prison in Barcelona already overflowing with five thousand refugees, many of them fellow Poles fleeing from the Nazis in Vichy.

One of the few contacts the prisoners were permitted was with the Polish Red Cross, which showed its own cryptographic imagination by sending secret messages to the prisoners in the guise of a list of names of Polish prisoners who were to receive food packages:

1. Zygmunt Przybylski
2. Komisja Przyjezdza   [Commission arrives]
3. Jutro Zestolicy   [tomorrow from the capital]

4. Bedzie Uwas   [itwill visityou]
5. Przygotujcie Uwagi   [prepare remarks]
6. Owarunkach Wobozie   [onconditions incamp]
7. Ispis Chorych   [and sicklist]
8. Trzymaj Ciesie   [Good luck]
9. Mikolaj Cieslak
10. Marian Wozniak

Rejewski and Zygalski were later moved to another prison camp; finally, after months of diplomatic pressure from the Allies, and a hunger strike by the prisoners, the Spanish authorities agreed to release all of the refugees. On July 21, 1943, Rejewski and Zygalski left Madrid for Portugal, where they were picked up by a British warship, transferred to an airplane at Gibraltar, and arrived in England at last on July 30.

GC&CS was unwilling to bring them to Bletchley Park, however; the Poles were sent to the Polish military command in London and then assigned to a small Polish intercept and cryptanalytic unit based at Boxmoor outside London, where they worked the rest of the war on SS and SD hand ciphers.

Langer and Ciezki were not so fortunate. Betrayed by their guide, they were surrounded by Gestapo agents on a dark, rocky path just inside the French–Spanish border on the night of March 10. They were held at Stalag 122 in Compiègne, then taken in September to SS concentration camp No. 4 at Schloss Eisenberg in Czechoslovakia. It took the Germans six months to realize who they had in their clutches. One day in early March 1944 Langer was summoned to appear before a joint Gestapo–Wehrmacht commission that had arrived at the camp. They had two questions for him: Would he work as a double agent? And how much had the Biuro Szyfrow, which he had commanded in 1939, and which had continued to operate in France in 1940, been able to read of the German Enigma machine?

Uncertain how much his interrogators knew, Langer took a huge gamble. With considerable self-possession and quick thinking, he coolly admitted that the Poles had indeed broken the Enigma before the war. But the procedural changes the Germans had introduced in late 1938, he explained, had made it impossible to continue the work. Langer then suggested that his interrogators call in Ciezki, who was the real cipher expert, to confirm his statement. Much to Langer's relief, Ciezki did. It was an act of incredible courage that, it is no exaggeration to say, saved the Allied code breaking operation at a crucial moment just months before D-Day.

Bertrand played an even riskier gambit. On instructions from the Resistance he remained in France, under a false name; then on January 5, 1944, he was arrested on a Paris street. The Abwehr told him they knew he was an agent in contact with London and offered to spare his life if he would switch sides. Bertrand pretended to go along with the scheme. He was released, whereupon he promptly fled with his wife, made contact with London, and was picked up from an improvised airstrip in the Massif Central late on the night of June 2 and flown to England.

———

On July 27, 1942, radio monitors in London picked up an astonishing broadcast to the German people by Admiral Dönitz. It was extraordinary for a Nazi official to present the public with anything but self-glorifying propaganda, but Dönitz began his speech by warning that "more difficult times lie ahead of us." The triumphal German U-boat successes reported in the last few months might, he cautioned, have left the German people unprepared for the "harsh realities of submarine warfare." People had to be prepared for bad news. It was, as far as the Admiralty was concerned, a "tip straight from the horse's mouth" that the convoy battles were about to begin again in the mid-Atlantic.

The "harsh realities" were harsh indeed. As much as Dönitz had tried to create an aura of heroism and chivalry about his U-boat captains, the Atlantic struggle was among the grimmest, most merciless, and most bloodthirsty of the entire war. The Second World War produced reams of mind-numbing statistics of death and destruction, yet the numbers in the U-boat war still have a capacity to shock. The Germans built 1,162 U-boats; 785 of them were sunk. Of the 41,000 men who served on U-boat crews, 28,000 were killed. U-boat crews who survived one patrol at sea were veterans. Early in the war they were volunteers, young men in their early twenties drawn by the image of glamour and swagger of an elite corps, and often propelled by an ardent Nazism they shared with their admiral. But that did not alter the appalling realities of life aboard the boats. It was not for the claustrophobic. The crew ate and slept in the same place; there was one toilet for forty-four officers and men, and even that could not be used when the submarine was submerged; there were no showers, no air-conditioning, and precious little fresh food or even fresh air much of the time.

Under orders from Hitler the U-boat captains had begun the war observing the international Law of Prize, which required that the crew of a merchant ship under attack be removed safely. Dönitz chafed at such restrictions from the start, arguing that by convoying merchantmen under the protection of warships the Allies had forfeited such protections. And arming merchant ships and running them without lights in fact made them warships in point of law, liable to attack without warning, he insisted. The debate over such legal niceties did not last for very long. On February 18, 1940, the Germans declared unrestricted U-boat warfare on Britain and France. There were occasional acts of chivalry by individual U-boat commanders, in which ships' crews were permitted to take to lifeboats provided they did not try to send a distress signal, but mercy was not the order of the day. And in September 1942 Dönitz issued new orders that left no doubt that the war had entered a new and more savage phase:

1. All attempts at rescuing members of ships that have sunk, including attempts to pick up persons swimming, or to place them in lifeboats, or attempts to upright capsized boats, or to supply provisions or water, are to cease. The rescue of survivors contradicts the elementary necessity of war for the destruction of enemy ships and crew.
2. The order for the seizure of commanding officers and chief engineers remains in force.
3. Survivors are to be picked up only in cases when their interrogation would be of value to the submarines.
4. Be severe. Remember that in his bombing attacks on German cities the enemy has no regard for women and children.

Service aboard the merchantmen and their escort vessels was as terrifying as it was aboard the U-boats, and without the glamour. The statistics here, too, were shocking; 2,452 merchant ships were sunk in the Atlantic; more than thirty thousand men of the British merchant navy were killed. If a U-boat got in among the convoy lines that stretched across miles of ocean, the merchant ships were under orders to keep moving; there was simply nothing else to do. Stopping to pick up survivors was sure to add to the Germans' toll, and there was one horrible image that those who crossed the Atlantic could never erase from their memories. "I saw it first in HMS *Alaunia* in 1940," recalled Hal Lawrence, as his ship passed within feet of men floating in the water, survivors from a torpedoed ship:

They shout, even cheer, as you approach; the red lights of their life jackets flicker when they are on the crest of a wave and are dowsed as they slip into the trough; their cries turn to incredulous despair as you glide by, unheeding, keeping a stoical face as best you can. But the cold logic of war is that these men in the water belong to a ship that has bought it and that a couple of dozen more ships survive and must be protected. . . . each time was as bad as the first. We *never* got used to it.

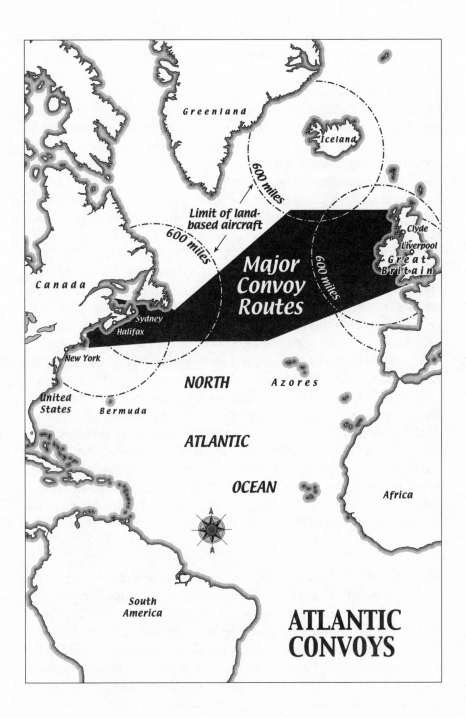

Greenland

Iceland

Limit of land-based aircraft

600 miles

600 miles

600 miles

Canada

Clyde

Liverpool

Great Britain

Sydney

Halifax

Major Convoy Routes

New York

NORTH

Azores

United States

Bermuda

ATLANTIC

OCEAN

Africa

South America

ATLANTIC CONVOYS

Some convoys had a rescue ship taking up the rear that could stop to pick up survivors; many convoys did not, and in fact so many of the rescue ships had themselves been sunk that by the end of 1942 there were only enough available for one convoy in four. Experienced sailors developed a fatalistic approach to their odds of survival. If you were on a ship carrying iron ore, you slept on deck because it would sink like a rock when hit and there would be only seconds to scramble overboard. If you were carrying an ordinary cargo, you could sleep below but you left your clothes on and the door open. If you were aboard a tanker, you got undressed, shut the door, and got a good night's sleep because if you were hit you would be blown to bits anyway in the resulting explosion and fireball.

A good many of the men who sailed on the convoys were *not* experienced. They were given a quick training and thrown aboard. An American seaman recalled his first watch when, approaching the chief mate on the bridge as he had been instructed at Merchant Marine school, he saluted snappily and barked out, "Relieving the watch, Sir!" The chief mate stood there for a moment dumbfounded, then said, "Oh my, how lovely." Many of the officers who manned the escort vessels of the Royal Canadian Navy, upon which a disproportionate burden for escort duty fell, were reservists from the prairies who had never seen the ocean before. "We were all badly trained, scared stiff, and most of the time wished to God we had joined the Air Force," said one. The Canadian Navy's "cheap and nasty" Flower-class escort ships, small corvettes with their open bridges and lookout positions, were "sheer unmitigated hell" when winter storms tore across the Atlantic.

The worst spot on the crossing was the place the Americans called "Torpedo Junction," a five-hundred-mile gap halfway between Newfoundland and Iceland where the convoys were beyond the range of Allied land-based bombers. It was there in late summer 1942 that Dönitz turned his attention, ordering his wolf packs to take up patrol lines. The Battle of the Atlantic was rejoined.

———

Meanwhile the blackout of Shark, the German U-boat key, continued. From August through November 1942, 147 ships were sunk in attacks on convoys. On November 22, the Admiralty's Operational Intelligence Center, with clenched teeth showing through stiff upper lip, sent Bletchley Park a memorandum asking if perhaps "a little more attention" might be paid to the Shark problem. The Battle of the Atlantic, it said, is "the one campaign which Bletchley Park are not at present influencing to any marked extent— and it is the only one in which the war can be lost unless BP *do* help."

It was not as if Bletchley Park were ignoring the problem. But the crypt-analytic impasse seemed insurmountable, at least until the four-wheel bombes were ready, which was not going to be any time soon. There was only one small chink in the four-wheel Enigma's armor that might be exploitable. The special short signals used by the U-boats to transmit weather and enemy-sighting reports were enciphered with the fourth rotor set to A—the "neutral" position that made the four-wheel machine fully compatible with the existing three-wheel Enigma. So in principle, short signals could be used to generate menus that could be run on the standard bombes. Even under the best circumstances it would be a stretch: The short signals were just that—short, sometimes consisting of but a single three-letter group. It was in theory possible to build a bombe menu by chaining together several short signals from different messages using their known relative rotor settings, and one heroic attempt was made in late September 1942 to do just that. A special indicating system for specifying rotor settings to be used with short signals was broken by GC&CS during 1942 by intensive study of such signals sent on the three-wheel Dolphin key, and using that data and a lot of guesswork, Hut 8 pieced together a plausible menu from Shark "B-bars," which were a variety of short signals that carried sighting reports and other operational messages. The bombes were pressed into service running it through all 336 possible wheel orders—but with no success.

Another form of short signals was known at Bletchley Park as "WWs" or "Willy Willys." These were weather reports, and each consisted of a dozen or so letters, each letter standing for a meteorological observation such as temperature, barometric pressure, and wind direction. This seemed more promising. Even better, weather reports sent by the U-boats were collected and rebroadcast several times a day for use by other German naval units by a powerful German transmitter at Norddeich—and Norddeich used a fairly simple meteorological cipher that had been broken by Hut 10, Bletchley Park's Met Section, over a year and a half earlier, in February 1941. This was the perfect cribbing opportunity. It was often possible to match up a specific Norddeich synoptic with a specific U-boat short signal by noting their times of origin, or by exploiting the fact that the U-boat short signals only permitted round numbers to be reported; thus the matching Norddeich synoptic would always have the location of the report given as a whole degree of latitude and longitude, and the wind direction would always be a number divisible by four.

It would have been the perfect cribbing opportunity, that is, but for one insurmountable obstacle. A new weather short signal book had gone into effect on January 20, 1942, and although Hut 8 had been able to work out some of the new short-signal codes before Shark switched to four wheels ten days later, its work was far from complete. More seriously, the WWs used their own indicator system in which a single letter stood for one of twenty-six rotor settings specified in a table book; these were completely different from the B-bar indicators, and only a few had been identified.

And so things stood in the fall of 1942, when, on the morning of October 29, a U-boat was spotted in the Mediterranean halfway between Port Said and Haifa. Four British destroyers were ordered out at once to hunt it down. Shortly after noon they made sonar contact; the destroyer *Petard* dropped a salvo of five depth charges and her crew immediately saw oil and air bubbles rise to the surface. The other destroyers joined the attack with dozens more depth charges. At one point, suspecting that the U-boat was lying at the bottom below the maximum depth at which the fuses on the depth charges could be set, *Petard*'s crew tried stuffing soap into the pressure sensors of the charges to delay their detonation. All afternoon and well into the night the attack continued; German survivors later said they counted 288 explosions. At 10:17 P.M., a depth charge ripped a hole in the U-boat's bow, and her captain decided he had had enough. As *U-559* broke the surface, the British destroyers opened fire, and a four-inch shell blew off the base of the sub's conning tower. That was the end of any resistance from the German crew. Within minutes, a boarding party from *Petard* was scrambling onto the U-boat's deck—not waiting for a boat to be lowered, Lieutenant Anthony Fasson, Able Seaman Colin Grazier, and Tommy Brown, a sixteen-year-old canteen assistant who had lied about his age to join the service, jumped into the water and swam across. The sub was already taking on water as the three began pulling documents out of the captain's cabin, at one point using a machine gun to smash open a locked cabinet. The water continued to rise dangerously. Another party from the destroyers arrived in a whaler and Brown made three trips up and down the ladder of the U-boat to hand over the recovered documents; finally one of the officers told Brown not to go down again and to tell the other two to come up at once. They had just started up when the U-boat, without warning, plunged beneath the waves. Brown managed to jump free and avoid being pulled under by the powerful suction of the sinking ship. Fasson and Grazier were carried down to their deaths.

But among the booty they had secured before that tragic end were the complete indicator list and code tables for the new weather short signals. The documents reached Bletchley Park on November 24. This was probably Bletchley Park's luckiest break yet. On December 2, the head of the Met Section noted in his diary: "A new development is that the Naval Section expect to be able to read the submarine traffic again shortly through the weather messages." The months of utter frustration gave way to a surge of hope, and Huts 8 and 10 were galvanized into action. The weather short signals were still far from ideal material to build a bombe menu from, and to allow for possible variations and turnovers as many as three different all-wheel-order runs were required for each one of the menus, over a thousand bombe runs in all. The menus were run nonstop for two weeks without success, but optimism was still high. The crucial problem now was perfecting the reading of the Norddeich signals, which underwent frequent changes in encipherment tables; one edge Hut 10 had was its discovery that sets of tables were reissued periodically, usually about a month after their first appearance, and would then remain in force for about six days. It took several days to solve a new table, but a reissued table could be identified and read at once, and when that happened Hut 10 was able to produce cribs with almost no delay.

On December 13 a message went from Bletchley to the Admiralty, and it was immediately relayed to OP-20-G in Washington:

> As a result month of most strenuous endeavor a few days of traffic will be readable in immediate future and this may lead to better result in near future. You will I am sure appreciate the care necessary in making use of this info to prevent suspicion being aroused as to its source. We have found this especially difficult when action by routing authority outside Admiralty is required. It would be a tragedy if we had to start all over again on what would undoubtedly be a more difficult problem.

In the early afternoon that same day, Hut 4 teleprinted to the Admiralty the positions of a dozen Atlantic U-boats, as determined from decrypts of fifteen weather short signals transmitted during the period of December 5 to 7. Traffic for the period of noon December 25 to noon December 26 was read at Bletchley beginning at 9:00 P.M. on the twenty-sixth, the first near-current decryption of Shark in almost a year. It came just hours too late to save ON154, a fast westbound convoy that was spotted by *U-664* that same afternoon. Dönitz sent a series of almost frantic messages to his U-boat commanders, urging them to be fearless and reassuring them, or so he insisted, that British anti-surface-vessel radar was ineffective. Two days later one of the wolf packs converged upon the convoy, sinking nineteen merchantmen and two escorts.

But though Dönitz did not yet know it, the tide had turned, this time for good.

After a long and rather petty wrangle within the Admiralty, Fasson and Grazier were posthumously awarded the George Cross, an honor intended primarily for acts of gallantry by civilians; the legalistic argument was that since Fasson and Grazier had died half an hour after the fighting stopped, they were not eligible for a combat medal. Brown received the lesser George Medal—and was then promptly discharged for being underage. It was all rather a grudging recognition for acts of true bravery in the face of extreme danger, acts that yielded perhaps the single most important crypto-logic breakthrough of the entire war.

———

There then began a period of shadow warfare between the Allied and Axis intelligence services.

All of the familiar difficulties in getting military authorities to take signals intelligence seriously were multiplied many times over in Nazi Germany. Regimes bent on aggression are by their nature less concerned with what the enemy is thinking than are their potential victims. Within the balkanized and hierarchical German military command there were no fewer than seven different code breaking bureaus, a fracturing of effort and a source of jeal-ousy that repeatedly impeded progress. The Axis cryptanalysts would never succeed in breaking the Allies' most secure codes, the Typex and SIGABA machines and one-time pads. Yet at least some of the warring factions within German signals intelligence were able to build up a considerable degree of technical competence, and, even more remarkably, some were able to com-mand the attention of German commanders—at least commanders like Rommel and Dönitz who defied the usual stereotypes. When the British army overran and captured a signals unit of Rommel's Afrika Korps in July 1942, they discovered that the German commander had built up a highly ef-ficient system for intercepting and exploiting British plain language and voice transmissions—a system that Rommel was able to reconstitute in time for the final battles in Tunisia in March 1943. German POWs and captured documents would reveal periodically throughout the war that German Army field units were also breaking the American M-209 cipher machine, though this was a fairly insecure device used for enciphering tactical signals, made all the more insecure by the fact that it was so difficult to reset the ro-tors that operators tended to reuse settings day after day.

But it was the B-Dienst that was scoring some of Germany's few success-ful attacks on high-level Allied codes, and it was Dönitz who more than any other German commander appreciated the crucial importance of this work. In the attack on high-level traffic, German cryptanalysts in Berlin were em-ploying equipment that was strikingly similar to that used at Arlington Hall and OP-20-G, IBM machines and even some RAM-type equipment they developed themselves. These included special-purpose calculators that used teletype tape to count digraphs and compute numerical differences between code groups or subtract trial additives. It was not quite as advanced as the American equipment, nor nearly so rapid, but it was not that far behind, either.

Dönitz's enthusiasm for signals intelligence was exceptional, but not entirely surprising. His chief problem was finding the Allied convoys at all amid the vast expanses of ocean. His tight, centralized, and highly personal system for controlling the U-boats under his command day by day and even hour by hour depended on timely information. From the beginning to the end of the war, information on convoys dominated the weekly reports of the B-Dienst. No detail about Allied shipping was considered too trivial to include, and Dönitz amassed a substantial library of data on the habits and patterns of convoy movements that was almost a mirror image of the files of the Submarine Tracking Room. Intercepted signals were invaluable, too, for verifying the often exaggerated claims by U-boat captains of ships sunk— the British and Americans obligingly transmitted reports that identified their losses by name—and even for helping Dönitz know where his own submarines were in the intervals between their transmissions. During its most successful periods it was not unusual for B-Dienst to read British naval signals with a delay of less than twenty-four hours. The breaking of the Allied convoy code in March 1942 became the determining factor in Dönitz's decision a few months later to resume the convoy war in the North Atlantic. From June to November virtually every order he sent to a U-boat group at sea was an attempt to exploit information he had obtained from decrypted Allied signals.

The heavy use of radio by both sides in the convoy battles, and the alacrity with which commanders on both sides put to use the information that their code breakers provided them, set the stage for the climactic battle of spring 1943. In one of the oddest quirks of the code war, cryptanalytic fortunes had once again shifted on both sides simultaneously. The breaking of Shark by GC&CS coincided almost to the day with B-Dienst's loss of current reading of Cypher No. 3, the Anglo-U.S Convoy Code. On December 15 the British had resumed enciphering indicators in all of their naval codes, including Cypher No. 3, and now, suddenly, the Germans were reading fewer than 10 percent of the Allied signals even within seven days of their transmission. Dönitz ordered immediate action, and B-Dienst, at a time when it was strapped for manpower, doubled the staff assigned to Cypher No. 3.

Reading each other's mail under these circumstances was more like looking in a hall of mirrors; it led to some strange instabilities as each side reacted not only to what the other side was saying about its own intentions but to what each was saying about the other. Upon resuming its reading of Shark, GC&CS was reassured to discover that the U-boat orders contained only "tardy and conjectural" information about the convoys. That false sense of security was amplified by one cryptanalytic blind spot that persisted. As Hut 8 worked through back traffic from November and December when time permitted, it was unable to break the special "officers-only" naval Enigma settings for those months—and it was these special signals that were most likely to relay any sensitive intelligence information that Berlin had discovered. The officers-only, or *Offizier*, messages were enciphered by first running the text through an Enigma at the day's basic setting but with a special stecker and initial rotor setting. The resulting cipher text was then *reenciphered* at the general setting of the day. When it was received by a U-boat, the radio operator would decipher a message that read something like OFFIZIER MAX XFGWO PWQNC MPDJQ. . . . He would then turn the text and the machine over to an officer to handle the second decipherment at the special officers-only setting that the code name (MAX in this case; there were twenty-five others, one for each letter of the alphabet) called for.

Breaking *Offiziers* posed special cryptanalytic challenges. On the one hand the wheel order and ring setting were known once the general settings of the day had been broken out from other traffic, so it was only a matter of finding the stecker and starting position; on the other hand, cribs were extremely difficult to develop given the limited use and restricted circulation of these messages; there was little in the way of routine phrases. The first-ever break of an *Offizier* came from the tried and true "FORT" method after one message was spotted as a probable continuation of another. But serendipity and opportunism were the order of the day in cribbing *Offiziers;* one of Hut 8's triumphs was the breaking of a sixteen-letter *Offizier* sent just after the final attack on the battleship *Tirpitz* on November 12, 1944, by British Lancaster bombers. The results of the attack were not yet known at Bletchley Park, but someone optimistically thought to try the crib TIRPITZ GEKENTERT—"*Tirpitz* capsized"—and the message broke.

If a successful *Offizier* crib was available the bombe was not needed at all, as hand methods could be used. When no cribs were available, one of the American film comparator machines, Hypo, proved especially useful once it went into service in November 1943. Hypo would encipher one or more high-frequency letters at every possible starting position and, by measuring the total amount of light that passed through superimposed films, identify which starting position yielded the highest number of hits against the actual cipher text. (Hypo was also used to recover "duds" by this statistical method when no cribs were available.)

The first Shark *Offizier* was not read until January 5, 1943—it was a signal on entry instructions for the port of Bordeaux—so GC&CS had no indication that their German counterparts had been reading Cypher No. 3 for an extended period before December 15, 1942, when No. 3 was tightened up. The Admiralty for its part was extremely careful to put only "appreciations" in the U-boat alerts it sent to convoys, and the information was as usual disguised by attributing it to other sources, mainly direction-finding fixes on U-boat radio transmissions. So B-Dienst was equally oblivious to British success in reading Shark. This double blindness would play a crucial role in the events of the next few critical months.

Actions speak louder than words, though, and even as Dönitz rushed more U-boats to the Atlantic—by January 1943 there were forty on patrol along the convoy lanes—they simply were not sinking ships. Dönitz's U-boats were unable to find a single one of the eight "expected convoys" that he had alerted them to between January 3 and 18. There were several reasons for this, luck and bad weather among them, but almost certainly the chief reason was that GC&CS had at last evened the score in the shadow war. For the first half of January, Shark was broken on about half the days with less than two or three days' delay. The teleprinters in the basement of the Citadel clattered away with decrypts from Bletchley Park, some just hours old, and within minutes the information was in the hands of Rodger Winn and the Submarine Tracking Room. Again and again the convoys were alerted to alter their routes to avoid the waiting U-boats.

The weakness of Dönitz's command system was that, although central control made it possible for him to act swiftly on intelligence and concentrate his forces to move in for the kill, it also made for a lot of chitchat on the airwaves. Short signals reduced air time, but still the U-boats had to go on the air to report enemy sightings, to request permission to move, to report when they had cleared the Bay of Biscay on their outbound journey, and to report their approach on the return so that escorts and minesweepers could be provided; every time a U-boat radioed a signal it was required to include a report of its position and fuel supplies, and if it did not report for a few days it would receive a radioed order to do so.

By mid-January the waiting became intolerable, and Dönitz ordered his U-boats to abandon their dispositions at either end of the convoy lanes in the eastern and western Atlantic and to concentrate at the "pinch point" south of Greenland.

He also sounded the first serious security alarm. It was becoming increasingly difficult to explain the convoy diversions on the basis of radio direction finding or agents' reports of U-boat departures from France. On January 28, Dönitz wrote in his log that there seemed to be only two possibilities: "that the enemy has succeeded (a) in temporarily breaking into our cipher data or (b) that in some place or other in our own ranks there is treason."

He immediately ordered the *Stichwort* procedure to be activated to alter the daily Enigma keys according to the set of secret sealed instructions. An investigation was launched by the Naval Communications Service. The next few days brought more alarms. B-Dienst's reading of Cypher No. 3 was still lagging but it contained some deeply disquieting discoveries. A convoy from the United States to Gibraltar had been heading directly for a German patrol line when it was sent a signal warning of the presence of U-boats and ordered to change course. Several of the appreciations sent to other convoys contained suspiciously accurate "direction-finding" fixes on German submarines. One reported the precise number of U-boats—twenty—on patrol in one area. How could the British have known that from direction-finding? And then there was the report on January 29 that warned that two U-boats were at 31° N 39° W, probably for a rendezvous with a resupply ship. The only problem was that this hadn't happened yet. The U-boats had indeed been ordered to that spot, but were still en route at the time the British report went out. Unless the enemy was prescient, no direction-finding fix could explain that one.

The investigators from the Naval Communications Service in Berlin lined up every intercepted Allied message in chronological order alongside every U-boat message to see if it was conceivable that the British had broken their codes. Everyone with access to U-boat orders was interrogated and placed under surveillance; only Dönitz and his U-boat chief of staff, Admiral Eberhard Godt, were excused from suspicion. That prompted Dönitz to remark to Godt, in his only recorded attempt at humor, "Now it can be only me or you."

The British had certainly been trying to convince the Germans that they were being betrayed at every turn, mainly through another stunt that Commander Ian Fleming had dreamed up. The *Atlantiksender* was the British counterpart of Tokyo Rose, a radio broadcast beamed to U-boat crews with a daily program of entertainment and propaganda, the highlight of which was exceedingly detailed and accurate information on what the U-boat crews were up to. Everything from parties that got out of hand to the scores of their football matches that had ended just hours earlier were dutifully noted. "This afternoon, Third Flotilla, La Rochelle, played Twelfth Flotilla, Bordeaux," the station reported one day. "The score was 5–3." It followed with the names of the players who had scored the goals on each side. The intention was not to cover ULTRA; it was rather "black propaganda" for its own end, but it did serve to deflect some suspicion during the Naval Communications Service investigation.

Still, treason seemed almost impossible, but then so too did a compromise of the impregnable Enigma cipher. A few days after the Allies had sent their January 29 message about the U-boat rendezvous they had followed up with other appreciations that correctly reported a change in the rendezvous location. Yet if their source was the German radio signals it meant that the British code breakers had weathered the *Stichwort* change without a hitch, which seemed utterly inconceivable. Then the hall of mirrors began to play its disorienting game. If the British *were* reading German messages, concluded Vizeadmiral Erhard Maertens, the Director of Naval Communications, surely they could not escape knowing that their own signals were being read. Yet they had taken no steps themselves to tighten up their own codes. Moreover, the British were not always right: A few of their U-boat appreciations were indeed uncannily accurate, but for the most part they were "monotonous" and vague, stumbling and erratic. "For such a practice there is no other explanation than this: the enemy cannot do it any better." He titled this chapter of his report, "The Enemy is Reading our Ciphers!?!"

That seemed a conclusive argument against a breach of the Enigma. Of course Maertens's mistake was to view cryptanalysis as an all-or-nothing proposition; his bigger mistake was not to realize that the British deliberately watered down their signals intelligence before distributing it to the field precisely in order to conceal their cryptanalytic success.

Throughout late January and February 1943 the tide of the battle swept back and forth day by day, sometimes hour by hour, according to the shifting successes of GC&CS and B-Dienst. In February the Germans were still only able to read about one to three current convoy messages a day, and very few were of much operational significance. But the accumulating back traffic was creating for Dönitz a huge store of intelligence on the patterns and behavior of convoys, and in particular it had caught a significant change from an eight-day cycle to a ten-day cycle in sailing schedules from New York. Current traffic read by Bletchley Park through noon of February 17 permitted the successful diversion of convoy HX226 from New York to England around the "Ritter" and "Neptun" patrol lines. But breaking the key for the eighteenth took Hut 8 nineteen hours, and it was not until early morning of the twentieth that GC&CS learned that Dönitz had ordered his U-boats to extend their patrol lines to the southeast. A second convoy, ON166, had also been diverted based on the original reports of Ritter's and Neptun's position; a very slight further diversion to the south would have saved the ships altogether. But the new information came just too late; ON166 was caught and fourteen of its sixty-three ships sunk.

The impish guiding spirit of cryptanalytic fortune struck again in mid-March. From the second week of the month B-Dienst sprang to life again and began reading the convoy signals rapidly and in large quantity. Simultaneously the most nerve-racking blackout of Shark yet struck Hut 8. With only a few days' warning, the weather short signal book was changed on March 10 to a new edition—which Hut 8 did not possess. On March 8, Admiral Godfrey warned the Admiralty that the short-lived success against Shark was about to come to an end; the impending loss of their only sure source of cribs would lead to a blackout of Shark for "some considerable period, perhaps extending to months."

There followed ten tense days while Hut 8 tried to do what had hitherto seemed impossible, namely break Shark by using B-bar short signals. The problem was that there was no neat source of cribs of the sort the Norddeich weather broadcasts had provided for the WWs; B-bars could contain any of hundreds of different sorts of reports. All available information on the U-boats was brought to bear. Direction-finding data might provide cribs for position reports; known positions of Allied convoys might suggest which U-boats were transmitting sighting reports; electronic methods for fingerprinting the individual background noise produced by different U-boats' radio transmitters could establish the signature that each message carried, and a related technique called TINA could identify Morse operators by their individual "fist"—the way they slightly elongated or shortened their dots and dashes and paused between letters and words. The cribs were so short that even when they could be ginned up, multiple messages had to be linked to produce an adequate bombe menu, and that was possible only if the messages had nearby rotor starting positions. Yet each crib had a fairly high chance of being wrong. Ten days passed without a single Shark signal read. Then on March 19 came a triumphal message from "C" to the Prime Minister: "We have been successful reading the U-boat telegrams for the 16, 17, and 18 instant." Churchill replied at once, "Congratulate yr splendid hens!"—the allusion was to Churchill's little joke of referring to Bletchley Park as his hens that laid eggs but never cackled.

Again hours determined a convoy's fate, but this was to be the B-Dienst's high-water mark. As Hut 8 struggled to break back into Shark, the German code breakers read a series of signals giving the exact location, speed, and course of two eastbound convoys, HX229 and SC122, which had left New York the first week of March. Diversion orders to the convoys were read by the Germans and in Dönitz's hands within hours of transmittal, and when forty U-boats converged on the ships on March 16 it was a slaughter. One escort and twenty-two merchantmen went down, 146,000 tons of shipping in a single action that saw one U-boat lost.

Dönitz was jubilant, but it was his last hurrah. Already, in February, there were signs of cracking morale among the U-boat service. Shark decrypts revealed a number of cases in which U-boat captains began to show timidity in the face of air attacks; the number of boats reporting "mechanical difficulties" while transiting the Bay of Biscay on their outward journeys increased dramatically. In late March the U-boat command sent a reproving message, chiding U-boat commanders for failing to exhibit "healthy warrior and hunter instincts." A few weeks later Dönitz was reduced to sending petulant threats: "Whoever is of the opinion that offensive action against convoys is no longer possible is a weakling and not a true U-boat commander."

By April the cryptanalysts had fought each other to a standstill; the Atlantic was so saturated with U-boats that evasive routing was no longer an option even with excellent and current intelligence. But other weapons were now turning the tide. The Allies had begun to equip patrol planes with new ten-centimeter-wavelength radar, against which the German radar warning receivers were ineffective; the withering Hedgehog, which could fire a salvo of small depth charges over the bow of a ship, made attacks far more deadly; escorts and escort carriers were being transferred from North Africa after Rommel was defeated in Tunisia; and most of all the Allied naval commands had finally broken the strategic air forces' jealous hold over the bomber force, and two hundred very-long-range aircraft were ordered transferred to the protection of the convoys. The air gap was closed. On May 24 Dönitz ordered his U-boats withdrawn from the North Atlantic. They would return again, but never to the success they had achieved.

The final cryptanalytic honors went to the Allies, too. In mid-May three *Offizier* messages were broken using the crib EIN ERWARTETER GELEITZUG—"an expected convoy." All revealed that the Germans had precise knowledge—latitude and longitude to a degree, speed to a tenth of a knot—of Allied convoy movements. The content and time of transmittal of the German messages coincided almost exactly with Allied signals sent in Cypher No. 3. OP-20-G had had to go to Admiral King to gain permission to actually look at the Allied signals—it proved to be a huge bureaucratic and security tangle—but once they did, the case was absolutely clear. GC&CS had meanwhile found evidence of its own. On June 10 a new convoy code, Cypher No. 5, went into service. The door was shut at last. Cypher No. 5 remained unbroken for the rest of the war.

Allied mastery led the Americans to take some chances that at first alarmed the British. The U.S. Navy lobbied vigorously in the spring of 1943 to attack U-boat refueling rendezvous, arguing—correctly—that knocking out the "milk cow" tanker submarines would play havoc with Dönitz's operations. The British vetoed the plan as too risky, pointing out—also correctly—that since rendezvous were carried out in radio silence according to instructions radioed as much as two weeks in advance, a successful attack would have to give away the fact that the German codes were being read. But in June, July, and August the U.S. Navy carried out a series of crushing attacks on U-boat refueling points, destroying three tankers and eleven operational U-boats. By July of the following summer, sixteen of the seventeen German tanker U-boats that had been built were sunk. Half of those were destroyed by U.S. forces acting on ULTRA intelligence. "The British were more clever in [ULTRA's] use," concluded Commander Kenneth Knowles, who directed the Atlantic tracking room at Navy headquarters in Washington, "we, more daring." Though he added: "But then they had so much more to lose."

Dönitz cast about feverishly to explain his mounting losses, at first blaming radar, then stray emissions from the radar warning receivers carried on the U-boats—which he ordered switched off. (To make sure they carried out this order, U-boat captains were instructed to remove an essential component from the receiver, place it under lock and key, and enter the action in their ship's log.) Enigma decrypts revealed at one point that Dönitz's suspicions had settled on infrared detection; the British immediately fanned those fears with double agent reports. The infrared-reducing paint subsequently applied to the U-boats actually increased their radar signature. So certain were the Germans of the inviolability of their codes that four decades later some veterans of B-Dienst were still insisting that all these stories of the British having broken the Enigma were so much nonsense: The British were incapable of *die geistige Arbeit*—the mental work—required for cryptanalysis.

———

By the time Dönitz pulled his U-boats from the North Atlantic, cooperation between OP-20-G and Bletchley had at last begun to work smoothly. On May 3, 1943, the first American four-wheel bombes were completed and underwent preliminary testing in Dayton at the National Cash Register factory. Although Hut 8 was clinging by its fingernails to the B-bar cribs and its three-wheel bombes, there were frequently long delays in breaking Shark with these methods, delays that would continue throughout the summer. The British were not getting far with their four-wheel machines. By summer 1943 only a single Cobra was in operation. The first High Speed Mammoth had been delivered in May (nickname: "Darwin"), and additional ones were being built at a rate of about two a month, but they were not working well at all. On June 3 Nigel de Grey sent Commander Wenger at OP-20-G an urgent inquiry: "Chances of breaking Shark on short signals are much less favorable than before . . . in view of this four wheel bombes become more than ever important. . . . How does your bombe production program stand?"

The answer, luckily, was quite well. Three weeks later, on June 22, the prototype machines at Dayton ("Adam" and "Eve") broke the *Offizier* settings for June 9 and 10, which were promptly teleprinted across the Atlantic; this was their first operational success. The American bombes had some teething problems, too, but by the time the first units arrived at the Naval Communication Annex from Dayton on August 31, the kinks had been worked out and new models were flying off the assembly line at a rate of six or more per week. By October 15, thirty-nine bombes were running; by December 15 the total had reached seventy-five.

OP-20-G, determined to run its own crib program, at first wasted thousands of hours of bombe time on cribs that the more experienced hands in Hut 8 knew were futile; the only really reliable cribs were ones obtained from "cross ruffing" general messages sent in the home-waters naval key, Dolphin. But it really didn't matter. America had done what America did best, mobilized industrial power on a mind-boggling scale. With a score of American bombes arriving each month, the need to rely on the jigsaw-puzzle-like process of piecing together B-bar cribs began to fade into a bad memory. Banburismus was abandoned at long last, too; with the American machines there was now so much bombe time to spare that husbanding mechanical resources at the expense of human labor was no longer necessary, or efficient. By the end of the summer of 1943 Shark was being broken continuously and currently, and except for occasional periods when a shortage of traffic caused problems, it would continue to be broken currently for the rest of the war.

In fact the problem henceforth would sometimes be to find enough work to keep all of the American machines busy. Almost as soon as the bombes arrived they were being put to use on "Hut 6 jobs" (known as "Bovrils")—that is, three-wheel Luftwaffe and Army keys, which could be run in about fifty seconds per wheel order on the high-speed four-wheel machines. Even after mid-1944, when OP-20-G completely took over the breaking of Shark using intercepts teleprinted from Bletchley, there was machine time in abundance to spare for Bovrils.

It was the way of Washington that the Navy conducted its own foreign relations and the Army did, too. The growing closeness of the U.S. Navy and the British in tackling naval Enigma from the fall of 1942 on had been threatening to leave the Army out in the cold. But the lesson of OP-20-G's success in forcing GC&CS to give it a piece of the Enigma action had not been lost on Arlington Hall. If building your own bombe was the way to make the British cry uncle, the Army could play that game, too.

The continuing obstacles to cooperation between the British and American cryptanalysts arose mainly from policy differences, but there were cultural differences, too, far more acute in 1942 than they would be after the war, when the communications and transportation revolutions (revolutions that the war itself did much to foster) would break down transatlantic differences in attitude, manner, and customs. In the 1940s the knowledge that even many well-educated Britons had of Americans was limited to the actors they had seen on movie screens. British suspicion over an American lack of security-mindedness certainly had its basis in reality, but it was multiplied by caricature many times over; likewise American resentment of British standoffishness and superiority was inflamed more by literary stereotypes than by any actual experience, for the simple reason that most Americans didn't have much experience with Englishmen. Captain Geoffrey Stevens, the British liaison officer at Arlington Hall, arrived in July 1942 and immediately ruffled feathers with what the Americans took to be his pompous manner, accentuated by his habitual costume of Sam Browne belt, polished black riding boots, and swagger stick. Much of this was just American touchiness and oversensitivity, and Stevens no doubt kept his true feelings to himself—though in point of fact his true feelings were everything the Americans feared. In his first few months Stevens sent a series of exasperated letters back to GC&CS complaining about everything from a picnic in Rock Creek Park ("did you ever get involved in one of these fantastic affairs?") to the "regimental gossip" of American women at a dinner party to "our fat friend Kully" and "his habit, which I am sure you must have observed in some degree, of shouting people down to contradict something they have no intention of saying." Worst of all was the "hopelessly over-organized" American operation and the immaturity of the Americans who ran it:

> Sometimes I think they are just a lot of kids playing at "Office." You must have noticed yourself how very many childish qualities the American male has: his taste in women, motor-cars, and drink, his demonstrative patriotism, his bullying assertion of his Rights, his complete pig-selfishness in public manners and his incredible friendliness and generosity when he likes you—Hell! Anyone would think I didn't like them. But perhaps it is as well I'm fond of children.

Jealous of the Navy's progress on the Enigma and concerned that the British monopoly on Army Enigma traffic would become a source of professional embarrassment to Arlington Hall as American commanders prepared to face the German Army in the field for the first time in North Africa, William Friedman strove to present GC&CS with a fait accompli of his own to match the Navy's. Within two weeks of the Navy's decision in September 1942 to build a bombe, the Army decided it would, too. OP-20-G said the Army could join its contract with NCR, but Friedman had what he thought was a better idea. In place of the rotating drums of the bombe, the Army wanted to use telephone switching relays. On December 15, 1942, the Army signed a contract with AT&T for $530,000 with an "AAA" priority rating to build just such a device.

But it was clear that Arlington Hall had little notion of what was actually involved in creating a signals intelligence operation of the scale needed to handle Enigma traffic. The single Army bombe, known as 003 or "Madame X," was the equivalent of 144 Enigmas, about four times the size of one standard British bombe. It was a highly innovative design and incorporated some features that greatly speeded the operations, including a system for automatically changing the wheel order. In the British bombes this required physically removing the rotors and replacing them, a time-consuming procedure. But its huge cost—which would double by the time the project was done—made it totally impracticable for the sort of mass production of traffic required for a serious attack on Enigma intercepts. Moreover, without its own intercept capability in the proximity of the Continent, the U.S. Army would be wholly dependent on traffic relayed from Britain in any case. It was inconceivable that the Army could truly create an independent capability of reading Enigma traffic, at least not soon.

Nonetheless Madame X proved an effective lever. The Army had already been making noises about British nonreciprocity and had shown its displeasure by initially blocking a security clearance for Alan Turing, who, during his visit to the United States that winter, was scheduled to visit AT&T's Bell Laboratories to consult on a speech-scrambling device. But Turing's visit provided the right dramatic opportunity to spring the surprise. On January 4, 1943, Arlington Hall received permission to reveal to Turing and Tiltman, who was in America at the time as well, "the fundamental principles and details" of the bombe. Dale Marston, who was then directing the development of the rapid analytical machinery for Arlington Hall, later recalled that when Tiltman was briefed on the work, he immediately said that the Army and GC&CS "had better get together." Turing was shown the actual prototype at Bell Laboratories in New York City on February 5.

Months of skirmishing along the well-established pattern ensued, as GC&CS fought the rearguard action it knew so well by now. Of course the Americans could take part in *research* on the Enigma, if that was what they wanted to do; it was just that *exploitation* of the traffic must remain under British control, for "security" and to "prevent duplication." The British imperiously delivered a draft treaty setting out those terms. Within B.P. the attitude was even more condescending. Hut 6 wrote a memorandum dismissing the American attempt to "participate in an already proven success, so that they may not appear to lag behind the British either in acumen or knowledge," and concluding that "the Americans have no contribution to make." A British liaison officer in Washington began disparaging even the American work on Purple and hinted that Britain was prepared to sever all signals intelligence cooperation if the Army refused to accept the British terms.

But cooler and wiser heads quickly intervened. Telford Taylor, who was at Bletchley Park as a liaison officer during the spring of 1943, cabled back urging sang-froid: The British threats to sever all contacts were not really worth taking seriously, since neither the British nor the American Chiefs of Staff would permit such a breach. Taylor advised rejecting the British proposal, but cautioned against making unreasonable demands:

> . . . we should not phrase [our proposal] so broadly that it seems to envision a *duplicate* operation at Arlington Hall, or to impose *undue* burdens on the British in supplying us with traffic and other aids. What we really want *at this time* is to gain a foothold in "Enigma" and develop technical competence, and gradually develop a supplementary operation so as to improve joint coverage. What we *ultimately* want is independence, but if we get the foothold and develop our technique, independence will come anyhow. As our position in Europe gets better established, we will be less dependent on the British for intercept assistance; as our skill in dealing with traffic grows, we will need less help in securing "cribs."

Colonel McCormack, who was also at Bletchley to help assess the situation for Arlington Hall, cabled back on May 13 urging compromise, too; he noted that it was indeed ridiculous to think of routing to Washington all raw traffic and attempting to duplicate all of the ancillary reference and index material, built up at GC&CS over three years, that was needed to tackle the German traffic. "If [Colonel Preston] Corderman [the head of Arlington Hall] wants his people to learn what makes this operation tick," McCormack wrote, "he had better send them here to learn it, because they never on God's green earth will learn it from anything that Arlington will be able to do in any foreseeable future." On the British side, Gordon Welchman urged moderation, too; while "the idea of a separate 'E' organization being built up from scratch elsewhere seems to us to be absurd . . . on the other hand it does seem logical that the Americans should take some hand in work on 'E' and we certainly need help."

On May 17 the "BRUSA" agreement was reached between GC&CS and the War Department, and it took precisely the line the moderates had been urging. The Americans pledged to follow the strict letter of the law in disseminating and putting ULTRA to use in the field. An American party would be sent to Bletchley Park, equipped with British bombes, and permitted to participate fully in solving Enigma traffic. The Americans could set up their own intercept station in Britain as well. A U.S. liaison officer would be able to examine all messages and summaries and pick out selected decrypts for transmission directly to Washington or American theater commanders.

———

In the event, the American invasion was a huge success. An advance contingent of about twenty American officers and men from Arlington Hall crossed the Atlantic in a troopship in August; the not very convincing cover story they were given to tell their fellow passengers was that they were carrier-pigeon handlers from the Signal Corps. Arriving at Bletchley, the Americans were pleasantly surprised to find that everything they had heard about the British was untrue, or almost everything.

Following what seemed to be by now an unshakable tradition for American cryptanalytical detachments, the U.S. contingent took over a girl's school near Bletchley as their headquarters; with only a minor amount of grumbling from Hut 3 and Hut 6 that there was no time to train a bunch of amateurs, the visitors were immediately put to work as translators and emenders, cryptanalysts, and bombe operators. Major William P. Bundy, who was appointed operations officer of the newly formed 6813th Signal Security Detachment, was told to get the best people he could find, and that helped ease their introduction into the somewhat rarefied atmospheres of the two huts. So did the fact that the commander of the American contingent was Major Roy Johnson, who headed Arlington Hall's Enigma section and who had already done a tour as a liaison officer at Bletchley Park. And so did the more material fact that the Americans were, for administrative purposes, a special detachment under the European Theater of Operations–United States, now headquartered in England and building up for the D-Day invasion. According to Army routine that meant they were entitled to ample stocks of Army provisions, which they freely shared with their new British colleagues. The 6813th had its own cooks, including one who had run a famous bakery in Philadelphia, and they turned out far above average meals—especially compared to the norm for wartime Britain, where Spam or reconstituted eggs or meat gravy was about as good as it got. (The 6813th also had as its executive officer the former deputy warden of the Georgia

State Penitentiary who, Bundy said, never encountered "a disciplinary problem worthy of his steel.")

Everyone was on his best behavior; Bundy went to lengths to prove that the Americans took security seriously, and the British went out of their way to be warm and friendly and steer clear of the cold and condescending stereotype. "They received us in the most extraordinarily open way," recalled Bundy. Even when they were tempted to be condescending, they wound up instead being charmed by the Americans' infectious exuberance that stifled any incipient raised eyebrows. Derek Taunt, a young mathematician from Cambridge who worked in Hut 6, recalled one American, Bill Bijur, who celebrated his promotion to first lieutenant by handing out cigars to everyone in the hut at midnight, "a vivid memory perhaps because it typified his generous, extrovert nature tinged with a touching naivety."

The 6813th had about seventy officers and men, and it was soon joined by the 6811th, which manned the American intercept station at Bexley, Kent, and the 6812th, which set up at Eastcote, near Harrow just northwest of London, to run its own bank of ten bombes. Eventually the Americans in Britain under Project BEECHNUT, as it was called, numbered about 290.

Being under ETOUSA led to some less welcome bureaucratic wrinkles; the 6813th headquarters at Little Brickhill happened to be right on the Old North Road, and there was a constant stream of American brass running up and down the road who would see the unit's discreet, though not discreet enough, "Signal Corps, 6813th Detachment" sign and come in for a look. "So we led a kind of double life," Bundy said. "At work we were mostly on first-name terms, informal, mixed-in higgledy-piggledy, behaving in effect like civilians in a very unheroic organization. Then back at the post we were saluting and behaving in the most correct GI fashion for understood defensive purposes"—to avoid being reported by some visiting general.

Another problem was that while B.P. had a strict rule requiring everyone to take a ten-day leave every three months in order to get away from the intense mental strain of the work, the Americans had to answer to ETOUSA's "Zone of the Interior." Its commander, General J. C. H. Lee—his initials were said to stand for "Jesus Christ Himself"—had laid down an equally strict rule that no one under him was going to goof off, and nothing more than a three-day pass was authorized. Bundy's creative solution was to send his men off on a three-day pass to some distant point in the British Isles like Edinburgh with instructions to send a telegram on the third day, "Have bad cold. Ask extension." Bundy would then cable back: "Extension granted."

The pressures of the work were subtly different in fall 1943 from what they had been earlier. Gone was the uncertainty and agony of days passing without a break in a crucial key, but gone too was the exhilaration and thrill of finding those breaks. There was no mistaking the gloomy humor with which Hut 8 announced in its Weekly Report in late July, "Dolphin is coming out with what some of us feel is lamentable regularity and we have had no Banburismus to do lately." As the postwar history of Hut 8 observed, "The problems which arose between September 1943 and the end of the war were . . . essentially of the type which could be tackled with known methods. The cryptographer thrives on stagnation and the activity of the last eighteen months of the war made our lives much more difficult." The work was still excruciatingly exacting but the pressure was now just not making mistakes, and not getting bored. Dilly Knox, working from his bed through the fall of 1942, succumbed at last to the cancer that had been wracking his body but not apparently his mind, and died on February 27, 1943; his death truly did seem to mark the end of an era for GC&CS, and indeed for cryptanalysis. The age of the lone eccentric scholar was over. By 1943 the transformation of Bletchley Park from an extension of an Oxford senior common room to a factory was complete. Dreary two-story steel and brick office blocks replaced the dreary wooden huts, the work force of four thousand men and women—the total would reach over ten thousand by 1945—came and went in shifts around the clock, transported to their billets and back in fleets of buses and vans; the watch in Hut 3, now ensconced in D Block (though forever known as Hut 3), sat around its horseshoe table sifting through the decrypts that arrived in a steady flow, the quiet interrupted only by the occasional debate over the exact meaning of a German word, or the sound of the glass window opening and shutting as the finished signals were passed through a hatch to the duty officer in the next room for final approval, or the footsteps of someone walking over to consult the index where hour by hour new cards were added to the hundreds of thousands that listed every personal name, every ship, every unit, every technical term, every weapon mentioned in every message. The bombe menus went out over teleprinter lines to the several outstations, where the bombes had been dispersed to minimize the risk of losing all the priceless machines if German bombers scored a lucky hit; the bombe crews efficiently plugged up the machines, ran the menus, and teleprinted the results back. It was about as exciting as working in an explosives factory: dull, monotonous, routine, but with no margin for error.

By this point in the war the toilers of Bletchley Park had worked out all sorts of ways of relieving the monotony and pressures and tensions, and a considerable cultural life had spontaneously arisen in their haphazardly formed community. There were four string quartets, a choir, and an orchestra, a music society that put on Purcell's opera *Dido and Aeneas,* a serious drama group that staged Shakespeare. And there was the annual musical revue at Christmastime that Herbert Marchant, the deputy head of Hut 3 and a former German and French master at Harrow, directed. "We had tremendous talent there," Marchant recalled: writers from *Punch,* professional actors, novelists, musicians. The revues were full of inside jokes (one was titled "Confidential Waste") but even so, William Bundy said, "could have played the West End or Off Broadway rather profitably."

A "B.P. Recreation Club" seems to have been less successful offering fencing, bridge, and "gramophone parties," but there were frequent dances, sometimes several a week, that became a fixture of the Bletchley Park social life. Bletchley Park had "emptied the Scottish universities of all their lady graduates and they emptied the squirearchy of England of their pink cheeked daughters," Bundy recalled, and there were more than a few well-to-do young women who liked to put on nice parties. The women's naval service, the WRNS, took over a huge Tudor mansion, Crawley Grange, along with several other stately homes in the vicinity as billets for the women assigned to Bletchley Park; the houses had beautiful ballrooms with paneled walls and gleaming wood floors, where dances were regularly held.

For the Americans who came over, life at Bletchley in the middle of nowhere was in many ways more interesting and diverting than life back in Washington would have been. Army recruiters shamelessly enticed prospective workers with photographs of prewar Arlington Hall, with its tennis courts and riding stables, and played up the glamour of working in the nation's capital. In fact, wartime Washington was a dull and lonely place for most of the people assigned to Arlington Hall and the Naval Communications Annex. The Navy's facility was in a residential area that lacked movie theaters, shops, restaurants, or just about anything else for that matter. Washington's F Street, the downtown commercial center, was a half-hour's ride on the streetcar each way. Naval officers at OP-20-G, most of whom rented rooms in private houses, would ride downtown at the end of their shift and try to find something to eat in one of the restaurants, always packed with other government workers. There wasn't much entertainment in the town, either. The museums that today crowd the Mall were not yet in existence. Washington's post-Prohibition liquor laws, drafted by FDR personally and designed to prevent a return of the old saloon culture, permitted only beer to be served at bars; if you wanted hard liquor you had to sit at a table and be served by a waiter, and no alcohol at all could be sold after midnight on Saturday. George McGinnis, a radio engineer at the Naval Communications Annex, remembers riding streetcars for two hours on his occasional free Sunday afternoon to play golf. At OP-20-G, unlike at Bletchley Park, no one got leave, period, except for a family emergency. At Arlington Hall one group of young men and women formed what they called "The

World's Greatest Cryptanalytic Bowling Team." That was about as exciting as things got.

———

But Washington was still "small town enough and Southern enough," as David Brinkley wrote in his history of the American capital in wartime, that it felt it needed to do something for all the young women who were flooding into the town from all over the country—some as young as sixteen, many away from home for the first time, homesick, and lonely. Every day, it seemed, some group in town "set up yet another committee to arrange trips, parties, picnics, boat rides on the Potomac, bicycle trips in Rock Creek Park, and hot dog roasts on somebody's lawn," Brinkley wrote.

Along with the hundreds of thousands of civilian women office workers was something new: hundreds of thousands of women in uniform. Both Britain and America had had small women's services in the First World War to free men from desk jobs and support functions. But the policies had been inconsistent among the services: Were they soldiers and sailors, or not? American nurses, physiotherapists, telephone operators, and dieticians had served in Europe in the First World War but without the benefits or protections soldiers received; they were not eligible for sick pay and had had to scrounge for living quarters.

In Britain, the need to recruit women for war work faced little controversy. Just before the outbreak of the Second World War, the British Army established the Auxiliary Territorial Service to accept women volunteers; the Royal Navy revived its Women's Royal Navy Service from the previous war; the RAF had its Women's Auxiliary Air Force. In 1941 the British women's services were declared to be a full part of the armed forces, though that did not alter the British practice of paying women two-thirds of what men of the same rank received.

In America, things were more complicated. Congresswoman Edith Nourse Rogers called on General Marshall early in the spring of 1941 to lobby for creating a women's service with full military status (and equal pay), and said she intended to submit a bill that would do just that. Marshall asked for a week to think about it. The week stretched on and on. When Rogers's bill was finally taken up by Congress after Pearl Harbor, the scene on the floor was pandemonium. "Think of the humiliation! What has become of the manhood of America?" thundered one congressman. "Are you planning to start a matrimonial agency?" asked a senator.

The bill finally passed in May 1942 only after Marshall lent his support. The rest of the Army still wasn't so sure about the idea. "What the hell are we supposed to do with them?" said Major Cannon R. Page of the First

Army Corps. "Young girls away from home for the first time and thrown in here with all these horny enlisted men? Does the Army want to send a girl home to her mother with the clap? Or pregnant? It's going to be a god-damned mess!"

In fact, it wasn't. The Army received 35,000 applications for the Women's Auxiliary Army Corps (it later became simply the Women's Army Corps) in the first week. When the Navy established its women's service, the WAVES, a few months later it received 25,000 applications for 900 officer commissions. Throughout the war the services would be beset with wild rumors: 250,000 pregnant WACs were being sent home from North Africa, Mrs. Roosevelt had personally issued "a supersecret order" to issue WACs condoms. The services traced some of the rumors of immoral conduct to the enterprising prostitutes in Harrisburg, Newport News, and Baltimore known as "Victory Girls" who had taken to wearing WAC-like outfits. In fact, pregnancy rates were lower among women's services than they were among unmarried women of the same age in the civilian population and there were surprisingly few problems.

Bletchley Park, Arlington Hall, and OP-20-G quickly became major employers of female labor, both civilian and military. By February 1944 the staff of OP-20-G totaled 3,722; of those, 2,813 were WAVES. Twenty-seven WAVES barracks were built across Nebraska Avenue from the Naval Communications Annex on a site that now, in an interesting quirk of history, includes part of the grounds of the Japanese Embassy. On V-J Day, the 10,371 people on the staff of the U.S. Army's SIS—now called the Signal Security Agency—were about evenly divided between military and civilian personnel; WACs made up about 20 percent of the military side (most were assigned to Arlington Hall and to the intercept stations at Vint Hill in Virginia and Two Rock Ranch in Petaluma, California) while more than 90 percent of the 5,661 civilian employees were women.

At Bletchley Park RAF mechanics had first been used to run the bombes, but "as an experiment" eight Wrens were brought in to help on March 24, 1941: "it was doubted if girls could do the work." Within a year the Wrens had taken near complete charge of the bombe operations. By the end of the war the bombe section consisted of 256 RAF mechanics who maintained the machines and 1,676 Wrens who ran them.

Bombe outstations Wavendon and Gayhurst were established in outbuildings on the grounds of the stately houses of those names where the Wrens were billeted in the countryside around Bletchley. Later, when the bombe outstations at Stanmore and Eastcote near London were established, the Wrens were put into barracks—complete with double decker bunks and strict military discipline, which was a bit of a shock after the rather informal country house life and dances they had enjoyed. Of course for many of the Wrens, the arrival at Bletchley had been a bit of a shock, too, in the first place. Diana Payne was "still dreaming of life at sea, with the romantic idea of marrying a sailor," when she was put on a train to Bletchley, then driven sixty minutes to be deposited into the middle of the country as far from the sea as one could be.

The strain of the work took its toll, and so did the fact that for thirty years afterward the Wrens were only permitted, when asked what they had done during the war, to say they were "writers." But it still was an extraordinary experience for many, to be out on their own, doing a vital job, and living a life that, if not romantic, was often far better than what they had left behind. Joan Clarke lied about her age (she was sixteen, a year short of the minimum) to get into the WAAF as a wireless operator: "It was awful to start with," she said, but home was worse, and soon she got used to things, learned Morse code quickly, and one day it was announced that volunteers were needed for the Y Service. What was the Y Service? "We can't tell you," was the answer, "except there's no transmitting." Clarke volunteered. She and the others selected for the special job were trained in German wireless procedure, given a thorough security investigation, even interviewed by a psychiatrist, and then posted to the RAF intercept station at Chicksands Priory. There was skepticism about women there, too, and the very first day Clarke and one other girl were thrown into the hardest job, clearly an attempt by the men to prove that women couldn't cut it. After that first day the men shut up.

But for the most part the women moved into the work with little resistance. Cryptanalysis was not traditional male work; it wasn't traditional *anything* work. Though as the official history of the WACs in World War II concluded, quoting a WAC staff director, "It was proven over and over again that women were far better equipped than men for routine but detailed work." The statement was referring directly to the performance of WACs assigned to the Second Signal Service Battalion at Arlington Hall, and "routine but detailed" was as good a description as any of what Allied cryptanalysis had mostly become by the end of 1943.

CHAPTER 10

Command of the Ether

AMONG the hundreds of newly hired clerks, typists, mathematicians, electricians, linguists, mechanics, and IBM engineers who arrived at Arlington Hall in the summer of 1943 were two men from opposite ends of the world, opposites indeed in just about every way. William Weisband was thirty-five years old, born of Russian parents, raised in Egypt. He had emigrated to the United States in the 1920s and become an American citizen in 1938. Weisband spoke fluent Russian; he was gregarious and popular at Arlington Hall, where he took a four-week Italian-language course in June 1943 before leaving for a tour of duty in Europe.

Cecil Phillips was a soft-spoken and seemingly unambitious eighteen-year-old from the mountains of North Carolina, a college dropout who had been rejected for the draft for flat feet. Phillips's mother told him to go back to college or go get a job, so he happened to turn up at the U.S. Employment Office in his home town of Ashville on the same day that an Army lieutenant from Washington showed up with a quota of clerk positions to fill. "How would you like to go to Washington and be a cryptographer?" the lieutenant asked. "That sounds interesting," Phillips replied. The lieutenant, obviously surprised that someone actually knew what he was talking about, said, "You mean you know what that means?"

Phillips did, having once possessed a Little Orphan Annie decoder ring, though that was about as far as his knowledge went. But he was promptly hired as a GS-2 clerk at $1,440 a year and told to report to Arlington Hall a week later, on June 22, 1943. Phillips's first job was stamping the date on incoming intercepts. Having proved his competence at that task he graduated to stapling. But his first boss at Arlington Hall, Lieutenant Bill Fleischman, who had just been through a cryptanalysis course at City College of New York himself, "wanted to teach in the worst way," Phillips recalled, and set aside a couple of hours each day to initiate Phillips and his half a dozen other new recruits in the mysteries of cryptanalysis. It was soon evident that Phillips's talents went far beyond stamping and stapling. On May 1, 1944, he was led to the back end of one of the wings of A Building, where an area was partitioned off with plywood dividers from the rest of the open wing. People walking down the corridor of the building might not even notice this area, and if they did they could not have gotten in there. People at Arlington Hall knew not to ask one another what they were working on, but if anyone had asked what went on behind the plywood walls it is certain they would have received no answer. At night the desks were draped with cloths. The usual tools of the trade, dictionaries and maps and such, were nowhere to be seen. This, Phillips was told, was "the Russian problem," and it was where he would now be working.

The Russian problem did not officially exist. On Arlington Hall's organizational charts it appeared only as "Special Problems" or "B-II-b-9," designating it one of the many subdivisions of the cryptanalytic branch. It was a secret even from the British. Two years earlier Geoffrey Stevens, the British liaison from GC&CS, had discovered that the Navy was gathering Russian diplomatic cables and speculated that "sooner or later they will inevitably try to break this since they do not trust the Russians further than they could throw a steam-roller." When SIS secretly began working on Russian diplomatic traffic in February 1943 under the direction of a slightly mysterious Ukrainian named Leonard Zubko, Stevens—who was extremely diligent in making the rounds of Arlington Hall—soon learned about it. A few weeks later the project was rather publicly shut down. Two months later it was quietly revived, this time under the charge of Ferdinand Coudert, a gifted linguist who spoke fluent Russian. The whole ruse seems to have been designed to mislead the British. A few months later Coudert drafted a request to the office at Arlington Hall that handled exchange of information with GC&CS: "Have the British any information they could make available to us concerning Russian Diplomatic or Commercial Codes?" But someone apparently intervened, and the request was never forwarded to London. Diplomatic traffic in any event remained the one extremely touchy area between the Americans and the British. In reality, GC&CS had begun its own work on Russian nonmilitary traffic at almost exactly the same time, around June 1943. Their project, code named ISCOT, focused primarily on signals passed over clandestine radio networks that the British had discovered operating between Moscow and Communist partisans in German-occupied Europe. Only with the end of the war did the British and Americans let each other know what they were up to, at which point full cooperation began against their former ally, who was by then rapidly becoming their new enemy.

The BRUSA agreement of May 1943 discreetly sidestepped the issue of diplomatic traffic; though it called for full cooperation and a complete exchange of cryptanalytic results and intercepts, it only specifically covered codes used by the military and air forces of the Axis powers and by the Abwehr. Informally, a good working relationship was forged between Arlington Hall and Denniston's diplomatic section at Berkeley Street. But the British drew a line at providing Washington with copies of neutral countries' diplomatic messages sent via cables that the British controlled, and were leery about American requests for code materials from countries such as Iraq that remained within the British sphere of influence. An American liaison officer from Arlington Hall reported in November 1944 that Denniston "frequently gets the impression that we are utilizing the war to exploit British cryptographic knowledge" in fields unrelated to actually winning the war. The United States for its part ceased sending Berkeley Street information about Latin American countries' codes in September 1944.

The extension of Lend–Lease to the Soviets had brought about a huge increase in the Soviet presence in the United States. There were trade missions and purchasing commissions and inspectors stationed at the factories that were filling Soviet orders. These Russian representatives all regularly communicated with Moscow via commercial cable. As Arlington Hall's Russian section began sorting through all this cable traffic, it soon became apparent that there were five different code systems in use. One, which made up about half of all the telegrams, clearly dealt with commercial matters, as they came from places where Soviet purchasing missions were located. The other four appeared to be purely diplomatic, originating at the embassy and consulates. It was also clear that, as had been known since the 1920s, the Soviets were using four- and five-digit codes enciphered with a one-time pad. If used correctly—and if the one-time pads were truly random—it was unbreakable.

Lacking anything better to try, however, one member of the team, Richard Hallock, who had been an archaeology professor at the University of Chicago, decided to begin a brute force search of the trade traffic. Starting in the fall of 1943, he had the machine branch punch the first five message groups from ten thousand of the trade messages onto IBM cards and list them in numerical order. When it was all done he had found seven double hits, which could well have been the product of pure chance. But as Hallock and his team began working through the seven pairs of overlapping messages using the standard technique of creating tables of differences, they found they could recover some of the additives from the one-time pads and the underlying code groups. It soon became clear that they were on to something. The one-time pads were not one-time; some were more like two-time.

Every message appeared to have been enciphered starting at the upper-left-hand corner of a page of one-time pad. There were sixty additive groups per page and every now and then an entire page had been reused.

Work progressed slowly through the winter and spring. In July 1944, while the principal effort continued on the trade code, Phillips was placed in charge of the rest of the Russian problem. It was both a vote of confidence in the nineteen-year-old cryptanalyst and a statement of how little hope for success there seemed to be in breaking the diplomatic codes. There were far fewer of these messages to begin with, and they were divided into four separate systems, which stretched the odds even further.

Phillips sifted through the traffic and was not getting much of anywhere when, in November, he decided to take a look at one batch of messages from New York to Moscow. And dimly through the sea of meaningless four-digit numbers a shape began to loom. An enciphered code should produce a purely random distribution of digits. The very first cipher group in the messages Phillips was looking at were not random, however; there was something ever so slightly odd about them. He began to count how often each digit occurred; each ought show up 10 percent of the time. The digit 6 appeared 20 percent of the time. It was at the point that he took the stack of messages over to Genevieve Grotjan—she of Purple fame, who had been assigned to the Russian problem—and she at once said, "That looks like clear key." She had recognized the four-digit groups that began each message as additive groups from the upper-left-hand corner of one-time pad pages that had been recovered from the trade code. The one-time pads, in other words, were being reused in the diplomatic codes, too. What was more, Phillips and Grotjan had just cracked the indicator system that would tell which page had been used: The indicator transmitted as the first group of each message was the actual additive group at the upper-left-hand corner of the key page that was used to encipher the following text.

The IBM machines were pressed into service again; so was Copperhead, the Navy's rapid optical comparator, to search for double hits in several thousand messages sent on the other Soviet diplomatic systems. The non-randomness in the first additive groups that Phillips had spotted was a pure fluke, probably the result of a printing machine that was used to generate the one-time pads mechanically, and that operated with a slight bias. In fact, only a small portion of the additive pages had this unevenness, and had Phillips chosen a different batch of messages he never would have spotted it. Nor, if the KGB's cryptographic center in Moscow had done its job right, would Arlington Hall have gotten anywhere. Over the next few years the Russian section would search through 750,000 telegrams, and find 30,000 pages of reused one-time pad. It would later become clear that for a few months in early 1942 the KGB, under pressure of wartime shortages, had printed duplicate copies of pages and bound them, often with different page numbers, into separate one-time pads.

It would also become clear that while one of the four diplomatic systems was being used for perfectly legitimate consular business, the other three were assigned to the KGB, Soviet military intelligence, and Soviet naval intelligence. And the twenty-nine hundred messages transmitted from 1940 to 1948 that were eventually read by Arlington Hall, some as late as the 1970s, revealed an extensive Soviet espionage operation in the United States.

It was one of the great ironies of the Cold War that for all the ill-informed, paranoid, and malicious ravings of the McCarthyites, and the harm that they did to countless perfectly innocent people, the truth about Communist infiltration of the United States was, in at least a few areas, even worse than they suspected. The cables revealed that the Soviets had penetrated the Manhattan Project at multiple points, with at least four agents at Los Alamos reporting to a network controlled from New York. The cables pointed to highly placed and well-informed sources in the War Department's General Staff, in the OSS, in the British Embassy in Washington. They chronicled day by day the work of members of the Communist Party–U.S.A. who were reporting directly to Moscow on military plants and technologies. They told of a ruthless KGB operation to hunt down, kidnap, and return to the Soviet Union deserters from the Soviet merchant fleet living on the West Coast of the United States. Details of KGB trade craft in the messages, including passwords and recognition signals, countersurveillance, recruiting, and forgery, showed a highly professional and experienced intelligence agency at work.

In 1945, Weisband, back at Arlington Hall, was assigned to the Russian section as a "linguistic adviser." One day he stood over the shoulder of Meredith Gardner, another linguist in the section, as Gardner extracted the text of a KGB telegram that contained a list of American atomic scientists working on the Manhattan Project. Five years later Weisband was fired from his job after an aeronautical engineer, James Orin York—who had worked at the Northrop and Douglas factories on the West Coast in the late 1930s and 1940s—told the FBI that Weisband had been his contact for passing secret information on aircraft and aircraft engines to the Soviets.

Weisband refused either to admit or deny he had been a spy for the Soviets, though at least one intercepted signal seemed to point pretty clearly to him. But there was another man, unquestionably a Soviet spy, who also learned all about Arlington Hall's Russian project (which would later be code named VENONA). Kim Philby arrived in Washington in the fall of 1949 as the British intelligence service's liaison officer, and he almost immediately became a regular visitor to Arlington Hall, where he was given complete run of the Russian section. It was not long thereafter that Moscow's American agents, especially those who had dealt with the atomic spy Klaus Fuchs, started receiving instructions that they had better be ready to flee the United States through Mexico. It was about this time, too, that the Soviets stopped using their duplicate pages of one-time pad, though this seems to have been unrelated to anything they had learned from spies: They had simply run through all of the repeated pages, and the wartime mistake that had led to their being printed in the first place was never to be repeated.

Bletchley Park had its Russian spy, too—or possibly spies. John Cairncross worked in Hut 3 for about a year in 1942 and 1943. He "kept his roots" in London, he explained to colleagues, and traveled up and down to the metropolis every weekend. What his colleagues did not know was that in his suitcase each week was a selection of decrypts about the Russian front, which he passed on to his Soviet handler in London. What Cairncross did not know was that most, if not all, of the decrypts were being sent to Moscow through official channels anyway. Apparently the constant fear of being caught proved too much of a strain, and Cairncross asked to be transferred to MI6 in the summer of 1943.

Reference to a possible second Soviet spy at Bletchley appears in the VENONA decrypts. His code name was BARON, and he apparently worked at Bletchley before Cairncross did; one VENONA message sent to Moscow in early 1941 contains the verbatim contents of an ULTRA decrypt that had been read by Hut 3. Remarkably, Weisband, Cairncross, and BARON appear to have been the only spies to have directly penetrated the British and American code breaking operations during the entire war—and they were working for a nominal ally. Axis agents never came close to learning what was, after the atomic bomb, the greatest secret of the Second World War.

———

The Allied cryptanalysts had proved by this point that the tiniest of flaws was enough to split the most perfect cryptographic edifices wide open. Even one-time pads were not immune, even one-time pads that were really (unlike the Russian ones) used only once. For several decades the German diplomatic codes had defied solution, but with the break in Floradora in 1943 attention turned to the even more secure GEE, the one-time-pad system. There was always the chance the Germans were using their pads more than once, and there were several thousand cribs now available from messages that had obviously been sent to different embassies and consulates simultaneously using both Floradora and GEE. There were also 3,651 pages of actual one-time pad key that had been obtained from the FBI's 1940 interception of the I. G. Farben courier. It was probably impossible nonetheless, and only a dozen people were put to work on the problem at Arlington Hall. They decided to do another huge IBM run; this would list every known or recovered additive group in numerical order along with its location by page number, row, and column in the one-time-pad pages. There was no obvious reuse to be found, but there was another one of those slight undulations in randomness that set a cryptanalyst's antennas vibrating. Individual additive groups were of course likely to appear more than once among the hundreds of thousands of five-digit additives by pure chance; that was no surprise. The surprise was that the groups that came before and after such repeated groups tended to have an unusual number of digits in common. What followed was one of Arlington Hall's most brilliant cryptanalytic feats, a digit-by-digit combing through of the additives in search of an underlying pattern. They found it in January 1945. One other clue came from a minute examination of small printing flaws in the numerals on the actual one-time-pad sheets. It became clear that each of the forty-eight five-digit groups on a page was printed with a different printing wheel; in other words, all 240 digits were printed at once by a gang of wheels that rotated in a complex pattern from page to page. The numerals on each wheel were placed in a nonsequential order. But once that trick was discovered, the cryptanalysts had a way to predict the key sequence of the one-time-pad sheets. An IBM gizmo was ginned up to test cipher text against different possible runs of additive. By June 1945, 110 people were working on GEE, and during the first half of 1945 the traffic yielded several crucial pieces of intelligence that were not yet known by other means, including the fact that the Japanese had begun to use a new medium tank and self-propelled gun.

The thinnest of wedges—later driven by the greatest concentration of mathematical might that Bletchley Park had yet assembled—was also used to split apart the German Army teleprinter ciphers. Standard teleprinter machines use a simple five-bit binary coding in which each character is represented by a series of five on/off pulses. The letter A, for example, is 11000, B is 10011, carriage return is 00010. The idea of adding coding wheels to a standard teletype machine in order to automatically encipher these streams of pulses was pursued by the German military services early in the war, and in 1941 British monitoring stations began to intercept "non-Morse" signals that had been generated with these experimental cipher machines. Bletchley Park gave these the code name "Fish," and it did not take very long to figure out what they were. It would later become clear that there were several different machines in use. The Lorenz company built machines known as the *Schlüsselzusatz* ("cipher attachments") SZ 40 and 42, which acquired the Bletchley Park name "Tunny"; Siemens built the T52 *Geheimschreiber* ("secret writer"), known to Bletchley Park as "Sturgeon."

Binary arithmetic is easy to realize in hardware, and that was part of the beauty of the German invention. The machines had two banks of five cipher wheels each, one wheel for each bit of the teletype code. At each position the wheel would cause an electrical circuit either to send a pulse or not—a 1 or a 0. For each bit, a bank of relays would add the pulses from the two banks of cipher wheels to the impulse generated by the teletype keyboard for the character that had been pressed (or, more often, punched in advance onto teletype tape). The whole process would be fully automatic, and the resulting stream of enciphered 1s and 0s, in the form of beeps and spaces, could be transmitted at very high speed over landlines or radio.

The immediate problem for GC&CS was to recover the sequence of five-bit additives used to mask the plain text characters, and here the ubiquitous talents of John Tiltman again came to the fore. On August 30, 1941, a German operator sent a four-thousand-character-long message. His recipient did not receive it properly and asked it to be resent. The operator then sent it again, using exactly the same wheel starting position, but abbreviating the first word, *Spruchnummer*—"message number." This was a fabulous piece of depth. Subtracting the two streams of cipher text from one another stripped the additive out of the equation entirely, producing a stream of five-bit groups that was the difference of two plain-text messages. Tiltman set to work on it, guessing at words as needed, and discovering that they were indeed simply the same text shifted a few characters over from one another. Subtracting the recovered plain text from the original cipher text then yielded a four-thousand-character-long key sequence.

The next problem was to figure out what wheel patterns would generate such a sequence. That task fell to William T. Tutte, a young Cambridge mathematician. Taking the first bit of each five-bit additive, he tried writing out the key sequence in tables with varying numbers of columns until a repeating pattern began to emerge in each row. He then repeated the process for each of the other bits. After four months he had figured out the number of positions on each wheel and the sequence of 1s and 0s on each. Finding messages in depth and cribbing by hand could in theory recover the initial wheel settings for each transmitted message, and that approach was favored by Major Ralph Tester, whose group in the "Testery" in F Block had first been assigned the problem. Max Newman, a mathematician who had taught Turing at Cambridge, lobbied for a more automated approach. Tutte's analysis had shown that one bank of wheels, which B.P. called the χ wheels, stepped with each character typed; the other bank, the ψ wheels, stayed put for long stretches, advancing at irregular intervals. That meant that most of the time, about 70 percent of the time in fact, a character enciphered with a 1 on the ψ wheel would be immediately followed by another character enciphered with a 1; or a 0 followed by another 0. And *that* meant that if the output of the ψ wheels for sequential characters was written out and then added to it-self shifted over by one character, a disproportionate number of 0s would occur, since in binary noncarrying addition $1 + 1 = 0$ and $0 + 0 = 0$. Plain text also had this bias (though not quite as pronounced) because of the occur-rence of doubled letters in German words, and also just the chance way 1s and 0s were assigned to the various letters of the alphabet in the standard teletype code. So again, adding plain text to itself with a shift of one charac-ter would also produce a slight preponderance of 0s. The upshot was that performing this same shift-and-add treatment on a string of actual cipher text would yield a sequence of 1s and 0s that tended mostly to track the pat-terns generated by the χ wheels alone—since both the plain text and the ψ wheels zeroed themselves out of the equation most of the time.

What was required, then, was to prepare two long loops of teletype tape. One would contain the cipher text added to itself with a shift of one; the other would contain, for every possible starting position of the χ wheels, the χ wheel output added to itself with a shift of one. The two tapes would then be compared, character by character, and the number of coincidences counted. The correct χ wheel starting position ought to result in the greatest number of hits between the two tapes.

The first machine built to do this was called Heath Robinson (the name was the English equivalent of "Rube Goldberg"), which arrived at the "Newmanry" in May 1943. Jack Good, a statistician who formed part of Newman's formidable team of mathematicians, said that one of the greatest secret inventions of the war was the discovery that ordinary teletype tape could be run at thirty miles per hour without tearing. It did, however, tend to stretch—especially the χ wheel tape that was used over and over. That was of course fatal, because absolute synchronization between the two tapes was essential for a correct count. Turing served as a more or less in-formal consultant to the Fish project, but he suggested that Newman sum-mon Tommy Flowers, an electronics specialist from the Post Office Research Station. Flowers, before the war, had designed and built experimental tele-phone switches that used vacuum tubes in place of any moving parts, and he at once suggested that the solution was to design circuits that would gener-ate the χ wheel patterns electronically. The resulting Colossus Mark I was completed and installed at the Newmanry in February 1944; it contained fif-teen hundred vacuum tubes (later models had twenty-five hundred) and proved far more reliable than Robinson, which, Good said, did have the one redeeming feature that it was usually possible to diagnose a fault by the type of noise it made—and sometimes by the smell it made, as there was one particular recurring problem that caused the machine to overheat and attempt to catch fire. The Colossus was the first computing device with a substantial electronic memory and it was programmable by switches and patch cords; it was even capable of some conditional logic, adjusting a calculation according to data accumulated in the course of a run. By the end of the war ten Colossi were in service. Like the America RAM equipment, the Colossus contained many innovative elements of the modern digital computer without itself being in the direct line of descent that led to the computer; other wartime machines, especially the American ENIAC, which had punch-card input and output, circuits to perform arithmetical operations and square roots, and some ability to store programs electronically, have a far better claim to having given birth to the computer age.

But there is no doubting that the Colossus spawned an intelligence bonanza. The German radioteletype circuits connected Berlin directly to the headquarters of theater commanders and army groups, and revealed German intentions and strategy at the highest levels. Although Fish never rivaled Enigma in the quantity of decrypts produced (during 1943 only about three hundred Fish decrypts were coming in a month versus about eighty thousand Enigma messages), it supplied a series of priceless reports on German appreciations of the Allied invasion threat in the spring of 1944—and of German preparations to counter it. The Allied deception plan that would prove crucial in the success of D-Day owes a huge debt to Bletchley Park's breaking of the German teletype machine.

————

ULTRA verified the effectiveness of the massive D-Day deception; it also in a very real sense gave birth to it. In 1939, the British SIS did not even know the name of its German counterpart, the Abwehr. Two years later, thanks in considerable measure to ULTRA, the British had either arrested or were running as double agents every single spy that the Abwehr had attempted to infiltrate into the country.

In December 1940, Oliver Strachey's ISOS operation at Bletchley Park succeeded in breaking a cipher used by the Abwehr to communicate with stations abroad. ISOS, depending on who you talked to, stood for "Illicit Signals Oliver Strachey" or "Illicit Series Oliver Strachey" or "Intelligence Series Oliver Strachey" or even "Intelligence Service Oliver Strachey," but whatever its name it dealt with the Abwehr's hand ciphers. ISOS decrypts soon yielded two huge dividends. First, they provided Dilly Knox's "ISK" section, which handled all nonsteckered versions of the Enigma, with the cribs it needed to break the Abwehr Enigma. Much Abwehr traffic was du-

plicated between the hand cipher and the Enigma channels, the Enigma being reserved for stations in occupied territory, where the risk of losing one
of its machines to enemy hands was far less. (The Abwehr's version of the
machine, while unsteckered, had the complication of employing rotors with
as many as seventeen turnover notches.) By December 1941, Knox's section
had begun breaking the Abwehr Enigma traffic regularly.

The other dividend that ISOS paid, and which ISK compounded once it
became available, was in providing advance warning of attempts by German
intelligence to infiltrate agents into Britain. In spring and summer 1941 the
Abwehr six times tried to smuggle agents in aboard refugee ships from Norway; five of those attempts ISOS disclosed in time for the new arrivals to be
arrested almost as soon as they set foot in Britain. In the sixth case the message which revealed the presence of three German agents aboard one ship
was not broken for several months, at which point it was discovered, alarm-
ingly, that the British Special Operations Executive had already recruited
the men for sabotage missions in Norway. Two of the men were promptly arrested, but there were a tense few months when the third, who had already
been dispatched to Norway, was arrested by the Germans—but he never betrayed the SOE.

Many of the arrested agents were promptly "turned around" by the
British intelligence services, sometimes reunited with their radios in a
prison cell where they were set to work establishing contact with their German handlers and following often elaborate and brilliant scripts prepared
for them by their British handlers. Reading the Abwehr codes not only
made it possible to carry out such elaborate deceptions, but even more important it made it possible to check on whether the Abwehr was falling for
it. They were. Even when the British "Double-Cross Committee" that ran
the controlled agents decided it was worth blowing an agent for the sake of
a particularly important deception operation, it seemed impossible to blow
him. Abwehr officers were always eager to seize on other plausible explanations for an agent's lapses, rather than admit that they had been had.

By the beginning of 1941, recalled J. C. Masterman, the Double-Cross
Committee's chairman, it began to dawn on the British authorities—"dimly,
very dimly"—that *every* German agent in Britain was under their control.
There simply were no others. By July 1942, that dim guess had become a certainty, in large measure as a result of confirmation now provided by the ISK
high-level decrypts.

The Double-Cross Committee's star double agent was a Spanish anti-
Fascist, Juan Pujol Garcia, code name GARBO. When the war began Garcia
had tried and failed to get himself recruited as a British agent and so conceived the daring idea of first offering his services to the Germans so he'd

have something more to offer the British. He marched up to the German embassy in Madrid; the Abwehr in due course signed him up; and, in July 1941, he left, supposedly for England, fully equipped with secret ink, money, and a list of questions. Instead of going to England, however, he went to Lisbon where for nine months, with nothing more to guide him than a map and tourist guide of Britain, a Portuguese publication on the British fleet, and whatever technical journals he could find at the public library, he proceeded to compose a series of highly inventive intelligence reports for his German masters. Despite a few wonderful gaffes—in one report, supposedly from Glasgow, he explained his success in prising out information from the locals by observing that "there are men here who will do anything for a litre of wine"—the Abwehr was convinced of his bona fides. Garcia "recruited" three imaginary subagents plus a courier, an airline employee who, he explained, carried his dispatches to Lisbon for posting to his handler in Madrid.

In January 1942 Garcia again approached the British, making contact with the SIS in Lisbon. The British were obviously skeptical; his story was fantastic. But the best voucher he had was ULTRA, and a few months later, on April 2, ISOS decrypts showed unmistakably that the Abwehr had no doubt about Garcia's reports. GC&CS broke a signal from Madrid to Berlin relaying Garcia's report of a nonexistent convoy from Liverpool to Malta. That was shortly followed by naval Enigma traffic betraying a flurry of German preparations to intercept the convoy. Garcia was flown to Britain later that month.

GARBO's reputation was then built up even further by one of the most brilliant deception schemes of the entire war. TORCH offered the opportunity that fall. On October 29, GARBO wrote a dispatch reporting the departure of a large convoy that had sailed from the Clyde on the twenty-sixth. The letter was carried to Lisbon, where the SIS waited until the Admiralty confirmed that the convoy had actually been spotted by Axis reconnaissance. Then the letter was mailed, on November 4. A second letter, bearing a November 2 postmark, but actually not mailed until the seventh—the day before the Allied landings, and thus sure to arrive too late—reported that troop transports and battleships camouflaged in Mediterranean colors had left the Clyde. His German handlers were duly appreciative: "Your last reports are all magnificent, but we are sorry they arrived late." GARBO's reputation was now unshakable, and on D-Day the ruse was repeated, but this time with a poison pill. By 1944 GARBO's network of imaginary subagents, all carefully charted and tracked by the Double-Cross Committee, had grown to thirty; they included a garrulous RAF officer, a Ministry of Information official with extreme left-wing views, a Venezuelan businessman in Glasgow, a Communist Greek sailor in eastern Scotland, a Gibraltese waiter in a service canteen, an Anglophobe American sergeant, and eight Welsh nationalists. In the early morning hours of June 6, GARBO sent an urgent radio dispatch to his Abwehr control: the invasion was imminent. Again, the message was perfectly timed; by the time it was in German hands the invasion had begun. Then three days later came the master stroke that the entire operation had been building up to. GARBO, having "met" again with his network, reported urgently that the attack on Normandy was a feint; the real blow was yet to fall at the Pas de Calais. This, of course, was what the Germans had been inclined to believe for months, and it was what the entire Allied deception operation had been aiming to reinforce, notably with the creation in Britain of an entire fictional army group, FUSAG—First United States Army Group, under the command of General George S. Patton. GARBO's June 9 dispatch pointedly noted that not a single FUSAG formation had taken part in the Normandy landings, proof that they were being held in reserve for the main thrust. The transcendent brilliance of the scheme was that it kept the deception going for days, indeed even weeks, *after* the attack had actually begun—it actually managed to make the failure of a nonexistent unit to materialize a source of grave anxiety to the enemy. In late June GARBO received a message that produced no end of satisfaction at the Double-Cross Committee: The Germans were awarding him the Iron Cross for his heroic work.

Fish decrypts in the spring of 1944 had confirmed that the Germans were convinced that the Pas de Calais was the Allied target; so did a series of invaluable Purple decrypts from Oshima, one, in November 1943, reporting in incredible detail his inspection tour of the German "Western Wall" defenses. In late December Marshall had urgently summoned Eisenhower to Washington (find "someone else to run the war for twenty minutes," he said), in part because he felt Eisenhower desperately needed a break, but in part to go over this new information with him. For weeks after D-Day, Oshima's reports continued to provide reassurance to Eisenhower that the deception was still working. On June 9 Oshima cabled to Tokyo that the Germans were on guard against a landing in the Calais and Saint Malo regions; a full month after D-Day, the Japanese ambassador reported that the Germans were still girding for a second attack in the Channel area, by Patton's nonexistent force.

In the final year of the war the double agents scored one last coup. As the V-1 and then V-2 weapons began to hit London, the Abwehr asked its agents in England to report their time and location of impact. Under the guidance of Britain's scientific experts, the agents supplied subtly skewed data that would lead the Germans to believe they were overshooting their target of central London. Under the influence of these false reports, the mean impact point of the V-2s began moving eastward at a rate of two miles a week, until by mid-February 1945 most of the rockets were falling well outside the boundaries of the heavily populated metropolis.

———

One by one the enemy code systems had fallen, and with them the Allied mastery of the airwaves was becoming lopsided. The Japanese naval codes continued to undergo regular changes and improvements, but OP–20–G had hit its stride and had little difficulty keeping up. Regular reading of the Japanese convoy codes gave American submarines an almost total mastery over the Japanese supply lines, and each day at 9:00 A.M., as if they were punching a clock in an office, the Pacific fleet's intelligence officers would sit down with top officers of the fleet's submarine force to divvy up the targets. By January 1944 U.S. submarines were sinking a third of a million tons a month.

JN-25 carried about 70 percent of all Japanese naval traffic, and it too was read almost continuously for the rest of the war. On April 14, 1943, a JN-25 decrypt landed on the desk of Edwin Layton, Nimitz's intelligence chief, and it at once electrified and disturbed him. The message began: ON APRIL 18 CINC COMBINED FLEET WILL VISIT RXZ, R–, AND RXP IN ACCORDANCE WITH THE FOLLOWING SCHEDULE: 1. DEPART RR AT 0600 IN A MEDIUM ATTACK PLANE ESCORTED BY 6 FIGHTERS. ARRIVE RXZ AT 0800. It would be Admiral Isoroku Yamamoto's death warrant. Layton and Nimitz held an uneasy conference. Nimitz asked Layton—who, during his time as a language student in Japan, had actually known the admiral—whether the Japanese Navy could replace Yamamoto. Layton confirmed that Yamamoto truly was irreplaceable. That settled the matter as far as Nimitz was concerned, and the orders went out. RXZ was identified as Ballale, a small island in the Solomons just south of Bougainville. Sixteen P-38 fighters from Henderson Field on Guadalcanal, flying at the extreme limit of their range, intercepted Yamamoto's airborne entourage flawlessly. As Captain Thomas Lanphier fired a long burst into one of the Japanese bombers, flames erupted from its engine, a wing shredded into pieces, and the plane plunged into the jungle below. Three Zeros and the second bomber also were shot down. The Japanese government waited until May 21 to break the news of Yamamoto's death to the fleet, and to the public.

———

There was still one last unsolved enemy code, and given its centrality to the entire war effort, it was becoming an acute embarrassment to Arlington Hall. OP–20–G had been working on Japanese naval codes for years before the war. But the Army's SIS, lacking the required intercept capability and diverted by the high payoff of the Japanese diplomatic codes, had put almost no effort into studying the similar Japanese Army codes. In April 1942 the SIS had a grand total of eleven people in its Japanese Military Codes Section, and only four of those were assigned to look specifically at the Japanese Army codes. They found a weltering confusion of different systems, and about all the cryptanalysts could do was sort and file the traffic as it came in. All of the high-level systems were four-digit, two-part codes enciphered with additives, and messages sent in each carried a four-digit "discriminant" that identified the system being used. Before Pearl Harbor there had been one major system used at the top echelons; it bore the discriminant 5678 and was designated by the American cryptanalysts as ATRW. But as the Japanese Army swarmed across the Pacific in 1942 and the burden of traffic grew rapidly, new additive books were issued for separate zones. By May 1942, the main system had been split into 7890, which SIS designated JEM, and 2345 or JEN. There was also a separate system, 2468 or JEK, that apparently was used by the Water Transport organization, which was in effect the Army's own Navy, used for moving troops around the Pacific. Further splits into regional and service zones continued, and just keeping it all straight was a huge clerical task at Arlington Hall. There seemed so little hope of achieving a cryptanalytic break that the traffic analysis section was given first dibs on all of the incoming messages, and since there were only two teletype lines linking SIS to Two Rock Ranch in California, which intercepted most of the traffic, nearly all the traffic came in by mail. Some was coming in by mail from Australia, too, where a contingent of U.S. Army personnel had arrived in mid-April to set up an intercept and cryptanalysis unit at MacArthur's request. It was already one month to as much as eight months old when it finally showed up in Washington; then the traffic analysis unit would pore over it for another three months; then the cryptanalysts would at last get to see it. There was a war on, but in the summer of 1942 still only two dozen people were working on the main military code of what was, at least for most Americans, the main enemy.

When the cryptanalysts finally got their shot, the only hope seemed to be a huge brute force IBM run. The indicators used to specify the starting point in the additive book for each message were enciphered in a totally impenetrable code of their own. Punching four thousand messages in all possible offsets consumed three to four million IBM cards and the job dragged on from April 1942 to the end of the year before the cryptanalysts finally threw up their hands in frustration and despair. There was just too much material to examine, too high a chance of random coincidence generating double hits. A few message pairs turned up with triple hits, but even those led nowhere. Statistical calculations showed that a single double hit in a system that used a fifty-thousand-group additive book had only a 1 in 45 chance of being causal; even a triple hit was only a 50–50 proposition.

Frank Lewis, who was at this point assigned to the Japanese Army section, had been unhappy about the brute force approach from the start—"brute force and bloody ignorance" was more like it, he felt—and thought it might be more profitable to start with the oldest back traffic and try to work forward to establish continuity. The prewar ATRW system used a fairly straightforward indicator system to designate the starting point in the additive book: At the beginning and end of the message were two four-digit code groups of the form BPPS RCRC, where B was the number of the additive book (1 to 3), PP was the page number (0 to 99), S was a sum-check obtained from adding the first three digits with noncarrying arithmetic, and RCRC was the row and column on the page, repeated twice as a check. The indicators were themselves enciphered with a table of one hundred additives. The question now was where, in the current systems, was the indicator hidden, and how was it enciphered?

One clue came in March 1943, when four captured message forms of 5678 were delivered to Arlington Hall. The notations on the forms suggested that certain of the message text groups themselves were somehow used as a key to specify the encipherment of the indicators. But the much more important break came just a few weeks later. A cable from GC&CS called attention to a curious phenomenon: The first digit of the third cipher group in each message was not completely random. The U.S. Army group in Australia, now known as Central Bureau Brisbane and now under the command of Abe Sinkov, who had been sent out in June 1942, weighed in with another discovery on April 1. For any given digit in the initial position of the second message group, there were only three possible digits that ever occurred in the first position of the third group. Whenever the second group began with, say, a 1, the third group began only with 3, 5, or 6. Whenever it was 2, the third group began with 2, 8, or 9. That was extremely telling, for, as Lewis's historical study of 5678 had shown, the B in BPPS could only have the values 1 to 3. Clearly the second text group of the message was being used as a key of some kind to encipher the BPPS indicator, which was then inserted into the message as the third group. Similarly, it became evident that the fourth message group was used as the key to encipher the RCRC indicator, which was then inserted as the fifth group of the message. Within a week both CBB and Arlington Hall had broken the system. It used a 10 × 10 conversion square; the plain text digits 0–9 ran across the top, the key digits down the side, and the table contained the cipher text digits. For example:

plain text

		0	1	2	3	4	5	6	7	8	9
	0	7	9	6	0	3	1	4	2	8	5
	1	3	1	0	7	9	6	2	8	4	5
	2	7	5	0	8	2	9	6	1	3	4
	3	8	4	9	5	2	6	1	3	0	7
key	4	7	8	2	9	4	1	3	0	5	6
	5	6	9	5	1	2	8	4	0	3	7
	6	5	8	4	9	1	3	0	7	2	6
	7	3	9	6	8	4	1	0	5	7	2
	8	7	4	1	0	6	2	9	8	5	3
	9	3	6	4	7	5	0	1	9	2	8

If the unenciphered BPPS indicator were 2697 and the second message text group, which serves as the key, were 0724, then the enciphered form of BPPS using this conversion square would be 6040.

CBB sent a cable to Washington on April 6 announcing they had cracked it, and giving their results. Arlington Hall the next day sent a cable back saying, in effect, "yeah, we did, too," which forever left the Brisbane team slightly suspicious.

Be that as it may, this was progress at last. From March 1 to April 30 the staff of the Japanese Army section was tripled, to 270, with more than a hundred assigned to JEK. Meanwhile a few weeks' work established that 7890 and 5678 used a similar but more complex encipherment system for the indicators, in which certain digits of the message groups referred to a separate printed key list that provided the key digits to be used with the conversion squares. The conversion squares were also changed often, an enormous headache.

But several things helped. One was the very difficulty of the Japanese language itself. Whenever a Japanese code clerk encountered a word or proper name that was not included in the code book, he had to spell it out using kana characters, which were each assigned their own four-digit code groups in the book for just such a need. But words spelled in kana are often ambiguous in Japanese, a language with many homonyms. For example the word *sentō* can mean public bath, residence of a retired emperor, the first scaling of the wall of a besieged castle, fighting together, or scissors. In written Japanese such words are rendered unambiguously by the use of ideographic characters known as kanji, which are derived from Chinese characters. With the beginnings of telegraphy a standard commercial telegraph code was developed that provided numerical equivalents for Chinese characters. (In 1935, John Tiltman once spent weeks working on what he thought was a batch of Japanese military intercepts, forwarded from the British intercept unit at Hong Kong. He soon established that it was an unenciphered code and began to identify the groups that stood for numbers; an interesting feature was that the numerical order of the code groups followed the stroke-order in which the corresponding Chinese characters were written. He had gotten quite far along when a friend of his, a Chinese interpreter for the War Office, took a look and explained to Tiltman that what he had "broken" was the publicly available Chinese Telegraph Code.)

When a Japanese code clerk got stuck, he would thus simply use the Chinese Telegraph Code value to clarify the meaning of the preceding word that had been spelled out in kana. The military code books assigned the code groups 1951 and 5734 to mean "Chinese Telegraph Code follows." Even though the Chinese Telegraph Code groups were disguised by an additive, the same as the rest of the text, it was an absurd security flaw. For one thing the code groups 1951 or 5734 appeared in messages so often that, along with the code groups for "stop" and other common words, they provided a reliable entry point for a code breaker. For another, since the kana groups that preceded 1951 and the Chinese Telegraph Code groups that followed stood for the same word, it wasn't too hard to guess one from the other.

With the indicators broken, the standard processes for placing messages in depth and beginning additive and code group recoveries could begin in earnest. By early June 1943, messages in 2468, the Water Transport system, were being read and forwarded to military intelligence. The decrypts were often not current, but since many announced in advance the planned sailing dates and routes of transports and supply ships, the information could still be put to use profitably even with a time lag. On August 20, a Water Transport message was broken from nine days earlier that announced that a supply convoy would be leaving Palau on the twenty-sixth and unload at Wewak on September 1. At Wewak the five cargo ships and two destroyers were greeted by a flight of U.S. medium bombers that came in at masthead height.

Progress on the other systems was slower, and continual cryptographic changes in all of the systems made it a week-after-week struggle. But another flaw appeared even as the schemes employed by the Japanese to encipher indicators and text grew more and more complex. The communications networks were broken into smaller and smaller zones; the additive books were changed more frequently; and beginning in February 1944 the encipherment system for 2468 was changed completely to the mind-numbingly tedious procedure of using conversion squares to encipher the actual message text. Instead of applying the additives to the code groups with noncarrying addition, the additives were now used as keys for conversion squares. But even this dramatic security measure contained the seed of its own undoing. With so many changes and so many separate networks—and now, so many Army units cut off in isolated islands of the Pacific—it was simply impossible for the Japanese Army to issue new code materials in writing. Instead new conversion squares were transmitted, in code, by radio, and in one of the supreme ironies of the cryptographic war, the highly stereotyped patterns of these cryptographic messages themselves provided reliable cribs for breaking. The streams of numbers were invariably followed by ten special sum-check groups and it was all so patterned that Arlington Hall later even developed a special-purpose IBM attachment, dubbed CAMEL, that could automatically spot any sequence of ten groups that could be the encipherment of ten sum-check groups.

In September 1943 work on the Japanese Army codes had progressed enough that it was split off into a separate branch, designated B-II, with Kullback in charge; all other work, which was mainly diplomatic traffic, was lumped together in B-III, the "General Cryptanalytic Branch," and Rowlett was named to take charge of that. (All of the bombes and Rapid Analytical Machinery also wound up in B-III. The third main branch, B-I, was devoted exclusively to translating Japanese decrypts.) The flow of Japanese Army decrypts improved, reaching almost two thousand for the month of January 1944, though most of those were still Water Transport signals. Then it became a tidal wave. Soldiers from the final echelon of the Japanese Twentieth Division, withdrawing from Sio, New Guinea, in a pouring rain, decided they had quite enough troubles without lugging a heavy steel trunk containing the unit's code books. Sweeping the area with a metal detector to search for Japanese booby traps, an Australian engineer froze as his headphones began to beep; a demolition squad was called in and unearthed the box. It was, as Sinkov later said, "the entire cryptographic library" of the Twentieth Division. The haul was forwarded to CBB at Brisbane and Sinkov's unit set to work at once, hanging up the sopping wet pages on clotheslines and training fans and heaters full blast on them. The pages were photographed and copies sent to Arlington Hall, and by March messages transmitted in the main Japanese ground systems were being read at Brisbane and Arlington Hall as quickly as the Japanese could read them. Some thirty-six thousand messages were broken in March. Actually, Arlington Hall was in some cases reading the messages even faster than the Japanese could. In October 1943, B-II had devised a fully automatic deciphering system using IBM machines, and now with the recovered additive books, that system for mass-producing decrypts really took off. Intercept stations sent the raw traffic over teletype directly to Arlington Hall; the teletype machines punched out paper tape that was fed into a machine that transferred the text to IBM cards; a card library of recovered additives and conversion squares stripped the encipherment, the resulting code groups were looked up in another stack of cards, and the decoded text printed out. By the end of June 1944 more that 4,300 messages a day were being punched onto IBM cards, and more than 2,500 a day were being automatically decoded. Work on the Japanese Army problem started slowly, but it ended with a bang.

Although Arlington Hall and CBB cooperated fully, there was one bone of contention that was never resolved. MacArthur insisted from the first that any unit in his area had to report to him, not the War Department. Colonel Carter Clarke, the head of Special Branch, tried several times to set up a secure channel by which CBB could send decrypts to Washington for processing through his office and then dissemination to commanders. Mac-Arthur thwarted it at every turn. In December 1943 General Strong, the head of military intelligence, sent two liaison officers to Australia to set up a secure link; part of the agreement with the British for handling ULTRA in fact explicitly required that such secure channels be established in all theaters. MacArthur had a subordinate issue an order that the two were not permitted to send radio messages without his permission. A year later Clarke, personally bearing a message from Marshall, flew out to see MacArthur and explain the requirements. Clarke got as far as Hawaii when MacArthur's headquarters ordered him back to Washington.

MacArthur's staff distributed ULTRA far more freely than the War Department, the Navy, or the British did, and the Navy more than once complained of the lack of security with which ULTRA intelligence was sent to commanders over radio channels within MacArthur's theater. But MacArthur was a force unto himself. Finally a copy of the new ULTRA regulations was delivered to the general. MacArthur had his staff rewrite it to suit him, signed it, and sent it back to Washington.

––––––

The rapid disintegration of the Japanese position in the Pacific by this point saved MacArthur's indiscretions from having much of a cost. By 1944 Japan had suffered irreparable losses in carriers and trained pilots; her defensive perimeter had collapsed in all directions, as first the Aleutians, then Guadalcanal, and then finally, after two years of horrific jungle fighting, New Guinea were wrested from her by the Allies. In truth, the subtleties of cryptanalysis were giving way in all theaters to the cruder realities of brute military force during the last year of the war. Code breaking had worked miracles when the chips were down. It had pulled through victory at Midway with the narrowest of odds, victory at a time when victory was indis-

pensable. It had halted Rommel in his tracks two hundred miles from Cairo
and its magic had galvanized a demoralized British army at Alam el Halfa.
It had helped turn the tide in the Atlantic at two critical junctures when
Britain faced unsustainable losses and was literally within months of starva-
tion. Yet it was the nature of things that attackers care less about subtlety
than defenders; they care less psychologically and they care less as a matter
of simple military calculation. The conventional wisdom is that an attacker
needs a 3 to 1 superiority in men and material when moving against a prepared defensive position, and an army that assembles that sort of force of
armor and ammunition and manpower is probably going to win by sheer
overwhelming weight no matter what. It won't necessarily win easily or
cheaply, however. And as the sheer overwhelming weight of American industrial power and manpower swept back the German armies in Europe,
the Allied generals paid one last terrible price for their slipping back into
the old familiar habits of spending too much time commanding armies and
too little time reading the enemy's intentions. The place was an old familiar
place for showdowns between Germany and the Western powers: the Ardennes forest of Belgium. Hitler's plan was his usual combination of genius
and raving. The Germans would cut through the Allied lines, seizing the port
of Antwerp and severing the front in two. The Allies would be thrown into
disarray. They could not of course be dislodged from France, but their
timetable would be so disrupted, Hitler fantasized, that Germany would be
able to race back across the eastern front and strike against the Russians,
whereupon the West would come to its senses and join Germany against
their true mutual enemy of Bolshevism.

The Allies' failure to anticipate the German counteroffensive, which began with a massive artillery bombardment against a weakly held American sector in the morning of December 16, 1944, was complex in its cause.
ULTRA had in fact provided ample warnings. But in fairness to the Allied
generals who looked at the situation as professional military men and concluded that the German Army was simply too battered to go on the attack,
that was a view held by the German generals, too. It was only Hitler who
dared to disagree with them both.

Forced to improvise, the Germans for once employed deception to the
full. An entire new panzer army, the Sixth SS, was formed under the guise of
"Resting and Refitting Staff 16." English-speaking commandos equipped
with captured American uniforms were to be dropped by parachute to sow
confusion. Yet ULTRA, and MAGIC, had penetrated most of the German ruses
in advance: A message from Oshima to Tokyo on September 4 had reported
on the ambassador's recent meeting with Hitler and Ribbentrop; Hitler was
forming a force of a million men, augmented by units being pulled back

from other fronts and a replenished air force, which would strike in the West probably in November. Luftwaffe Enigma messages read throughout November and early December revealed that fighters and fighter-bombers being ordered to the west were to be fitted with equipment for close air support operations. Orders emphasizing the need for strict security were themselves a strong hint that an offensive rather than a defensive operation was being planned. On November 1 a Führer Order called for volunteers for a special force; knowledge of English and American idioms was essential. Fighter aircraft were requested to protect large troop movements along railroads leading into the area behind the Ardennes region. On December 10 orders were intercepted imposing radio silence on all SS panzer units.

The fact of an offensive and even the time when the attack would begin were signaled in advance by these decrypts, though its place was less clear to see, and there is always the danger of distortion in retrospect; indications of where the offensive might fall were part of a kaleidoscope of conflicting indications that also pointed to the Aachen region farther north, which in many ways was a more logical spot. Yet that does not alter the fact that Allied euphoria and the old habits of ignoring intelligence bore much of the blame for the misadventure that ensued. The Allied command sifted the evidence and confidently concluded that the Sixth SS Panzer Army was merely a "fire brigade" that might scurry about trying to put out the flames of the latest Allied advance, nothing more. It took six weeks of heavy fighting to halt and push back the Germans. At one point the Germans had the U.S. 101st Airborne Division entirely surrounded; called upon to surrender, the division's commander General Anthony C. MacAuliffe gave his famous one-word reply, "Nuts!" and fought on. The English-speaking Germans in American uniform were dealt with by the improvised battlefield security measures immortalized in World War II movies: German commandos might know American idiom, but not many knew what to say when challenged to name the first baseman for the Brooklyn Dodgers. The GIs who came up with these inventive tactics showed considerably more subtlety and imagination than their commanders who had ignored Bletchley Park's evidence of the German plans.

The Battle of the Bulge cost Hitler two hundred thousand men, six hundred tanks, most of his remaining aircraft, and his entire strategic reserve. Hitler was defeated but fought on; for the remaining four months the war in Europe was one of grinding inevitability as the Allied war machine advanced, tanks, artillery, infantry, aircraft hammering away together in overwhelming force against an ever dwindling resistance still put up by fanatical SS units and terrified young boys and old men hastily drafted into service. There was little finesse to the Allied assault, but then there was little need

for finesse by this point. It was more like bludgeoning a dead but still twitch-
ing rattlesnake than stalking a wily bobcat, and you didn't need a mathematician from Oxford or Cambridge to wield a cudgel. The code war, at
least for this war, was over.

———

The code breakers at Bletchley Park and Washington most of the time did
not read the messages they broke; that was someone else's job. Even in Huts
3 and 4 the flood of traffic, the thousands of messages a day that came
through, most of them routine administrative details whose larger significance would emerge only through the patient piecing together of multiple
clues, induced a certain professional detachment. But there were a few signals that the code breakers of Bletchley Park and Arlington Hall and OP-
20-G would never forget. Late in 1943 or early in 1944, Walter Eytan, one
of the very few Jews who worked at Bletchley Park, was on duty in Hut 4
when a signal came in that had been intercepted from a small German-
commissioned ship in the Aegean. The ship reported that it was transport-
ing Jews *zur Endlösung*—"for the final solution." "I had never seen or heard
this expression before," Eytan recalled a half-century later, "but instinc-
tively I knew what it must mean, and I have never forgotten that moment. I
did not remark on it particularly to the others who were on duty at the time
and of course never referred to it outside BP, but it left its mark—down to
the present day." On May 1, 1945, Alec Dakin was sorting decrypts as they
came into a small wire tray of Hut 4 when he picked up one message from
the bottom of the stack. It was an *Offizier* sent to all U-boats and it began
with these words: DER FUEHRER ADOLF HITLER IST TOT. Hitler is dead. Five
days later came a flurry of signals ordering the U-boats to cease hostilities,
to make their way to British ports to surrender, and to strictly obey instructions to avoid "demonstrations of any sort." The next texts that came through
were in the clear. The bombes and decoding machines fell silent; within a
few months many of the machines, including all of the bombes, would be dismantled to ensure that Bletchley Park's secrets remained safely buried.

The code breakers at Arlington Hall had their experience of holding a
moment of history in their hands on August 13, the day the Japanese government transmitted a message to its ambassador in Switzerland with instructions that it be turned over to the Swiss government for delivery to the
government of the United States. Two hours before the Swiss Minister presented Japan's surrender to the State Department in Washington, it was broken and read at Arlington Hall.

The war was indeed over.

The signals intelligence war had, in truth, just begun.

Legacy

WITHIN six months Churchill was warning of "an iron curtain" descending over Europe; two years later Stalin had seized Czechoslovakia and blockaded Berlin. The Cold War replaced the hot war so swiftly that few in the West were ready for it, at least not psychologically. War weary troops wanted to go home, and in Manila, Paris, and Frankfurt, thousands of GIs rioted when demobilization didn't proceed swiftly enough to satisfy them. At Bletchley Park the Wrens packed up and went home leaving a skeleton staff to poke through the huts searching for scraps of confidential waste that had slipped behind a filing cabinet or gotten stuck in the back of a desk drawer. The civilians in uniform at Arlington Hall and the Naval Communications Annex took off their uniforms and went back to their lives.

They had won the war, they had changed the world, and that was that. What they perhaps did not realize was that they had also accomplished an even more remarkable feat: They had changed the Army and the Navy. Never again would generals and admirals ignore intelligence. Intelligence now *was* signals intelligence, and signals intelligence was the real McCoy. It was also a weapon perfectly suited to the new kind of war, the Cold War, that lay ahead—a war of subversion and espionage, of intentions and bluffs, of threats and deterrence. Reading an adversary's thoughts and anticipating his plans mattered more than ever in the nuclear chess game. And so even as the wartime code breakers went home, the organizations they had thrown together to defeat Hitler and Tojo quietly remained, quietly growing into huge permanent bureaucracies. The Army and Navy code breakers did the unthinkable and merged, forming the Armed Forces Security Agency, later the National Security Agency, establishing themselves at what would become a veritable city at Fort George G. Meade, Maryland, halfway between Baltimore and Washington. GC&CS became the Government Communications Headquarters, GCHQ, at Cheltenham. NSA and GCHQ listened in on everything—not just radio messages from Soviet bombers but transatlantic telephone calls, too, and even when the targets of the eavesdroppers were now British and American citizens there were no Secretaries of State to be heard complaining about other gentlemen's mail being read.

The veil of absolute secrecy remained, in fact drew even tighter, and that posed unique strains to a democracy in peacetime. Prosecuting a war was one thing, prosecuting citizens in a court of law was another, and when VENONA decrypts provided the decisive proof of Soviet espionage activities by Alger Hiss, Harry Dexter White, and Julius Rosenberg, the evidence was kept out of court and out of public knowledge. Astonishingly, even President Truman was kept in the dark by his subordinates. In 1949, the new head of the Armed Forces Security Agency, Admiral Earl E. Stone, was astonished to learn of VENONA and decided that the President and the CIA should be informed at once. Carter Clarke, now a general and head of Arlington Hall, vehemently disagreed. The President of the United States, he insisted, was not among those "entitled to know anything about this source." Clarke appealed to General Omar Bradley, the Army Chief of Staff; Bradley agreed that he would take personal responsibility for informing the President "if the contents of any of this material so demanded." It apparently never did, in the Army's view. If war was too important to be left to the generals, signals intelligence was too important to be left to the politicians.

With peace, the cryptanalyst's culture of paranoia reached a new ascendancy. In war there was always pressure from military commanders to balance the undeniable damage done by disclosure with the undeniable gains of use. In peacetime secrecy became an end in itself, and NSA and GCHQ became a force unto themselves, and sometimes a law unto themselves. The absurdity of it all was that in refusing to tell the American public the truth of the evidence against Hiss and others, the government was concealing cryptologic secrets that, while undeniably a secret from the American public, were hardly a secret to the Soviets, thanks to the betrayals of Philby and Weisband. The McCarthyites sensed the whole story was not being told, which encouraged their belief that monsters were being hidden in the closet, that there was actually a massive Communist conspiracy to infiltrate and seize control of the United States government; at the same time it set the American left in its belief, which it would cling to for decades, that *all* of the charges of Communist espionage were inventions of the red baiters. Hiss, White, and Rosenberg became martyrs of the left, whose complete innocence would become an article of faith. In fact, they were guilty as hell—and had the truth been told at the time, it might have spared the nation untold agony. It might even have put a stopper on some of the more ludicrous charges that Hiss and others were part of a Communist master plan that sold out China to the Reds through fiendish inside manipulation of American policy. Hiss and the others were spies, but they weren't the Manchurian Candidate.

———

Wartime necessity and wartime improvisation had made Arlington Hall and Bletchley Park places where rules mattered less than results, where an ability to do the work counted far more than being a company man. As meritocracy gave way to bureaucracy and the ring of peacetime secrecy was drawn ever tighter, the tolerance for personal quirks and intellectual innovation that had made the cryptanalytic bureaus seem more like universities than government agencies began to fade. To be sure, NSA continued to forge innovations in computing, building the first magnetic drum memory immediately after the war and, a few years later, the first computer in the United States to use magnetic core memory. One of the enduring legacies of code breaking in the Second World War and after was the huge boost it gave to the computer industry, and even today NSA remains a major consumer, and innovator, in high-end computing. And NSA continued to show at least a certain tolerance toward individuality; there were no political litmus tests imposed on the professional staff. If there had been any political cast to the staff of Arlington Hall it was, if anything, slightly left of center, and many of the top cryptanalysts who stayed on, rising to senior management positions at NSA in the fifties, sixties, and seventies, were political liberals. Frank Lewis blithely spent the years of Communist-hunting and loyalty investigations of the 1950s working as a senior NSA official while cranking out cryptic crossword puzzles for the far-left publication *The Nation,* and no one ever raised an eyebrow.

The case of Alan Turing was quite another matter. Turing was almost completely apolitical; but he was also homosexual and thus, to the minds of MI5 in postwar Britain, a security risk. Turing had never particularly made a point of his homosexuality, but neither was he ashamed or self-conscious about it; it was completely characteristic of him that he worked out things in his own mind and then it didn't matter what the world thought about it. And so when he went to the police in January 1952 to report that his house had been broken into and some things stolen—and that the suspect was a friend of a working-class boy Turing had in fact picked up and spent several nights with—and when the police started asking him some pointed questions about certain inconsistencies in his story and how exactly a Fellow of the Royal Society happened to be entertaining a nineteen-year-old printer's assistant in his house, Turing naively told them. Tracking down burglars was hard work, but prosecuting people who wandered into police stations and offered confessions was easy, and Turing had just confessed to a serious crime. "Gross indecency" was the legal term for consensual homosexual acts under an 1885 statute still on the books, and it was punishable by two years' imprisonment. Turing pleaded guilty and was spared a prison term if he accepted "treatment" for his disorder, including compulsory hormone injections.

At one point Turing had been offered a salary of £5,000 by Hugh Alexander to come back to GCHQ for a year to apply his postwar work on computers to the new problems of code breaking. In October 1952 Turing confided to a friend that he knew it was now impossible for him to be permitted to work at GCHQ. Two years later he killed himself by eating an apple dipped in cyanide.

———

It was perhaps inevitable that, with the zeal of converts, the military and political authorities that had so long neglected signals intelligence would come in time to overrate it. The remarkable successes scored in the war and the technical wizardry of it all was so persuasive—and the continuing fiascoes in attempting to run actual human agents so discouraging—that "national technical means" (which would later include spy satellites and other remote sensors) came to dominate intelligence gathering. There were times when the United States would pay for that lopsided emphasis. The dearth of "humint"—human intelligence—would become a recurring theme in the fight against insurgent and terrorist groups in the sixties, seventies, and eighties.

With the passing of a half-century it is possible to look back on the accomplishments of code breaking in the Second World War with a more level gaze. As a triumph of the human intellect they were extraordinary. At some critical junctures—Midway, Cape Matapan, Alam el Halfa, the first battles of the Atlantic convoys—the timely reading of enemy dispatches averted disasters that would have been terrible setbacks to the Allied cause. Particularly at sea, where simply knowing where the enemy is amid a vast expanse of water is often more than half the battle, signals intelligence played an unsurpassed role. So, too, in carrying out effective deception, where knowing what the enemy is thinking is practically *all* the battle.

But did code breaking "win" the war, or even shorten the war, as is so often claimed? There the answer is more equivocal. In the Battle of the Atlantic in the summer of 1941, ULTRA intelligence clearly helped tip the scales at a time when sinkings were reaching alarming proportions, yet many other factors were at work as well, not least the withdrawal of some U-boats from the Atlantic in support of the German attack on Russia, and the strengthening of Allied escorts. In the climax of the battle in the spring of 1943 ULTRA played an even greater role, yet so did new depth charges and radar and air power. Simple calculations that compare monthly tonnage losses before and after the breaking of Shark, and award all the credit to Bletchley Park for the tonnage thereby saved, constitute a classic post hoc ergo propter hoc fallacy.

Even in the cases where one can single out one battle in which ULTRA was clearly decisive, it is heading onto dangerous ground to assume that absent ULTRA the battle would have been lost. The whole game of "counterfactual" history, of what-ifs, is a descent into never-never land. History is lived forward, not in retrospect, and it is folly to assume that if one factor had been changed then all other factors left to determine the course of history would have stayed the same, too. If there had been no ULTRA how can we say that necessity would not have driven the Allies to take other measures that would have been as effective, or even more effective, in countering the challenges they faced? For example, had there been no ULTRA at all, and had the Atlantic sinkings continued and Britain's lifeline been threatened, pressures would have become irresistible to divert more bombers from the strategic air campaign and bring them to bear on the U-boat menace—as was in fact finally done in any event. It is not even inconceivable that that much-needed step could have come earlier had Bletchley Park never existed. In war, as in life, as Winston Churchill observed, fortune is all of a piece; one never knows when a seeming misfortune in fact has saved one from something even worse.

Arguments about ULTRA having shortened the war by several years bump up against the one weapon that defies the strictures against unitary explanations of history: the atomic bomb, which, it should never be forgotten, was built from the start with the intention and expectation that it would be used against Germany.

That said, there were a few junctures in the war where a British setback might have so weakened resolve to continue that negotiations for a cessation of hostilities might have begun. There were certainly those, such as Lord Halifax, who favored such a course in the summer of 1940 after the fall of France. ULTRA contributed only marginally to the Battle of Britain, but arguably more significantly to the safeguarding of Britain's ocean supply lines in the crucial months of 1940 and 1941 before America's entry into the war, a time when invasion and collapse were not inconceivable. Churchill was deadly serious when he contemplated plans for the government to flee to Canada to carry on resistance should Britain be overwhelmed. But again, the what-ifs soon cloud all rational speculation: Would a collapse of Britain have hastened Japan's entry into the war, and thus America's? Quite possibly; there is no end to such a chain of speculation.

Nor is it really necessary. Churchill and Stalin probably were both closest to the truth when they recognized that what really won the war was American industrial might, an unstoppable force once unleashed. But wars, again, are fought through time in a forward direction. Victories that are inevitable in retrospect are anything but in prospect, and in celebrating victories it is not just the result we celebrate but the sacrifice, brilliance, and courage displayed along the way. In the story of the breaking of the Axis codes, all were displayed in abundance. To question the more glib and extravagant claims for the effect of the code breakers' work on the war does not, and should not, take away one iota from the magnificence or the importance of their accomplishments.

———

The other lingering question about the code war was why it was in the end so lopsided. On more than one occasion, the answer was painfully simple; as Gordon Welchman said, "we were lucky." Had the design of the Enigma machine been even slightly different in a few of its particulars, particulars that were almost arbitrary design decisions, the Allies' task would have been immensely more difficult.

But there seemed to be more than chance behind many of the mistakes the Germans made, and in retrospect some historians have found it tempting to argue that German rigidity, and Nazi ideology, prevented the recruitment of the intellect required for effective cryptanalysis; there were certainly no Alan Turings in the B-Dienst. It is also easy to speculate that the multiple warring factions within the Nazi bureaucracy hindered progress both in breaking Allied codes and in detecting flaws in their own codes.

Yet things were not so black and white. However much the fatuous Göring disdained science and technology, so did many Allied commanders. And Dönitz certainly did not share their folly; even Hitler was almost as enamored of "secret weapons" as Churchill was. Nazi Germany never had any difficulty in mobilizing the nation's scientific and engineering talent in the production of jet engines and rockets; during the war Germany led the world in both fields. And when it came to bureaucratic internecine warfare it was hard to find the equals of the U.S. Army and the U.S. Navy: At one point in the war the Army discovered that the Navy had for six months refused to pass on to Arlington Hall decrypts of Japanese naval codes that would have been invaluable in breaking a Japanese Army code that contained frequent reencipherments of those same texts. At Bletchley Park, the U.S. Navy had even sought to bar U.S. Army officers from entering areas where work on the naval Enigma was taking place.

The German code breakers were certainly not backward when it came to technical understanding. They were fully familiar with IBM methods, and as early as 1943 were pointing out that it was theoretically possible to recover the settings of the steckered Enigma if a crib were available. In 1944 Lieutenant R. Hans-Joachim Frowein of the Naval High Command's communications security branch worked out a method that recorded a catalogue of Enigma rotor transformations on seventy thousand IBM cards; he showed how the cards could then be employed (in a manner analogous to Knox's baton method) to partially encipher plain text and matching cipher text and scrutinize the resulting pairs of encipherments for logical contradictions. (This method of "scritching" was applied late in the war by Arlington Hall's cryptanalytic computers known as the Autoscritcher and Superscritcher.) So at least some in the German code bureaus were perfectly aware of the Enigma's inherent flaws.

Improvements to the Enigma that would have overcome its major weaknesses were proposed throughout the war. In 1944 the Luftwaffe began to introduce a pluggable reflector that could be reconfigured at will, and that indeed caused no end of trouble to the Allied code breakers. Several special-purpose machines, including the Arlington Hall scritchers, were developed to try to cope with the problem; the standard method of attack required a hundred-letter crib, and even then there was no assurance of success. Another change came on July 10, 1944, when without warning some Luftwaffe keys suddenly became unreadable. A decrypt that day referred to certain messages being enciphered with a previously unknown device called "Enigma Uhr." A Hut 6 team worked forty-eight hours nonstop and determined that the Uhr—which means "clock" in German—was probably an attachment that would plug in to the stecker board and, by means of a rotating switch, generate forty different permutations among the ten selected steckerings for the day. Some of these permutations were nonreciprocal, and this was something entirely new: B might be steckered to E while E was steckered to H. For the code breakers, this meant it was now impossible to use the bombe with the diagonal board. The bombe could still be used, but with the diagonal board disconnected, much longer than normal cribs were necessary.

Yet these improvements were introduced halfheartedly, belatedly, and piecemeal, and throughout the war the German cryptographic services showed a confusion over what the Enigma's security really depended upon.

The Enigma as a cryptographic device occupied a strip of ground halfway between the traditional and modern worlds of cryptography, and therein the confusion lay. The Enigma was modern in conception in the way its exceedingly long key period was designed to foil a statistical attack; yet it was conventional in the way it also relied upon secrecy of design and in the means it provided (such as the ring setting) to disguise from prying eyes its daily setting. Many of the security scares the Germans faced were answered with improvements in the physical and "secrecy" security of the machine while doing nothing about its statistical insecurity. Indeed many of these measures actually made matters worse. Rules that forbade the use of the same wheel in the same position on successive days, or that required that the same starting-position letter not be used more than once a month in *Offizier* keys, made sense if what one feared were some spy or traitor glancing at a machine's settings one day, but all they did as far as Bletchley Park was concerned was to make their job easier by eliminating the number of possibilities that they needed to try as a month wore on. Today, encryption schemes depend soley on statistical security; the basic notion is that one can even make the encryption method public, and just make sure that there are too many combinations for any computer to run through in a reasonable length of time. So in some ways the Germans actually paid a price for being pioneers; they adopted machine ciphers too early, committing themselves to a design that even as it strode into the new cryptographic age kept one foot dragging behind in the old.

And yet there remains the incontestable fact that, in the code war, the Allies triumphed and the Axis faltered. The Allies made mistakes and recovered; the Axis did not. The Allies began the war with a neglect of and a disdain for signals intelligence every bit as great as Germany's and Japan's; within a few years they had repaired their past folly, while the Germans and Japanese largely failed to. Dönitz was the exception rather than the rule among Axis commanders in the importance he attached to intelligence; most German and nearly all Japanese commanders never even seemed to grasp the concept. The Japanese Navy disbanded its intelligence service in 1932, and even after it was revived five years later there was never any evidence that Japanese commanders paid it much heed. Japanese code breaking operations throughout the war remained decentralized and fragmented. Japanese cryptanlaysts did succeed in reading Chinese military and diplomatic codes, some low-level British weather and merchant-ship codes, and—probably their most significant success—American aircraft movement codes, which reported on the numbers and types of planes departing from and arriving at airbases throughout the Southwest Pacific Theater. Most of the work was carried out by separate Special Intelligence Sections attached to each Area Army and Air Army, though, and they never achieved the level of coordination or the critical mass of concentrated talent and effort necessary to break the high-level Allied codes.

It was hardly a coincidence that when the Allies faltered, as they did in failing to heed the warnings of the Ardennes offensive, they did so when swollen-headed with success and overconfident of their purely military prowess. But that is a failing that totalitarian and militarist regimes are far more liable to. In the end, totalitarian nations on the offensive tend to believe their own propaganda of invincibility and national superiority. Cryptanalysis is an act of supreme arrogance tempered with supreme humility: the belief that one's enemies' most closely guarded secrets can be prised open by an exercise of pure brain power, and the belief that those secrets are worth knowing.

Chronology

	Cryptanalysis	The War
PREWAR		
1923	Enigma machine exhibited at International Postal Union Congress, Bern, Switzerland	
January 1924	Laurance Safford appointed first head of U.S. Navy's Research Desk	
1924	First U.S. Navy intercept station established, in Shanghai	
1925	Joseph Rochefort assigned to Research Desk	
1928	"On-the-Roof Gang" classes in kana Morse code begin	
May 1929	"Black Chamber" shut down	
December 1932	Marian Rejewski recovers wiring of German military Enigma rotors	
30 January 1933		Hitler becomes Chancellor
1936	Japanese Red machine broken by Agnes Driscoll	
15 September 1938	New indicator system for German Army Enigma introduced	
29–30 September 1938		Munich conference
15 December 1938	Two new rotors for Army Enigma introduced	

	Cryptanalysis	*The War*
1939		
15 March		Germans invade Czechoslovakia
1 June	New Japanese naval code, JN-25, introduced	
20 February	Purple machine begins to replace Red on Japanese diplomatic circuits	
24 July	Meeting at Pyry between Polish, French, and British code breakers	
1 September		Germans invade Poland
3 September		France and Britain declare war
1940		
17 January	Poles break first wartime Enigma setting, using hand methods (Zygalski sheets)	
12 February	Enigma rotors VI and VII recovered from *U-33*	
14 March	First bombe operational at Bletchley Park	
9 April		Germans invade Norway
10 April	Yellow Enigma key appears and is broken	
26 April	Trawler *Polares* captured	
1 May	Double encipherment of Enigma indicators ceases	
May–July	Naval Enigma traffic for 22–27 April read using *Polares* materials and first bombe	
10 May		Germany attacks in West; Churchill becomes PM
21 May	Traffic in main Luftwaffe Enigma key (Red) read currently	
22 June		France capitulates
Summer		Battle of Britain
August	Rotor VIII captured	

	Cryptanalysis	The War
8 August	Bombe No. 2 (with diagonal board) installed	
September	Purple broken by U.S. Army. First JN-25 decrypts by OP-20-G	
1 December	JN-25B appears	
December	GC&CS breaks main Abwehr hand cipher (ISOS)	
1941		
January	Sinkov mission to GC&CS	
12 February		Rommel arrives in Tripoli
28 February	Luftwaffe Enigma key for North Africa (Light Blue) broken	
4 March	Trawler *Krebs* captured off Lofoten Islands; naval Enigma settings for February taken	
13 March	Hut 3 begins sending decrypts to Cairo	
24 March	First Wrens assigned to work on bombes	
28 March		Battle of Cape Matapan
March	Naval Enigma messages read for February using Lofoten material	
3 April		Churchill writes to Stalin warning of Hitler's intentions
7 May	Trawler *München* captured	
9 May	*U-110* captured	
May	Naval Enigma traffic read for 21 days with 3–7 days' delay	
15 May	Rochefort assumes command of Station Hypo	
27 May		*Bismarck* sunk
June	Naval Cypher No. 3 introduced. Naval Enigma traffic read currently	

	Cryptanalysis	*The War*
22 June		Germany attacks Russia
28 June	Trawler *Lauenburg* captured	
July–August	SS and Police decrypts reveal German atrocities on the Eastern front	
16–22 August	Denniston visits Washington	
6 September	Churchill visits Bletchley Park	
17 September	Chaffinch, Army Enigma key for North Africa, broken	
5 October	Separate Triton ("Shark") Enigma key introduced for Atlantic U-boats	
21 October	Hut 4 and Hut 8 appeal to Churchill	
19 November	Dönitz's log reviews possible source of leaks	
December	Abwehr Enigma (ISK) broken	
7 December		Pearl Harbor
1942		
January	Germans obtain key to U.S. military attaché cipher in Cairo	
1 February	U-boats begin using four-wheel (M4) Enigma on Atlantic key (Shark)	
15 February		Singapore falls
18 March	Current reading of JN-25B begins. Brazil rounds up German agents	
6 May		Corregidor surrenders
15 May	Kullback arrives at GC&CS as observer	
26 May		Rommel launches attack at Gazala
4 June		Battle of Midway
30 June		Rommel halted at El Alamein

	Cryptanalysis	*The War*
11 July	GC&CS permits Enigma decryption in Cairo	
Summer	Arlington Hall begins traffic analysis on Japanese Army messages	
8 August		Montgomery appointed commander of Eighth Army
31 August		Battle of Alam el Halfa
4 September	U.S. Navy approves construction of bombes	
October	Agreement on Enigma cooperation reached between GC&CS and OP-20-G	
23 October		Battle of El Alamein begins
29 October	U-559 captured	
8 November		TORCH, Allied landings in North Africa
13 December	Initial break of M4 Shark, using short weather code book from *U-559*	
1943		
14 January		Casablanca conference
30 January		Dönitz becomes C-in-C of German Navy
31 January		Germans surrender at Stalingrad
15 February	Final Floradora additives recovered	
April	Initial break in Japanese Army Water Transport code	
18 April	Yamamoto killed	
May	Heath Robinson delivered	
3 May	First two U.S. bombes at Dayton begin tests	
17 May	BRUSA agreement signed	

	Cryptanalysis	The War
24 May		Dönitz withdraws U-boats from North Atlantic
22 June	First result from U.S. Navy bombes sent to GC&CS	
10 July		Allied landing on Sicily
31 August	First U.S. Navy bombes arrive in Washington	
1944		
13 January		First V-1 hits London
19 January	Japanese code books captured in New Guinea	
February	Colossus Mark I delivered	
25 February	Additional fifty bombes ordered by U.S. Navy	
May	First use of RAM on Japanese Army problems	
6 June		D-Day
20 July		Failed attempt on Hitler's life
1 September	Remaining twenty-five Navy bombes on order canceled	
8 September		First V-2 hits London
November	Initial VENONA break	
16 December		Ardennes offensive
1945		
January	German one-time pad (GEE) broken	
12 April		Roosevelt dies
30 April		Hitler commits suicide
7 May		Germany surrenders
6 and 9 August		Atomic bombs dropped
14 August		Japan surrenders

APPENDIX B

Naval Enigma:
Its Indicating System and the
Method of "Banburismus"

The difficulties that Hut 8 faced in attempting to break the German naval Enigma in 1939 and 1940 were primarily due to its much more secure indicating system. Unlike the Luftwaffe and Army systems, the naval Enigma used a procedure that effectively prevented any information about the machine's setting from being revealed in the indicators themselves—unless one possessed a set of external coding tables used specifically for the indicators. After the first message on a Luftwaffe or Army key was broken each day, all other messages on the key were instantly readable. That was not the case with the main naval keys.

In the system adopted by the German Navy on May 1, 1937, the operator would select two three-letter groups from a printed list—a "procedure indicator group" and a "cipher indicator group." The two groups (in this example, VFN and HYU) would be written one above the other and a random letter of padding added to each:

X H Y U
V F N K

The letters were then split into vertical bigrams:

XV HF YN UK

and each bigram replaced with the one specified in a bigram substitution table that was in force for that day:

BM OG PY UD

BMOG and PYUD would then be transmitted at the start of the message. To determine the starting position of the rotors to be used for enciphering the actual message text, the procedure indicator group, VFN, would then be typed into the Enigma with its rotors set at the *Grundstellung*—the ground setting, or "Grund" specified on a daily key list. The three-letter group that emerged from this encipherment, say SPL, would be the position that the rotors were then set to for enciphering the text. The robust security of the system was a result of two different encipherment systems being employed: the bigram tables to encipher the indicator, and the Enigma to transform the unenciphered indicator into the actual message setting:

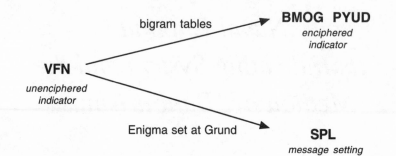

The bombe, when a successful crib was available, allowed Hut 8 to recover the stecker, wheel order, and message setting for an individual message. But that alone did not unlock the key to the indicating system that would allow the code breakers to directly decipher further messages from the same day. Without both the bigram tables and the Grund, the indicators transmitted with each message remained indecipherable.

Turing's "EINS" catalogue provided a means to recover further messages without having to start from scratch with a new bombe run; since the stecker and wheel order remained in force for a full day, only the message setting had to be found. At first the EINS catalogue was created by hand, by setting the rotors of the Enigma replicas to each of the 17,576 successive starting positions and typing in the word "EINS," using the stecker and wheel order of the day that had been located by the bombe run. Later a machine known as "Baby" automated the task. Baby was a mini-bombe of four Enigma machines set to consecutive positions; it was automatically rotated through all possible start positions and, in a later version, punched the encipherment of "EINS" at each position directly onto IBM cards. The cards were then sorted to created a printed catalogue in alphabetical order of all the tetragrams representing the 17,576 possible encipherments of EINS. The comparison of the cipher text of a message with the catalogue still had to be carried out by eye.

If the bigram tables were known and if enough messages were EINSed from a single day, it was possible to recover the Grund either with the bombe or by hand cribbing. The procedure was this: The bigram tables provided the unenciphered procedure indicator group for each message; the EINS method revealed the actual message setting used to encipher the message; thus for each message one had a matching piece of plain text (the unenciphered procedure indicator—VFN in the above example) and its corresponding cipher text (the actual message setting—SPL in the above example) for the Enigma set at the Grund. Several such cribs would provide enough data for a menu that could be used to establish at what position— the Grund—an Enigma would effect such plain-text–cipher-text substitutions.

Even before the first message of the day had been broken, knowing the bigram tables could be of enormous benefit in reducing the number of wheel orders that would have to be run on the bombe. This remarkable discovery, which Turing made in 1939, was the basis of the "Banburismus" procedure he devised, which would play a central role in the breaking of the naval Enigma until the U.S. Navy's production of a large number of bombes in 1943 made it no longer necessary to husband bombe time. Turing recognized that if two unenciphered procedure indicators were of the form ABX and ABY, then although one did not know what message settings these would be transformed to by the Enigma set at the (unknown) Grund, there was one thing that *was* certain: A in each case would be transformed to the same letter α and B would be transformed to the same letter β; thus the actual message settings for the two messages would have to be something like ZYA and ZYQ. That meant that their rotor starting positions were within twenty-six positions of one another.

The trick was to find two messages with such nearby indicators and then attempt to establish precisely how far apart their starting positions were by using the principle of "depth": The cipher texts of two properly aligned messages would have a higher index of coincidence than two randomly aligned messages—that is, the odds would be better than the merely random, one in twenty-six chance that the same letter of cipher text would appear simultaneously in the same position in both messages when properly aligned. The procedure was to punch messages that possessed nearby starting positions on long sheets of paper. The special paper used for this came from the town of Banbury, hence the pseudo-Latin term "Banburismus." An A to Z alphabet ran vertically on the sheet, and the message itself ran horizontally; a hole was punched at the proper spot to indicate which letter was in which position. By superimposing two Banbury sheets and sliding them relative to one another, overlaps would be immediately apparent wherever light shone through. If a series of two or three or more successive positions matched up, that was even more encouraging confirmation of proper depth; it suggested the same combination of letters or even the same entire word had been enciphered with the same stretch of underlying key. The search was extended by combing through all messages, whether they shared obviously nearby procedure indicators or not, for more certain signs of depth, such as repeated tetragraphs, pentagraphs, and higher hits; this was done using IBM cards and sorters.

The number of letters by which one message had to be shifted relative to another in order to place it in depth gave a true reading of the difference in their actual rotor starting positions. Thus if a message with the unenciphered indicator ABX had to be slid five positions to the left of a message with indicator ABY to align them in depth, it was clear that the actual message settings (produced by enciphering ABX and ABY at the Grund) had their third letters spaced five apart in the alphabet—something like ZYA and ZYF, or ZYB and ZYG, or ZYC and ZYH, and so on. Whatever the absolute message settings were, they had to be five letters apart.

This made it possible to begin constructing the cipher alphabet of the Enigma at position three of the Grund—that is, the plain-text–cipher-text substitutions the Enigma performed upon the third letter of each unenciphered indicator. If several interlocking depths were obtained in messages via the Banbury method, then the relative positions of several letters in the cipher alphabet were established, and they could be slid together along the plain-text alphabet until a position with no "crashes" occurred—that is, where there were no contradictions between reciprocal pairs of plain-text and cipher-text letters, such as a letter being enciphered by itself. For instance, if the Banbury sheets determined the following relative positions of pairs of messages:

message pair	*relative distance apart*
(unenciphered indicators)	*(determined by Banbury sheets)*
QWR, QWX	5
BBC, BBE	2
RWC, RWL	13
PNX, PIC	5

then the relative positions of R, E, C, L, and X could be placed in the cipher alphabet like so:

A B C D E F G H I J K L M N O P Q R S T U V W X Y Z
R . . . **X** . . . . . **C** . **E** . . . . **L** . .

The reciprocal nature of the Enigma meant that if X was transformed into F, then F had to be transformed into X; thus the position above was not possible—it has a crash (F goes to X but X goes to L). Sliding the two alphabets until a noncrashing position occurred yielded this possible solution:

A B C D E F G H I J K L M N O P Q R S T U V W X Y Z
. . . **C** . **E** . . . . **L** . **R** . . . . **X** .

which could be then further completed and tested for consistency by adding the reciprocals implied by the letter pairs:

A B C D E F G H I J K L M N O P Q R S T U V W X Y Z
. . **D** **C** **F** **E** . . . . . **Q** . . . **L** T U **R** S . . Y **X** .

The true identity of the third letter of each message's starting position was thus established: The message with unenciphered indicator RWC, for example, had an actual starting position of the form αβD (since, according to the reconstructed cipher alphabet above, C is transformed to D by the Enigma at the third position of the Grund); RWL had an actual starting position of αβQ (since L is transformed to Q).

The brilliant part of Turing's method was the next leap. The turnover notch was located at a different position on the various rotors. On rotor IV, for example, it occurred between J and K. That meant that if rotor IV was in the rightmost slot, then when the message that began at rotor setting αβD was typed in, the middle rotor would turn over after six letters of the message had been entered. If, for example, the actual rotor starting positions for the two overlapping messages αβD and αβQ were ABD and ABQ, then the rotor settings at successive positions in the two messages would be as follows:

```
1  2  3  4  5  6  7  8  9  10 11 12 13 14 15 16 17

A  A  A  A  A  A  A  A  A  A  A  A  A  A  A  A  A
B  B  B  B  B  B  B  B  C  C  C  C  C  C  C  C  C. . .
D  E  F  G  H  I  J  K  L  M  N  O  P  Q  R  S  T

                                    A  A  A  A
                                    B  B  B  B. . .
                                    Q  R  S  T
```

But this is impossible, since it means the turnover of the middle wheel would have occurred in the first message before it reaches its point of overlap with the second message, thirteen letters ahead: When the first is at setting ACQ, the second is at ABQ. The coincidence in cipher text discovered by the Banbury sheets would thus not have occurred, since the overlapping text would have been enciphered at a different rotor setting in the two messages. So rotor IV can be eliminated from consideration as the rightmost rotor.

The variation in the location of the turnover notch among the various rotors was obviously intended by the Germans to be a further source of cryptographic bafflement. But in fact it proved to be a terribly exploitable weakness. Had all eight rotors been made with the same turnover position, Turing's scheme would have been useless. In fact, the three rotors used exclusively by the German Navy, rotors VI, VII, and VIII, all were given the same turnovers (actually, they each had *two* turnover notches, between M and N and between Z and A). The notches for rotors I through V fell at Q–R, E–F, V–W, J–K, and Z–A, for which the official Hut 8 mnemonic was Royal Flags Wave Kings Above.

With a lot of luck, it was possible to repeat the process of elimination by analyzing turnovers of the left wheel, thereby establishing the identity of the middle wheel. Thus the 336 possible wheel orders could be reduced to as few as 6 that actually needed to be tested on the bombe. More usually, the procedure narrowed the right wheel down to one or three possibilities, leaving $7 \times 6 = 42$ or $3 \times 7 \times 6 = 126$ possibilities to be tested, still a substantial reduction from 336.

APPENDIX C

Cryptanalysis of the Purple Machine

It took Frank Rowlett and his coworkers at the U.S. Army's Signal Intelligence Service only a few months to determine that the Japanese Purple machine, like its predecessor the Red machine, employed two separate encryption channels. Six selected letters were connected via a plugboard to one scrambler; the remaining twenty letters went to another. The plugging was changed daily. But the "sixes" could usually be identified quite directly by making a frequency count of the cipher text. Since the sixes were scrambled among themselves, the frequency of each of these letters in the cipher text was equal to the average frequency of the six letters in the underlying plain text. Likewise, the frequency of each of the "twenties" was equal to the average plain-text frequencies of those letters. Picking six letters at random out of the alphabet almost always resulted in a set whose average plain-text frequency was either greater or less than that of the remaining twenty; thus the frequency of letters in Purple cipher text tended to fall into two distinct distributions.

Using a large amount of matching plain and cipher text, the SIS cryptanalysts were able to recover the underlying pattern by which the letters of the sixes were scrambled at each successive key position. This proved to be a 6 × 25 substitution table. Once this scrambling pattern was determined, it became a straightforward matter to identify which of each of the day's sixes corresponded to which letters in the substitution table.

For example, if the scrambling pattern were:

		plain text					
		a	b	c	d	e	f
key position	1	F	A	E	B	C	D
	2	D	A	C	F	E	B
	3	D	E	F	C	B	A
	4	C	E	B	F	A	D
	.	.	.	.	.	.	.

and if the frequency count of the cipher text had identified A, B, D, H, O, and R as the "sixes," the next step would be to make a separate frequency count of these letters at each key position of the message. The first, twenty-sixth, fifty-first, and

seventy-sixth letters were all enciphered at the same key position; likewise the second, twenty-seventh, fifty-second, and seventy-seventh; and so forth. Multiple messages from a single day (which all used the same plugging) could be grouped
together once the indicator system, which revealed which key position each message
had been started with, had been cracked. If the frequency count for key positions 1
through 4 shows the following:

key position	cipher text character					
	A	B	D	H	O	R
1	2	7	15	10	8	1
2	2	0	6	14	9	3
3	4	19	7	3	14	1
4	9	18	0	25	11	1

it is possible to begin making assignments using a knowledge of the actual frequency
of the letters *a, b, d, h, o, r* in romanized Japanese text. The letter *o* is by far the most
common of these six, with *a* next. Thus a likely solution is that for key position 1, *o* =
D; for position 2, *o* = H; for 3, *o* = B; and for 4, *o* = H. The substitution table has only
one column that fits this pattern, namely the fourth (in which d is transformed in sequence to BFCF; this is isomorphic to *o* being transformed to DHBH). Likewise, *a*
would appear by the frequency counts to be HOOB in these four successive key positions, which matches the pattern of the first column of the substitution table (*a* goes
to FDDC). With a few more frequency assignments, the full identities of the letters
in the substitution table can be nailed down, thus:

key position	plain text					
	a	d	b	o	r	h
1	H	A	R	D	B	O
2	O	A	B	H	R	D
3	O	R	H	B	D	A
4	B	R	D	H	A	O
.	.	.	.	.	.	.

The scrambling pattern of successive key positions in the "sixes" substitution table
was not one that could have been produced by a rotor, as in the Enigma machine.
Each key position was totally unrelated to the previous one. The hardware that
could produce such a scrambling pattern remained a puzzle until Leo Rosen's
chance discovery of the telephone "uniselector" switches provided the answer.
These consisted of six switches ganged together. At each of twenty-five steps they
connected the six input letters to the six output letters in a different arrangement.

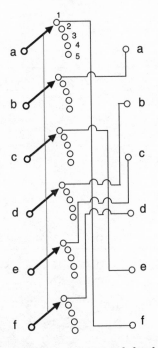

The wiring of a small section of the stepping switch for the "sixes" channel of the Purple machine, illustrating the interconnections for key position 1 in the substitution table above. All six switches move in tandem through twenty-five positions, then repeat the cycle.

Recovering the substitution pattern for the twenties proved far more difficult. The key length was considerably longer than twenty-five. Genevieve Grotjan's remarkable discovery was that although the substitution tables changed after every cycle of twenty-five letters, the tables formed by each cycle were all interrelated by certain patterns.

Eventually two such patterns of interrelationships emerged. In one, the identities of the letters in the table would change after each cycle of twenty-five key positions, but the way the letters hopped around from one line to the next within each table was consistent:

cycle 1

```
         a b c d e f g h i j k l m n o p q r s t u

      1  U A F T P B F Q R L O C M D G K S H N J I
key
position 2  B E D T I L P A C U S K N F H R G J M O Q

      3  A I K P O U T R E Q S D F G B H N J M C L
      .  . . . . . . . . . . . . . . . . . . . . .
      25 M N B C L H D P I T E R Q A F O G K U S J
```

cycle 2

```
         a b c d e f g h i j k l m n o p q r s t u

      1  H C A I G J D O M U F P T R Q E L S N B K

      2  J A R I K U G C F H L E N D S M Q B T F O

      3  C K E G F H I M A O L R D Q J S N B T P U
      .  . . . . . . . . . . . . . . . . . . . . .
      25 T N J P U S R G K I A M O C D F Q E H L B
```

The circled and squared letters in both cases followed the same sequence of movements from one line to the next; so did the letters in all the other positions. These tables, in other words, were *isomorphs* of one another.

In the other pattern, entire columns of the encipherment tables were found to transpose in each subsequent cycle:

cycle 1

key position	a	b	c	d	e	f	g	h	i	j	k	l	m	n	o	p	q	r	s	t	u
1	U	A	E	T	P	B	F	Q	R	L	O	C	M	D	G	K	S	H	N	J	I
2	B	E	D	T	I	L	P	A	C	U	S	K	N	F	H	R	G	J	M	O	Q
3	A	I	K	P	O	U	T	R	E	Q	S	D	F	G	B	H	N	J	M	C	L
.	.	.	.	.	.	.	.	.	.	.	.	.	.	.	.	.	.	.	.	.	.
25	M	N	B	C	L	H	D	P	I	T	E	R	Q	A	F	O	G	K	U	S	J

cycle 2

	a	b	c	d	e	f	g	h	i	j	k	l	m	n	o	p	q	r	s	t	u
1	A	P	R	I	J	E	L	D	F	U	T	S	O	Q	M	K	B	N	G	C	H
2	E	I	C	Q	O	D	U	F	P	B	T	G	S	A	N	R	L	M	H	K	J
3	I	O	E	L	C	K	Q	G	T	A	P	N	S	R	F	H	U	M	B	D	J
.	.	.	.	.	.	.	.	.	.	.	.	.	.	.	.	.	.	.	.	.	.
25	N	L	I	J	S	B	T	A	D	M	C	G	E	P	Q	O	H	U	F	R	K

Such patterns are exactly what one would expect when two or more alphabetic substitution tables are applied sequentially to plain text in a cyclic fashion. That is, the first letter of text is enciphered by line 1 of table 1, and the result enciphered by line 1 of table 2; the next letter by line 1 of table 1 and line 2 of table 2; and so on through lines 1 and 25; then through all twenty-five lines of table 2 again but using line 2 of table 1. When table 1 is the "slow" cycle, the isomorph pattern results. When table 2 is the slow cycle, the columnar transposition results.

Further analysis showed that the Purple machine used a cascade of three substitution tables for the "twenties." In the actual machine, these substitutions were effected by three banks of four uniselectors each; the four provided 6 × 4 = 24 ganged switches, allowing for the twenty letters, plus four extra switches that were used as control circuits. Any of the three banks could be set to be the fast, medium, or slow cycle. Although that added to the number of permutations the machine as a whole generated, it was a weakness akin to the changeable wheel order in the Enigma machine, for it provided a cryptanalyst further diagnoses of the wiring of each uniselector bank.

The Intercept Network

Shifting priorities and fortunes of war led to continual changes in the British and American intercept networks. In late 1943 OP-20-G was operating about 445 receivers for intercepting high-level Pacific traffic (the number would grow to about 775 by the end of the war) at four fixed sites:

Bainbridge Island, Washington	120 receivers
Imperial Beach, California	75
Wahiawa, Hawaii	200
Australia	50

German U-boat traffic was being intercepted at Chatham, on Cape Cod, Massachusetts; in addition, a network of direction-finding stations ran all the way from Greenland to Brazil. Other D/F stations were located on Pacific islands.

The U.S. Army network at this same time consisted of six fixed stations that concentrated on Japanese military signals and Axis diplomatic traffic:

Vint Hill, Warrenton, Virginia
Two Rock Ranch, Petaluma, California
Asmara, Eritrea
Fort Shafter, Territory of Hawaii
Fairbanks, Alaska
New Delhi, India

GC&CS in mid-1943 was receiving traffic from a far-flung network in the United Kingdom.

Royal Navy	
Scarbora	72 receivers
Flowerdown	45
Chicksands	16
Cupar	15
Shetlands	2

Army

Beaumanor	140
Harpenden	23
Keddleston Hall	36
Mobile units	6

RAF

Chicksands	105
Cheadle	75
Kingsdown	15
Waddington	24
Tean	19
Wick	14

Foreign Office

Brora	14
Cupar	13
Denmark Hill	23
Sandridge	54
Whitchurch	40
Knockholt	35

Post Office

St. Albans	8
Coastal stations	9

This was supplemented by overseas stations in Canada (Ottawa, Winnipeg, Grande Prairie, Point Grey, Victoria), Malta, Gibraltar, Alexandria, Cairo, Baghdad, and other sites in Egypt, South Africa, West Africa, and India.

Rapid Analytical Machinery (RAM)

British and American cryptanalysts made extensive use of IBM punch-card equipment during the Second World War; one of the chief applications of IBM machines was to create numerically ordered lists of sequences of message groups that had appeared in enciphered codes as a preliminary step toward placing messages in "depth."

The first general-purpose computers did not appear until after the war, and the IBM card methods, although surprisingly powerful, had distinct limitations; in particular they could not compare or tally large numbers of different variables simultaneously. A variety of single-purpose computing machines were thus devised to handle some of the special cryptanalytic problems that arose. None of these were "computers" in the modern sense, nor were they even lineal ancestors of the computer, although they did pioneer some component technologies (such as electronic counting and memory storage circuits) that would play a part in the first digital computers.

BOMBES

Standard:
Contained thirty-six three-wheel Enigmas (thirty in the first models)
142 built, by British Tabulating Machine Company

Test Plate ("Baby"):
Four three-wheel Enigmas set to consecutive rotor positions
Used to compile "EINS" catalogue for recovering the settings of naval Enigma messages after the first message of the day was broken: The machine would encipher the word EINS at all 17,576 initial rotor positions for a given wheel order, stecker, and ring setting and punch the results on IBM cards

Jumbo:
Standard bombe with the addition of the "machine gun," which automatically checked each stop for consistent steckers and printed out the results; allowed weaker menus to be run effectively
Fourteen built

Cobra ("Wynn-Williams" type):
A high-speed fourth wheel plus a vacuum-tube sensing unit plugged to a Standard
   bombe; a stopgap intended to handle the four-wheel naval Enigma problem
Designed by General Post Office engineers; proved unreliable in use
Twelve built

High-Speed Mammoth ("Keen" type):
A complete four-wheel bombe for the naval problem; used high-speed relays to
   sense stops
Fifty-seven built

U.S. Navy N-530 and N-1530:
Sixteen four-wheel Enigmas; high-speed bombe with electronic stop sensor
About 125 built, by National Cash Register

003 ("Madame X"):
144 three-wheel Enigmas; used telephone switching relays in place of rotors
One built, by AT&T Bell Laboratories for U.S. Army; mainly used for research

Autoscritcher and Superscritcher:
Electromechanical (Autoscritcher) and fully electronic (Superscritcher) computers
   used to attack Enigma with pluggable reflector
Built by U.S. Army; Autoscritcher was extremely slow; Superscritcher, completed in
   fall 1945, contained thirty-five hundred vacuum tubes

Duenna:
Small bombe unit to recover unknown settings of pluggable reflector
Several built for U.S. Navy

Bulldozer:
Thirty-two four-wheel Enigmas; "statistical bombe" for use when no crib was available, analyzed plain-text frequencies and stopped when a settable threshold was
   reached
One built, for U.S. Navy

Grenades:
Various attachments to bombes for finding settings when wheel order, stecker, and
   ring setting were known

FILM AND PAPER TAPE COMPARATORS

I.C. Machine:
Used to place cipher text in depth; $3\frac{1}{2}$ x $1\frac{1}{2}$ inch photosensitized glass plates
   recorded six hundred characters of message text; used single photocell to measure
   amount of light emerging from superimposed plates as their relative position was
   shifted
Built by Eastman Kodak

70 mm Comparator:

Messages were punched on wide (70 mm) paper tape, formed into loops up to twenty-four feet long containing up to seventeen hundred characters of message text, and fed through two reader heads at all possible relative offsets; electronic counters and a printer tallied coincidences at various intervals specified by plug-board

Built by National Cash Register and Gray Manufacturing

Hypo:

Used mainly for breaking duds and *Offizier* messages in naval Enigma traffic; 35 mm photographic film recorded the encipherment of one to five high-frequency letters (such as E, R, N) at all positions for a given stecker, wheel order, and ring setting and then compared with a message-text film

Copperhead:

Punched 70 mm tape comparator; optically compared a sequence of one hundred code groups at a time by testing for "blackout" of superimposed message tapes prepared with complementary coding; used to place enciphered codes in depth

Tessie:

Used to search for symmetrical sequences in cipher text; 35 mm film for two messages was automatically scanned to produce punched tape that recorded incidents of repeated letters within twenty positions of one another; the resulting punched tapes, which recorded the interval between repeated letters within each message, were then superimposed and examined visually for certain patterns

5202:

35 mm film comparator used to slide enciphered German teletype text ("Tunny") against known key sequences; capable of complex comparisons and statistical tests

Built by Eastman Kodak for U.S. Army; one delivered to GC&CS in April 1945

OTHER PAPER TAPE EQUIPMENT

Matthew:

Relay-based machine to perform rapid noncarrying addition and subtraction for additive stripping; Letterwriter paper tape for input and output

Mike:

Digraph frequency counter; input from two Letterwriter tapes; relay matrix directed signals to one of 676 electric counters that tallied all 26 × 26 possible digraphs

COLOSSI

Robinson:

Built by GC&CS for attack on Tunny; teletype tapes for message text and key sequences were fed simultaneously through two readers in continuous loops that ran through all possible offsets; electronic circuits counted coincidences

Colossus:

Advanced version of Robinson in which the key sequence was encoded internally by
electronic circuits; contained twenty-five hundred vacuum tubes to perform memory and Boolean arithmetic operations

Notes

Abbreviations Used in Notes
AI   Author's interview
BI   *British Intelligence in the Second World War* (Hinsley et al.)
CAC   Churchill Archives Centre, Cambridge University
GC&CS   Government Code and Cypher School Official Histories of World
    War II, National Cryptologic Museum
HCC   Historic Cryptographic Collection, National Archives at College Park
NACP   National Archives at College Park
OH   Oral history
PRO   Public Record Office, Kew, U.K.

Full references to published and unpublished sources cited in short form in the Notes below may be found in the Bibliography.

PROLOGUE: MIDWAY

Page
1   Leading the attack: Morison, *U.S. Naval Operations,* III: 389–98.
2   "Shangri-La": Burns, *Roosevelt: Soldier,* 224.
2   Yamamoto had had his eyes on Midway: Prange, *Miracle at Midway,* 21–25.
3   "We can accomplish anything": Layton, *I Was There,* 453.
4   When Rochefort showed up: Layton, *I Was There,* 32–34; Rochefort, OH, 5–7.
4   stomach in knots: Rochefort, OH, 45.
4   naval intelligence slush fund: Laurance Safford, "The Undeclared War, History of R.I.," November 15, 1943, SRH-305, NACP, 1–2; Layton, *I Was There,* 29–31; "Naval Security Group History to World War II," SRH-355, NACP, 40.
6   "Rowlett, do you know what this means?" Rowlett, *Story of MAGIC,* 3–5.
6   aide interrupted the President: United States, *Pearl Harbor Attack,* 10: 4659–65.
7   signals were intercepted by U.S. Navy operators: "History of GYP-1," CNSG 5750/202, Crane Files, NACP, 5, 14. This document provides a complete account of the breaking of JN-25 from 1939 through the war.

8 first real break: "History of GYP-1," CNSG 5750/202, Crane Files NACP, 6, 11, 17.

8 desperately short of help: "History of GYP-1," CNSG 5750/202, Crane Files, NACP, 21–23.

8 nowhere near enough to read current traffic: "OP-20-GY," CNSG 5750/198, Crane Files, NACP, 7, 14, 20. Other contemporaneous or nearly contemporaneous documentary sources are in total agreement that no JN-25 traffic was being read in the months leading up to Pearl Harbor; see, for example, "Activities and Accomplishments of GY-1 During 1941, 1942 and 1943," CNSG 5750/197, Crane Files, NACP, which lists the number of JN-25 messages read in 1941 as "none." The two other high-level Japanese naval codes were not being successfully read at this time, either. In the months before the Japanese attack OP-20-G was also attempting to break the "MAT," or materials code; Station Hypo was assigned the dead-end task of tackling AD, the little-used administrative code. No progress on either of these systems was made. Station Cast in Cavite made some progress against the Japanese Navy's merchant shipping liaison code and did apparently extract a few fragments of naval ship movement reports from JN-25 messages but these did not yield any significant intelligence.

8 Conspiracy theorists: Claims that OP-20-G was reading the AN code before Pearl Harbor have been made by several authors, notably Robert B. Stinnett in *Day of Deceit.* This book, however, misinterprets and misunderstands the process of cryptanalysis—and disregards the most important parts of the documentary evidence relating to the breaking of the Japanese naval codes: notably the fact that the Japanese introduced a completely new AN code book in December 1940. Stinnett reaches the incorrect conclusion that, because OP-20-G was close to reading the AN code in October 1940, it must have been reading the code throughout 1941 as well. See Budiansky, "Too Late for Pearl Harbor."

8 month-by-month progress reports: "OP-20-GY," CNSG 5750/198, Crane Files, NACP.

8 orders to the Japanese fleet: "Pre–Pearl Harbor Japanese Naval Dispatches," CNSG 5830/115, Crane Files, NACP.

9 mail service from the Pacific: "History of GYP-1," CNSG 5750/202, Crane Files, NACP, 27.

9 "Too bad it didn't kill me": Miller, *War at Sea,* 206.

10 "this can't be true": Burns, *Roosevelt: Soldier,* 162.

10 MacArthur, apparently frozen: Manchester, *American Caesar,* 206–7.

10 "on the ground, by God": Burns, *Roosevelt: Soldier,* 165.

10 "best we had": United States, *Pearl Harbor Attack,* 26: 37.

10 worse blow than Pearl Harbor: Prange, *Miracle at Midway,* 5.

10 Summoned to the Dungeon: Prados, *Combined Fleet,* 187–88.

11 "I can offer a lot of excuses": Rochefort, OH, 110–11.

11 rent a few precious IBM machines: "Military Study: Communication Intelligence Research Activities," 30 June 1937, J. N. Wenger, SRH-151, NACP, 22–23.

11 hacking and coughing: Rochefort, OH, 123–27, 132–33; Layton, *I Was There,* 422.

11 "ideal of an ideal man": Webb, AI.

12 took anyone he could get: "Naval Security Group History to World War II," SRH-355, NACP, 445–47.

12 "current decryption": "History of GYP-1," CNSG 5750/202, Crane Files, NACP, 29.

13 would change it to six: Prange, *Miracle at Midway,* 19.

13 "burning the top of my desk": Layton, *I Was There,* 411–12.

13 believe enemy will attack: "The Role of Comint in the Battle of Midway," SRH-230, NACP, 7.

15 cold water: Memo for Comdr. Redman, April 18, 1942, "OP-20-GI Files R.I. Disseminated Externally Mar 42–May 42," CNSG 5750/171, Crane Files, NACP; "History of GYP-1," CNSG 5750/202, Crane Files, NACP, 39–42; Layton, *I Was There,* 413.

15 a Japanese deception: "History of GYP-1," CNSG 5750/202, Crane Files, NACP, 42; Prange, *Miracle at Midway,* 47.

16 Estimate of Situation: Miller, *War at Sea,* 245–46; Prange, *Miracle at Midway,* 181.

16 two weeks' supply of fresh water: "OP-20-G File of Memoranda and Reports Relating to the Battle of Midway," SRMN-005, NACP, 235; Layton, *I Was There,* 421–22.

16 intercept unit in Melbourne, Australia: Station Cast, which moved from Cavite Naval Base in Manila Bay to the fortress at Corregidor in December 1941, was evacuated to Australia before the American surrender in May 1942. The unit was redesignated FRUMEL—Fleet Radio Unit, Melbourne.

16 a message dated May 26: "History of GYP-1," CNSG 5750/202, Crane Files, NACP, 45.

17 air attacks against Midway: "Midway and Yamamoto: Properly Revisited," NR 4632, HCC, 19–20. Despite this intelligence suggesting June 4 as the day that the Japanese carrier attacks would begin, Nimitz, perhaps out of caution, sent orders to be prepared for preliminary attacks on the night of June 2/3 or the morning of June 3. I am indebted to John Lundstrom for pointing this out to me and for supplying me copies of messages from the Cincpac Secret & Confidential Message File, RG 313, NACP, that confirm this.

17 Tokyo Rose: Prange, *Miracle at Midway,* 77.

18 "five miles out": Layton, *I Was There,* 438. I am grateful to John Lundstrom for pointing out to me that Layton's prediction was made early on the morning of June 4, not on May 27 as recounted in Prange, *Miracle at Midway,* 102. Prange acknowledges in the endnotes to his book that Layton in fact told him that this incident took place on June 4, but Prange decided that Layton's memory was incorrect. Yet Layton's statement is corroborated by the fact that Cincpac messages continued to point to June 3 as the attack date right up until June 3. The incorrect date of Layton's famous prediction has been propagated by the odd fact that Layton's posthumous coauthors of *I Was There* decided to accept Prange's version rather than Layton's.

20 Shimatta: Prange, *Miracle at Midway,* 265.

20 sank into a chair stunned and speechless: Miller, *War at Sea,* 256–57.

22 "drunken brawl": Rochefort, OH, 265–66.

22 The honors for breaking: "History of GYP-1," CNSG 5750/202, Crane Files, NACP, 51–52.

23 "by virtue of seniority": "OP-20-G File, Communications Intelligence Activities, 1942 to 1946," SRH-279, NACP, 9–14.

23 denied Rochefort a medal: Layton, *I Was There,* 451; "OP-20-G Exploits and Commendations, World War II," SRH-306, NACP, 5.

23 took Holtwick's breath away: "The Inside Story of the Battle of Midway and the Ousting of Commander Rochefort," CNSG 5750/577, Crane Files, NACP, 4–5. This is the somewhat mysterious document that Layton refers to in *I Was There;* its origins and authorship are unclear, but internal evidence suggests it was written around 1944 by a naval officer closely involved in the matter. Some of the accusations made in the document, apparently based on hearsay, are clearly incorrect. But several specific incidents the paper recounts, including the story of Holtwick's meeting with Redman, bear the signs of firsthand knowledge; they ring true and are consistent with other established facts of the case. Layton says that marginal comments on the document correcting some of its errors are in Rochefort's handwriting. See Layton, *I Was There,* 581.

23 "one unforgivable sin": Layton, *I Was There,* 452.

24 club to beat Rochefort with: Layton, *I Was There,* 467–68.

24 "a very stupid thing": Rochefort, OH, 258–61, 268–70.

1. "NO GOOD, NOT EVEN FOR INTELLIGENCE"
Page

25 "so-so cryptanalyst": Rochefort, OH, 36.

25 "hanged at the yardarm": Dorwart, *Conflict of Duty,* 54.

25 talked the War Department into creating: Kahn, *Codebreakers,* 351–54.

26 staff of sixteen: The Cipher Bureau in New York: Administrative Problems, "Yardley, The Friedman–Yardley Correspondence," Document 205, Herbert O. Yardley Collection, NACP.

26 greatest coup: Yardley, *Black Chamber,* 312–13.

27 "highly unethical": For a full account of Stimson and the "gentlemen's mail" quotation, see Kruh, "Stimson."

27 James Thurber: Thurber, "Exhibit X."

28 late into the night: Andrew, *Secret Service,* 108.

28 read American diplomatic codes: DENN 1/4, A. G. Denniston Papers, CAC, 7.

28 "lots of fun": Memorandum for Colonel Clarke, June 15, 1943, "Army and Navy Comint Regs & Papers," NR 4632, HCC.

28 pro-American official: Layton, *I Was There,* 117.

29 70 percent of its funding: Kahn, *Codebreakers,* 360.

29 $7,500 a year: Kruh, "Stimson," 85.

29 offered a job in Washington: United States, *Pioneers in U.S. Cryptology,* 4.

29 resigning his commission as a major: Yardley to Captain M. F. Shepherd, February 24, 1931, "Officer Resignations (Herbert O. Yardley and Robert Anderson)," Document 206, Herbert O. Yardley Collection, NACP.

29 indignant headlines: "The Role of Communications Intelligence in the American–Japanese Naval War, Volume II," SRH-012, NACP, 103.

29 cloud of suspicion: Layton, *I Was There,* 41.

30   Subject Yardley: Matters Pertaining to Publication Authorized by Former Chief of Cryptanalysis of U.S. War Department's Intelligence Bureau, 10 June 1931, Chief of Telegraph Section, Foreign Ministry, "Docs re Yardley & Japan, ca 1930," Document 207, Herbert O. Yardley Collection, NACP.

31   William F. Friedman: Kahn, *Codebreakers,* 370–71.

31   "training" and "research": Code, Ciphers, Secret Inks, Radio Interception and Goniometry, memorandum from the Adjutant General to the Chief Signal Officer, April 22, 1930, "Radio Intelligence (R/I) Detachments File, 1930–1940," NR 2142, HCC.

32   greater threat: Rowlett, *Story of MAGIC,* 38.

32   They were three schoolteachers: "Notes on History of Signal Intelligence Service," NR 3245, HCC, 88–89; Kullback, OH; Rowlett, *Story of MAGIC,* 6–33.

32   deny they were working on anything: Rowlett, *Story of MAGIC,* 16; Code, Ciphers, Secret Inks, Radio Interception and Goniometry, memorandum from the Adjutant General to the Chief Signal Officer, April 22, 1930, "Radio Intelligence (R/I) Detachments File, 1930–1940," NR 2142, HCC.

32   only major employer of college-educated professionals: Dinnerstein, *Antisemitism in America,* 87–90.

32   editor of the school's yearbook: Dinnerstein, *Antisemitism in America,* 87.

33   Civil Service notice: Kullback, OH.

33   tasks of pure drudgery: Kullback, OH; Rowlett, *Story of MAGIC,* 52–54.

34   "experimental" monitoring station: "Notes on History of Signal Intelligence Service," NR 3245, HCC, 67–68.

34   plaintive appeals: Memorandum for G-2, Ninth Corps Area, J. O. Mauborgne, September 12, 1932, "Intercept/Crypto Correspondence 1927–1941," NR 2123, HCC; High Speed Recording Apparatus, memorandum from J. O. Mauborgne to Chief Signal Officer, January 19, 1933, "Intercept/Crypto Correspondence 1927–1941," NR 2123, HCC; Radio Intercept, memorandum from Wm. H. Wilson to Assistant Chief of Staff, G-2, 9th Corps Area, February 3, 1933, "Intercept/Crypto Correspondence 1927–1941," NR 2123, HCC; Rowlett, *Story of MAGIC,* 84.

35   full staffing: Coordination of the Activities of the Radio Intelligence Detachment, memorandum from Adjutant General to Commanding General, Ninth Corps Area, September 15, 1936, "Intercept/Crypto Correspondence 1927–1941," NR 2123, HCC.

35   small intercept units: "The Achievements of the Signal Security Agency in World War II," SRH-349, NACP, 11.

35   "in the near future": Monitoring Activities, memorandum from S. B. Akin to Signal Officer, Eighth Corps Area, October 17, 1939, "Intercept/Crypto Correspondence 1927–1941," NR 2123, HCC.

35   "these penalties apply": "Notes on History of Signal Intelligence Service," NR 3245, HCC, 76–77.

35   One naval officer was sent to Tokyo: Prados, *Combined Fleet,* 8.

36   The Navy set up: Layton, *I Was There,* 56.

36   typewriters that printed kana: "The Undeclared War, History of R.I.," SRH-305, NACP, 1.

37   "On-the-Roof Gang": "Naval Security Group History to World War II," SRH, 355, NACP, 40–41; Leek, "On the Roof Gang," 34–35.

37 five other naval officers: Prados, *Combined Fleet,* 77–78.

37 Agnes Meyer Driscoll: Layton, *I Was There,* 33; Kahn, *Codebreakers,* 415–18; United States, *Pioneers in U.S. Cryptology,* 22.

37 Kullback was sent to Hawaii: Kullback, OH.

38 Promotion up to the rank of colonel: Kahn, "United States Views Germany and Japan," 501.

38 smacked too much of Prussian militarism: Cline, *Washington Command Post,* 16–17.

38 Eisenhower . . . Patton: Manchester, *Glory and Dream,* 6.

38 Munitions Building, and the companion: Brinkley, *Washington Goes to War,* 53–54, 61, 74; Frank Lewis, AI.

39 single mechanized regiment: Manchester, *Glory and Dream,* 6; Millett and Maslowski, *Common Defense,* 382.

39 "Mfg USA 1863": Brinkley, *Washington Goes to War,* 57.

39 American Tank Corps: Millett and Maslowski, *Common Defense,* 381–82.

39 "construction worker": Brinkley, *Washington Goes to War,* 59.

40 eat hay and whinny: The officer was Arthur "Bomber" Harris, who later became Marshal of the RAF; see Bennett, *Behind the Battle,* 363.

40 "Bonus Army": Manchester, *Glory and Dream,* 12–18; Eisenhower, *At Ease,* 213.

40 Washington and London conferences: Millett and Maslowski, *Common Defense,* 364–65, 372–73.

41 had established intelligence divisions: Millett and Maslowski, *Common Defense,* 261; Dorwart, *Conflict of Duty,* 10–11, 89.

41 "stepchild": Kahn, "United States Views Germany and Japan," 501.

41 Parker's more able successors: Dorwart, *Conflict of Duty,* 37.

42 "I don't want to see you": Layton, *I Was There,* 39; Rochefort, OH, 61–62.

42 "good sense to get out of it": Layton, *I Was There,* 49.

43 "warts": Graves, *Good-bye to All That,* 88.

43 not-quite-right color of khaki: Sassoon, *Fox-Hunting Man,* 261.

43 unseamanlike bearing: DENN 1/3, A. G. Denniston Papers, CAC, 7; Andrew, *Secret Service,* 102.

43 to mount standing patrols: Andrew, *Secret Service,* 86.

44 "Thank God": Andrew, *Secret Service,* 102–3.

44 "very man I wanted": Andrew, *Secret Service,* 86.

44 Sir Alfred Ewing: Ewing, "Special War Work, Part I," 194–95; DENN 1/2, A.G. Denniston Papers, CAC, 1.

44 contact with an amateur radio enthusiast: DENN 1/2, A. G. Denniston Papers, CAC, 1.

45 *Telconia* began pulling up: Kahn, *Codebreakers,* 266; Andrew, *Secret Service,* 107–8.

45 "Ewing, Admiralty": Ewing, "Special War Work, Part I," 195.

45 sole qualifications: DENN 1/3, A. G. Denniston Papers, CAC, 1

45 "boxlike room": DENN 1/3, A. G. Denniston Papers, CAC, 4.

45 truth was more prosaic: Andrew, *Secret Service,* 89.

46 "entirely misdirected": Andrew, *Secret Service,* 89–90.

47 "the dreadful news": Fitzgerald, *Knox Brothers,* 91–92.

47 Knox eventually took on: Fitzgerald, *Knox Brothers.* 66–67, 72, 76–78, 109; Andrew, *Secret Service,* 94.

47 The sailor in Room 53: "Alice in I.D. 25," DENN 3/3, A. G. Denniston Papers, CAC, 20.

48 "with dripping hand": Churchill, *World Crisis,* I: 466: Andrew, *Secret Service,* 97–99.

48 "news of the kind that never fails": Ewing, "Special War Work, Part I," 198–99.

48 "see the Intercepts": Kahn, "Churchill Pleads for the Intercepts."

49 changed every night at midnight: DENN 1/3, A. G. Denniston Papers, CAC, 12.

49 copies of the book were salvaged: Kahn, *Codebreakers,* 273–74.

49 "Notbywit": Andrew, *Secret Service,* 124.

49 flagged chart: DENN 1/3, A. G. Denniston Papers, CAC, 11.

49 forbidden to communicate: Andrew, *Secret Service,* 121.

50 Battle of Jutland: Andrew, *Secret Service,* 104–6; Beesly, *Room 40,* 155.

50 "none too pleasant" tasks: "Naval Section of G.C. and C.S.," HW 3/16, PRO, 11. A copy of this document, which consists of Clarke's reminiscences, is also apparently among Clarke's papers at CAC and parts appear reprinted in Clarke, "Government Code and Cypher School," and Clarke, "The Years Between."

51 Courtenay Forbes: DENN 1/4, A. G. Denniston Papers, CAC, 2.

51 the ones to take charge: "Naval Section of G.C. & C.S.," HW 3/16, PRO, 10–11.

51 "little value": Historical Notes, HW 3/33, PRO.

51 Lord Curzon: "Naval Section of G.C. & C.S.," HW 3/16, PRO, 12.

52 Dilly Knox: Fitzgerald, *Knox Brothers,* 169–72.

52 American diplomatic codes: Clarke, "Government Code and Cypher School," 221.

52 "Public Benefactors": "Naval Section of G.C. & C.S.," HW 3/16, PRO, 9–10.

53 will go on for ever!: "Alice in I.D. 25," DENN 3/3, A. G. Denniston Papers, CAC, 49.

53 "violent" game of tennis: Letter, William F. Friedman to Denniston, 6 June 1944, DENN 6/6, A. G. Denniston Papers, CAC.

53 report their successes: Filby, "Bletchley Park and Berkeley Street," 277.

53 Hugh "Quex" Sinclair: Andrew, *Secret Service,* 294–95.

53 crocodile skin cigar case: "Naval Section of G.C. & C.S.," HW 3/16, PRO, 12–13.

54 every cable sent to or from the United Kingdom: DENN 1/4, A.G. Denniston Papers, CAC, 17–20.

54 spy on former allies: DENN 1/4, A. G. Denniston Papers, CAC, 6–12.

55 "Floradora": Filby, "Floradora and a Unique Break."

55 "all German codes were unbreakable": "History of Hut Eight," NR 4685, HCC, 14.

55 radio direction finding had been used: Ewing, "Special War Work, Part II," 35.

56 "sobering influence": Andrew, *Secret Service,* 243–45.

56 "which will pay me the most" . . . draw him into conversation: Filby, "Bletchley Park and Berkeley Street," 280.

56 huge ruby ring: "The Friedman Lectures on Cryptology," SRH-004, NACP, 18.

57 That swine Lloyd George: Andrew, *Secret Service,* 262.

57 Curzon and Sinclair: Andrew, *Secret Service,* 268–69, 292–93.

57 compromise our work: DENN 1/4, A. G. Denniston Papers, CAC, 8.

58 Fifth Column panics: Andrew, *Secret Service,* 41–43.

58 "Jews and Bolshevists": Andrew, *Secret Service,* 415–16.

59 Sir Barry Domville: Andrew, *Secret Service,* 341.

59 primary task of ONI: Dorwart, *Conflict of Duty,* 53, 80–84.

59 "knew no Italian": Naval Section of G.C. and C.S., HW 3/16, PRO, 23.

59 "swept out of sight": Bennett, *Behind the Battle,* 19.

59 abolish its Military Intelligence Division: Bennett, *Behind the Battle,* 15.

59 Air Ministry was so committed: Watt, "British Intelligence," 267–68.

60 "bloody little man!": Andrew, *Secret Service,* 353.

60 "Situation Report Centre": Watt, "British Intelligence," 264–65.

60 William D. Puleston: Dorwart, *Conflict of Duty,* 88.

60 Japanese naval aviators: Kahn, "United States Views Germany and Japan," 476–77.

60 Albert C. Wedemeyer: Brinkley, *Washington Goes to War,* 60; Wedemeyer, *War and Peace,* 3–4.

61 "No good": Calvocoressi, *Top Secret Ultra,* 8.

2. NATURE OF THE BEAST

Page

62 medieval Arab treatise: Kahn, *Codebreakers,* 97–98.

64 Al-Khalil: Kahn, *Codebreakers,* 97.

64 6 million telegrams: Kahn, *Codebreakers,* 220.

66 definitive mathematical analysis: Solomon Kullback, "Statistical Methods in Cryptanalysis," SRMA-013, NACP.

68 "vilest of the vile": Churchill, *Second World War,* I: 323.

68 hated the Russians: Stokesbury, *World War II,* 64.

69 "must and will go": Shirer, *Rise and Fall,* 212.

69 messages became unreadable: Bloch, "Enigma Before Ultra: Polish Work," 145.

70 string of companies: Kahn, *Codebreakers,* 421–22; Deavours and Kruh, *Machine Cryptography,* 94.

72 about seventeen thousand letters: There are $26 \times 26 \times 26$, or 17,576, different possible combinations of the three rotor settings. However, in actual operation of the Enigma, the turnover mechanism causes a "double stepping" to occur in the middle rotor: each time the middle rotor advances to the position where it will trigger a turnover of the left rotor, it then immediately advances again (along with the left rotor) as the next letter is typed in. If, for example, the turnover occurs between E and F on the middle rotor and V and W on the right rotor, then an actual rotor sequence would be as follows:

ADU

ADV

AEW

BFX

BFY

BFZ

BFA

Thus the key length of the normal Enigma is actually $26 \times 25 \times 26$, or 16,900. When rotors with multiple turnover notches were later introduced, the key length was shortened even further. See Hamer, "ENIGMA."

73   150 million million such permutations: Miller, *Mathematics of Enigma*, 9.

74   simply to spoof: Kozaczuk, *Enigma*, 6.

74   German interest in the Enigma: Welchman, *Hut Six Story*, 207–8; Woytak, "Conversation with Marian Rejewski," 58.

74   course in cryptology: Kozaczuk, *Enigma*, 1–4.

75   "don't tell your colleagues": Woytak, "Conversation with Marian Rejewski," 52–53. There are conflicting accounts of when Rejewski began work on Enigma. Rejewski recalled it was late November 1932, but according to the convincing reconstruction by Bloch, "Enigma Before Ultra: Polish Work," 148, it must have been around October 15.

77   "hatted" codes: "A Cryptographic Dictionary," NR 4559, HCC, 43.

77   FROM KAGA: This illustrative example is adapted from Hinsley and Stripp, eds., *Codebreakers*, 278.

81   encoded at least five different ways: "History of GYP-1," CNSG 5750/202, Crane Files, NACP, 2.

82   "Welcome, gentlemen": Rowlett, *Story of MAGIC*, 19, 34–35.

82   a series of five-letter groups: Rowlett, *Story of MAGIC*, 87, 111.

83   copies of the Red code book: Laurance Safford, "The Undeclared War, History of R.I.," November 15, 1943, SRH-305, NACP, 11–12; Layton, *I Was There*, 31–32, 45–46.

83   "semblance of discretion": J. W. McClaran, Memorandum for the Director, April 10, 1933, "OP-20-G File on Army/Navy Collaboration, 1931 to 1945," SRH-200, NACP, 006.

84   Seeman Gaddis: Layton, *I Was There*, 109–10; "Japanese Suspicions," NR 1812, HCC.

84   resist its cloak and dagger urges: Rowlett, *Story of MAGIC*, 189.

85   Rowlett lay sleepless: Rowlett, *Story of MAGIC*, 112–22.

85   bulldozer of a New Yorker: Phillips, AI; Lewis, AI; "Second Monthly Letter," Captain Geoffrey Stevens, August 17, 1942, HW 14/49, PRO, 2.

85   Rowlett was far more reserved: Phillips, AI; Lewis, AI; Rowlett, *Story of MAGIC*, 252.

85   decades-long battle: Friedman was awarded $100,000 by Congress in 1956 for the development of the SIGABA/Electric Cipher Machine and other inventions that he patented but, because of their secrecy, could not profit from directly. Rowlett, with much justice, insisted that the crucial ideas in SIGABA had been his. Congress finally awarded him $100,000 in 1964.

86   "isomorphs" then began to be evident: Deavours and Kruh, *Machine Cryptography*, 212–20.

87   coax out of Wenger: Kullback, OH; Rowlett, *Story of MAGIC*, 112–13.

87   beat the Army to the punch: Rowlett, *Story of MAGIC*, 129–30.

88   "a thousand times over": "The Undeclared War, History of R.I.," SRH-305, NACP, 13.

88   "AD" code: Layton, *I Was There,* 77.

88   the new "B" machine: "Naval Security Group History to World War II," SRH-355, NACP, 8.

88   June 1, 1939: "History of GYP-1," CNSG 5750/202, Crane Files, NACP, 1.

3. "IL Y A DU NOUVEAU"

Page

89   a waste of time and money: W. F. Clarke, "Naval Section of G.C. and C.S.," HW 3/16, PRO, 19.

89   obtain approval for a substantial increase: DENN 1/4, A. G. Denniston Papers, CAC, 3–10.

90   "We want no Czechs!": Shirer, *Rise and Fall,* 397.

90   "last territorial claim": Churchill, *Second World War,* I: 308.

90   "left to themselves": Churchill, *Second World War,* I: 327.

90   courier in royal livery: Manchester, *Last Lion,* II: 359–60.

90   "Do not suppose that this is the end": Churchill, *Second World War,* I: 327.

91   After twenty-five years': Churchill, *Second World War,* I: 352.

91   "here was a man who could be relied upon": Shirer, *Rise and Fall,* 387.

91   threw out his prepared text: Shirer, *Rise and Fall,* 453–54; Manchester, *Last Lion,* II: 395–96.

92   British military expenditure: Churchill, *Second World War,* I: 336.

92   "weekends in the country": Manchester, *Last Lion,* II: 483.

92   did not like to be bothered: Bennett, *Behind the Battle,* 3–4.

93   Leonard Woolf: Woolf, *Downhill All the Way,* 254; Woolf, *Journey Not the Arrival,* 15.

93   On June 30: Kahn, *Seizing the Enigma,* 79.

93   *Il y a du nouveau:* Woytak, "Interview with Marian Rejewski," 80.

94   Knox had worked out: *BI,* 3(2): 950–51.

94   "regarded as practically impossible": Kozaczuk, *Enigma,* 56.

94   Humphrey Sandwith: Beesly, "Who Was the Third Man at Pyry?"

94   to conceal its activities: Kozaczuk, *Enigma,* 43–44.

94   "Where did you get these?": Bertrand, *Enigma,* 59–60.

95   Rejewski's answer infuriated him: Fitzgerald, *Knox Brothers,* 234.

95   "very derogatory remarks": Telford Taylor, Memorandum for Colonels Clarke and Corderman, January 22, 1944, "Early 'E' History," NR 4246, HCC.

95   "Nous avons le QWERTZU": Peter Twinn, in Hinsley and Stripp, eds., *Codebreakers,* 126–27.

95   gracious thank you: Kozaczuk, *Enigma,* 60.

96   Menzies was in black tie: Bertrand, *Enigma,* 61.

97   second observation: Rejewski, "Mathematical Solution of the Enigma," 3–8; Welchman, *Hut Six Story,* 208–9.

100   commercial Enigma machine: Deavours, "La Methode des Batons."

100   ordered him to drop all of his other work: Woytak, "Conversation with Marian Rejewski," 53.

100   remains an unanswered question in permutation theory: Rejewski, "Mathematical Solution of the Enigma," 7.

101   mid-November: Rejewski recollected the date as mid-December 1932, but

Bloch's careful reconstruction points to an earlier date. There is also considerable confusion about when Rejewski received the first two documents, but again Bloch makes a strong case that Rejewski must have had them from the start of his work that fall. They certainly would have been required to know that the machine had three wheels, a fixed reflector, and a stecker plug board. See Bloch, "Enigma Before Ultra: Polish Work," 148–50.

101 began work on the Enigma in October 1932: Bloch, "Enigma Before Ultra: Polish Work," 147.

101 "was fond of money": Andrew, *Secret Service,* 449.

101 filed them away on the shelf: *BI,* 3(2): 950–51.

101 Verviers and Liège: Bloch, "Enigma Before Ultra: Polish Work," 153.

102 four variables: Rejewski, "Comments on Appendix 1," 77.

102 As if "by magic": Kozaczuk, *Enigma,* 20.

102 Had they transmitted the starting position in plain text: Woytak, "Conversation with Marian Rejewski," 53.

103 general intelligence summaries: Kozaczuk, *Enigma,* 58.

103 Schmidt was arrested, and shot: Kahn, *Seizing the Enigma,* 115.

103 entered on index cards: Welchman, *Hut Six Story,* 211–13.

104 they called a cyclometer: Rejewski, "Mathematical Solution of the Enigma," 12–15.

104 raw and bloody: Kozaczuk, *Enigma,* 29.

104 "scrap of paper": Rejewski, "Comments on Appendix 1," 77.

104 ten or twenty minutes': Woytak, "Conversation with Marian Rejewski," 55.

104 75 percent of all intercepted messages: Kozaczuk, *Enigma,* 45.

105 vividly recall one intercepted signal: Kozaczuk, *Enigma,* 29.

105 was caught by SS troops: Shirer, *Rise and Fall,* 222.

105 tightened up their operating procedures: Welchman, *Hut Six Story,* 208, 213.

107 about a hundred messages: Welchman, *Hut Six Story,* 64.

107 For reasons that will probably forever remain obscure: Some accounts say the name was derived from the ticking sound the *bomba* made as it rotated; others that it was named in honor of the eponymous ice-cream concoction that Rozycki happened to be eating at the café where the mathematicians came up with the idea. The first seems slightly more convincing, though Kozaczuk, who interviewed Rejewski and other participants in the early Polish work, endorses the latter (Kozaczuk, *Enigma,* 63, note 1).

109 test was performed automatically: This plausible suggestion is made in Davies, "The Bombe," 133–36.

110 At Pyry, the Poles: *BI,* 3(2): 957.

110 noted in his diary: Shirer, *Rise and Fall,* 564.

110 Berlin radio interrupted its musical program: Manchester, *Last Lion,* II: 485.

111 To obtain such men and women: DENN 1/4, A. G. Denniston Papers, CAC, 5.

112 We dined very well: Vincent, quoted in Andrew, *Secret Service,* 452.

112 "All you ever say here is 'Third' ": Henry Dryden, in Hinsley and Stripp, eds., *Codebreakers,* 196.

112 "men of the professor type": Erskine, "GC and CS Mobilizes."

113 "strange fellows notoriously unpractical": Twinn, quoted in Andrew, *Secret Service,* 453.

113 "a polite note": Welchman, *Hut Six Story,* 9–11.

113   began showing up regularly at the Broadway: Hodges, *Turing,* 148, 151.

113   "Not very good": Hodges, *Turing,* 26.

114   "A. M. Turing showed an unusual aptitude": Hodges, *Turing,* 38.

114   Anti-War Council: Hodges, *Turing,* 71.

114   "whenever I found the doors wide open": Filby, "Bletchley Park and Berkeley Street," 274.

114   twice severely injured himself: Fitzgerald, *Knox Brothers,* 73; Welchman, *Hut Six Story,* 33.

114   maximum possible speed: Smith, *Station X,* 20.

114   "It's amazing how people smile": Fitzgerald, *Knox Brothers,* 235.

114   "Turing Machine": Newman, *World of Mathematics,* 2092–95.

115   "To our surprise and delight": Hodges, *Turing,* 138.

115   elaborate—and beautiful—assemblage of gears: Hodges, *Turing,* 156–57.

115   leading proponent of the doctrine of strategic bombing: Overy, *Air War,* 12–13.

118   simply bought it himself: DENN 1/4, A. G. Denniston Papers, CAC, 21; Smith, *Station X,* 20.

118   But then in 1882: Hinsley and Stripp, eds., *Codebreakers,* 306.

118   "ugly" and "hideous": See for example William F. Clarke, "B. P. Reminiscences," HW 3/16, PRO, 48; Diana Payne, in Hinsley and Stripp, eds., *Codebreakers,* 133.

118   During the Munich crisis: William F. Clarke, "B. P. Reminiscences," HW 3/16, PRO, 49–51.

119   On August 1, 1939: DENN 1/4, A. G. Denniston Papers, CAC, 21.

119   barely controlled chaos: Denniston to Menzies, September 12, 1939, HW 14/1, PRO; Denniston to Sinclair, September 16, 1939, HW 14/1, PRO.

119   lavish midday meal: Smith, *Station X,* 5.

120   section led by Lieutenant Colonel John Tiltman: Hinsley and Stripp, eds., *Codebreakers,* 209.

120   The French Army: Manchester, *Last Lion,* II: 589.

120   nightmarish journey: Kozaczuk, *Enigma,* 70–73.

121   *mañana:* Kasparek and Woytak, "Marian Rejewski," 21.

4. FIGHTING BACK

Page

122   "endlessly waiting": Woolf, *Journey Not the Arrival,* 9–11.

122   Gallup poll reported: Manchester, *Last Lion,* II: 608–10.

122   bundled in coats and mittens: Smith, *Station X,* 28–29.

123   adapt British Typex cipher machines: *BI,* 3(2): 952.

123   "sex-cyclometer": "Enigma–Position," November 11, 1939, HW 14/2, PRO.

123   "stayed so long in the bathroom": Professor E. R. Vincent, quoted in Andrew, *Secret Service,* 94.

123   an extremely difficult man: Welchman, *Hut Six Story,* 34–35.

124   sent an urgent request to America: Welchman, *Hut Six Story,* 87.

124   Knox furiously told him: Welchman, *Hut Six Story,* 71–72.

124   Turing eventually wrote up his analyses: "Turing's Treatise on the Enigma," NR 964, HCC.

125   "at any moment be canceled": *BI,* 3(2): 954.

125   Knox had already developed a scheme: Deavours, "La Methode des Batons."

127   "Jeffreys sheets": "Turing's Treatise on the Enigma," NR 964, HCC, 80–89, 95–96; *BI,* 3(2): 952.

128   geometric property: Derek Taunt's chapter in Hinsley and Stripp, eds., *Code-breakers,* 100–112, explains this concept with admirable clarity. See also "Cryptanalytic Report on Yellow Machine," NR 3175, HCC, which is a very complete American liaison report from 1943 describing the Enigma and the bombe method; also, "Turing's Treatise on the Enigma," NR 964, HCC, 97–104.

129   four Enigmas could be plugged together: Two excellent explanations of the bombe's wiring and theory of operation are Ellsbury, "Breaking the Enigma," and Davies, "The Bombe."

131   contracted with British Tabulating Machine Company: *BI,* 3(2): 954.

132   dispatched part of the second set: *BI,* 3(2): 952.

132   "I shall tender my resignation": Knox to Denniston, January 7, 1940, HW 14/3, PRO.

132   The French demurred: Memorandum to C.S.S. (personal), January 25, 1940, HW 14/3, PRO.

133   had been interchanged: *BI,* 3(2): 952.

133   Knox had broken three more days: Memorandum to CSS (personal), January 25, 1940, HW 14/3, PRO; *BI,* 3(2): 952–53.

133   "enigmatic scraps": Lucas, quoted in Smith, *Station X,* 35.

133   "traffic will vanish": Memorandum, Interception of Enigma Traffic, March 1, 1940, HW 14/4, PRO.

134   "arrive in a muddle": Memorandum, Five-letter Enigma Traffic with repeated three-letter indicator at end of heading, November 28, 1939, HW 14/2, PRO.

134   "Heil Hitler": Kozaczuk, *Enigma,* 87.

134   forwarded to Denniston an outline: Memorandum to Commander Denniston, November 18, 1939, HW 14/2, PRO; Welchman, *Hut Six Story,* 76–77.

135   "Which way does a clock go": *Dictionary of National Biography,* "Missing Persons" volume, s.v. "Knox, Alfred Dillwyn."

135   psychological study of successful scientists: Eiduson, *Scientists,* 105–6.

136   "shamelessly recruited friends": Welchman, *Hut Six Story,* 84.

136   "I was not a mathematician": Milner-Barry, OH.

136   "visions of London in flames": Milner-Barry, "C. H. O'D. Alexander," 3.

137   "I knew Gordon well enough": Milner-Barry, OH.

137   himself a chess fanatic: Kahn, *Seizing the Enigma,* 102.

137   "not cut out to be a businessman": Milner-Barry, "C.H. O'D. Alexander," 3–5.

137   took place in a showroom: Hinsley and Stripp, eds., *Codebreakers,* 265; "The British GC & CS," August 1, 1942, NR 3661, HCC, 2–3.

137   Tiltman proved the experts wrong: Hinsley and Stripp, eds., *Codebreakers,* 266; Some Reminiscences by Brigadier John H. Tiltman, "Army and Navy Comint Regs & Paper," NR 4632, HCC, 4.

137   asked a few innocuous questions: Marks, *Silk and Cyanide,* 2–3; Hinsley and Stripp, eds., *Codebreakers,* 264–65.

137   The "A" and "A–" students: Memorandum, September 23, 1942, HW 14/53, PRO.

138 holding a competition: Smith, *Station X,* 81–82.
138 circulated cryptograms and puzzles throughout the Navy: Layton, *I Was There,* 33.
138 received letters signed by William F. Friedman: Lutwiniak, OH.
138 "Everybody who had no visible talent": Bundy, "Some Wartime Experiences," 67.
138 University of Illinois: Marston, AI.
139 "no particular background of training": "Administrative History of the Military Cryptanalytical Branch (to 30 June 1944)," NR 2719, HCC, II, IIA, 5.
139 a week to several weeks late: Welchman, *Hut Six Story,* 229.
139 handed an office tray: Hinsley and Stripp, eds., *Codebreakers,* 310.
139 Menzies, who had been: Andrew, *Secret Service,* 343–44, 439.
139 Two other high officials at the SIS: Bennett, *Behind the Battle,* 29; Andrew, *Secret Service,* 439–50.
140 "were of course troopships": Christopher Morris, in Hinsley and Stripp, eds., *Codebreakers,* 238.
140 "surprise, ruthlessness, and precision": Churchill, *Second World War,* I: 591.
141 "Wrong Way Forbes": Miller, *War at Sea,* 58–59.
141 first break came April 10: *BI,* 2: 660; *BI,* 3(2): 953.
141 "unprecedented in the history of warfare": Bennett, *Behind the Battle,* 42–43.
142 Lloyd George took to the floor: Churchill, *Second World War,* I: 659–60.
142 "My warnings": Churchill, *Second World War,* I: 667.
142 Enigma indicators disappeared: Bloch and Erskine, "Dropping of Double Encipherment"; *BI,* 3(2): 953n.
142 first bombe: "Squadron Leader Jones' Section," HW 3/164, PRO, 1.
143 "Herivel tip": Welchman, *Hut Six Story,* 98–99, 230; *BI,* 3(2): 953–54.
143 "cillis": Welchman, *Hut Six Story,* 99–110.
144 "We are beaten": Churchill, *Second World War,* II: 42.
145 *"Où est la masse de manoeuvre?":* Churchill, *Second World War,* II: 46–47.
145 first day the Red key was read currently: Welchman, *Hut Six Story,* 229.
145 "I give the gift of myself": Stokesbury, *World War II,* 102.
145 began demolishing the walls: Shirer, *Rise and Fall,* 741–42.
145 Hitler ordered it blown up: Shirer, *Rise and Fall,* 742; Stokesbury, *World War II,* 103.
146 Bertrand replied on June 26: Bertrand, signal CXG 4, June 26, 1940, HW 14/5, PRO.
146 secure three airplanes: Kozaczuk, *Enigma,* 109.
146 "current plan": Bertrand, signal CXG 2465, June 25, 1940, HW 14/5, PRO. ("Plan actuel est continuer travailler dans [?]pareilles conditions.")
146 Denniston replied: Denniston to Bertrand, June 26, 1940, HW 14/5, PRO.
146 "P.C. Cadix": Kozaczuk, *Enigma,* 113–14.
146 "resume cooperation": Memorandum for The Director, March 5, 1941, HW 14/3, PRO; see also *BI,* 3(2): 954.
147 "the leading expert outside Germany": Hinsley and Stripp, eds., *Codebreakers,* 77–78.
148 miscalculation, incompetence, and misconduct: Churchill, *Second World War,* I: 654–55; Thompson, *War at Sea,* 35–37; Miller, *War at Sea,* 73–75.
148 "What is your source?": Hinsley and Stripp, eds., *Codebreakers,* 78–79.

149   letter from Churchill's top aide: HW 1/1, PRO.
149   "at least five-fold": *BI,* 2: 4.
149   defend themselves against Churchill's proddings: *BI,* 2: 4–5.

5. IMPOSSIBLE PROBLEMS

Page
150   *U-33:* "The History and Sinking of 'U.33,'" ADM 199/2057, PRO, 110–11.
151   "I forgot to throw the wheels away": Kahn, *Seizing the Enigma,* 111.
151   numbers VI and VII: *BI,* 3(2): 955.
151   "have it to myself": "History of Hut Eight," NR 4685, HCC, 14.
151   as early as May 1, 1937: "History of Hut Eight," NR 4685, HCC, 13.
152   procedure manual: "Enigma General Procedure," NR 1679, HCC, 4, 6.
153   FORTYWEEPY: "History of Hut Eight," NR 4685, HCC, 13.
153   Each would have to be run separately on the bombe: The thirty Enigmas
       (later thirty-six) contained in each bombe often meant that three separate
       menus could be run at once, using ten or twelve Enigmas for each menu. But
       because of the complication that turnovers of the middle wheel introduced
       into matching a crib to the cipher text, it was usually necessary to run several
       variations of the menu for each crib at each wheel order.
154   not even sure it was practical: "History of Hut Eight," NR 4685, HCC, 14, 21–
       22.
154   "EINS" catalogue: "History of Hut Eight," NR 4685, HCC, 21.
154   reading the 1938 traffic: "Turing's Treatise on the Enigma," NR 964, HCC,
       140–41.
154   creative and facetious names: "Squadron Leader Jones' Section," HW 3/164,
       PRO.
155   tangle of movable plugged cables: "Report on E Operations of the GC & CS
       at Bletchley Park," NR 3620, HCC, 60.
156   a particular horror: Whitehead, "Cobra and Other Bombes," 303; Hinsley and
       Stripp, eds., *Codebreakers,* 134.
156   "strictly forbidden weapon": Whitehead, "Cobra and Other Bombes," 303.
156   results had been far from promising: "Squadron Leader Jones' Section," HW
       3/164, PRO, 2.
156   bag was on the verge of sinking: Report of H.M.S. "Griffin," May 5, 1940,
       ADM 199/476, PRO, 252–54.
157   break a message for April 23: "History of Hut Eight," NR 4685, HCC, 22; Er-
       skine, "First Naval Decrypts," 43; Hinsley and Stripp, eds., *Codebreakers,* 113.
       Each of these accounts slightly contradicts the others. Erskine suggests that
       no Enigma settings could have been recovered from VP 26 because of the
       long delay that subsequently occurred in Hut 8's breaking of traffic for April
       22, 26, and 27. However "History of Hut Eight" states that some settings were
       found in the VP 26 material, but only for April 23 and 24, and that "the scrap
       of paper on which they were written was for some time ignored." That could
       explain both the initial delay of a few days in breaking traffic for April 23 and
       24 and the much longer delay in breaking the other dates—which required
       more involved hand methods to recover steckers for paired days, and bombe
       runs for the days on which only cribs were available.

157 solid triumph in late June: This date is well established by Erskine, "First Naval Decrypts," who thoroughly searched the records of decrypts teleprinted from Bletchley to the Admiralty, found in ADM 223/620, PRO. This corrects the account in *BI*, 1: 163, 336.

157 months of frustration ensued: "History of Hut Eight," NR 4685, HCC, 22.

157 "take away souvenirs": ADM 199/476, PRO, 258–61.

158 like people waiting for a miracle: "History of Hut Eight," NR 4685, HCC, 23–24.

158 "total inability make himself understood": "History of Hut Eight," NR 4685, HCC, 24.

158 I suggest we obtain the loot: "Operation Ruthless," ADM 223/463, PRO.

159 "very ingenious plot": "Operation Ruthless," ADM 223/463, PRO.

160 In a 1937 opinion poll: Burns, *Roosevelt: Lion*, 399.

160 Fewer than 10 percent of Americans: Manchester, *Glory and Dream*, 201.

160 FDR said in a radio address: Burns, *Roosevelt: Lion*, 395.

160 "Well, Captain, we may as well face the facts": Manchester, *Glory and Dream*, 202.

161 "a few nice boys with BB guns": Manchester, *Glory and Dream*, 202.

161 Friedman asked for a budget increase: "Correspondence on Expansion of the Signal Intelligence Service 1939," NR 4532, HCC, NACP; "Notes on History of Signal Intelligence Service," NR 3245, HCC, 79.

162 employee No. 10: Snyder, AI.

162 taught himself Japanese: Rowlett, *Story of MAGIC*, 28–29.

162 Hurt wandered over . . . "He had a theory": Kullback, OH.

162 official SIS plan: "Notes on History of Signal Intelligence Service," NR 3245, HCC, 80.

163 future wife, Elizebeth Smith: United States, *Pioneers in Cryptology*, 18–20.

163 four top-rank hires: Lewis, AI.

164 very long cribs: Deavours and Kruh, *Machine Cryptography*, 233.

164 predict the first few words of each message: Rowlett, *Story of MAGIC*, 146–47.

165 the building shared its power: Marston, AI.

166 round of Cokes: Rowlett, *Story of MAGIC*, 153.

166 shook with the banging of hammers: Rowlett, *Story of MAGIC*, 156.

167 only 240 combinations: Japanese Machines, March 29, 1945, "Enigma (Conferences, Theory, Related Inf.)," NR 1737, HCC. The key list at first contained only 120 combinations, and later was doubled in number.

167 first twenty-five letters of message text: "General History of OP-20-3 GYP," CNSG 5750/201, Crane Files, NACP, 15.

167 Safford also quite sensibly said: L. F. Safford, Memorandum for Admiral Noyes, July 27, 1940, "Army–Navy Collaboration, 1931–1945, Part I," CNSG 5750/225, Crane Files, NACP.

168 No one location could pick up all Japanese traffic: A Study of the Radio Intercept Activities of the Army and the Navy with Respect to Coverage of Foreign Diplomatic Traffic, August 24, 1940, "Study of Intercept Activities—1940," NR 2130, HCC.

168 oddest of compromises: Benson, *U.S. Communications Intelligence*, 13–14.

168   neither OP-20-G nor SIS was happy: Report by A. G. Denniston, October 31, 1941, HW 14/45, PRO, paras. 21, 23.

168   "dirty politics": Burns, *Roosevelt: Lion,* 424.

169   Stimson need not be told: Kruh, "Stimson," 85.

169   expressed nothing but satisfaction: Rowlett, *Story of MAGIC,* 172–73.

169   saw it as fundamentally dishonest: Kruh, "Stimson," 80–82.

169   If we can do this to the Japanese: Rowlett, *Story of MAGIC,* 162, 172.

169   received a letter from the Adjutant General: Cook, OH.

170   seeing Cook's thumbs: Rowlett, OH.

170    $37 billion defense appropriation: Stokesbury, *World War II,* 118.

171   "the old red blood": Burns, *Roosevelt: Lion,* 421.

171   two-thirds of Americans were now in favor: Manchester, *Glory and Dream,* 224.

171   "we shall fight on the beaches": Churchill, *Second World War,* II: 117–18.

171   began to blame Jewish ownership: Manchester, *Glory and Dream,* 220.

171   "greatest and cheapest victory": Burns, *Roosevelt: Lion,* 439.

172   running out of cash: Churchill, *Second World War,* II: 555–56.

172   "somewhat mixed up together": Burns, *Roosevelt: Lion,* 441.

172   Roosevelt overrode Secretary of State Cordell Hull's: Bath, *Axis Enemy,* 33–34.

173   "must not be understood to commit your government": Bath, *Axis Enemy,* 42.

173   "gold mine": Benson, *U.S. Communications Intelligence,* 16.

173   Britain's Super-Secret ASV: van der Vat, *Atlantic Campaign,* 210.

173   "plunked down on my desk": Bath, *Axis Enemy,* 35.

173   a dyed-in-the-wool anglophobe: van der Vat, *Atlantic Campaign,* 332–33.

173   dictating to a junior partner: Bath, *Axis Enemy,* 8–9.

174   shaved with a blowtorch: Cohen and Gooch, *Military Misfortunes,* 62–65.

174   British were astonished: Benson, *U.S. Communications Intelligence,* 17.

174   Friedman at once saw the advantage: Co-operation with GCCS, "Chronology of Cooperation," NR 2738, HCC, 2.

174   proposed to provide the British: J. O. Mauborgne, Memorandum to Assistant Chief of Staff, G-2, October 25, 1940, "Chronology of Cooperation," NR 2738, HCC.

174   "As regards German and Italian": A. G. Denniston, Memorandum to the Director (Personal), November 15, 1940, HW 14/8, PRO.

175   Should this expert make a favourable impression: Letter, C/5392, 22 November 1940, HW 14/45, PRO.

175   admitted to Walter Reed: Kahn, *Codebreakers,* 389.

175   promoted to director of communications research: "Notes on History of Signal Intelligence Service," NR 3245, HCC, 78.

175   Canadian delegation: Benson, *U.S. Communications Intelligence,* 19.

175   passage in a warship: Cable No. 3154, December 18, 1940, HW 14/45, PRO.

176   anchored at Annapolis: Kahn, *Seizing the Enigma,* 236.

176   The wooden crates were pockmarked: Currier, "My 'Purple' Trip," 195–96.

176   Shenley Park: Currier, OH.

176   were warmly welcomed: Smith, *Station X,* 125–26.

176   "Americans on a secret mission": Currier, "My 'Purple' Trip," 199; Currier, OH.

177   "most important letter of his life": Burns, *Roosevelt: Soldier,* 12–13.

177   "the whole program": Burns, *Roosevelt: Soldier,* 25.

177   picketed the British Embassy: Burns, *Roosevelt: Soldier,* 48.

177   "whether they are mothers": Manchester, *Glory and Dream,* 230.

178   working as a Nazi propagandist: Brinkley, *Washington Goes to War,* 29.

178   like lending chewing gum: Brinkley, *Washington Goes to War,* 51.

178   a few hours later: Burns, *Roosevelt: Soldier,* 49.

178   "C" reported to Churchill: C to Prime Minister, C/5906, 26 February 1941, HW 1/2, PRO; Erskine, "Churchill and the Start of the Ultra–Magic Deals."

178   "by word of mouth only": Weeks to Denniston, March 3, 1941, HW 14/45, PRO.

178   Army representatives gave a similar pledge: Benson, *U.S. Communications Intelligence,* 20.

178   not permitted to make written notes: Briefs for Field Marshal by Colonel Tiltman, Ref: General Marshall's letter to Field Marshal of 23/12/42, "Copies of Letters Between the Field Marshal and General Marshall, etc.," HW 14/60, PRO. Tiltman explicitly states in this 1942 document that the Americans were not permitted to take notes on the British methods for solving the Enigma. In a talk given fifty years later, Prescott Currier, one of the American participants, said that "we brought back all of the information we really wanted and there was never any question that anyone was holding anything from us" (Currier, "My 'Purple' Trip," 198). But the contemporary documents, including Weeks's own handwritten agreement with Denniston that specifies "word of mouth only," clearly show the strict limitations that were placed on what the Americans were allowed to have in writing. Currier in his talk was in part attempting to correct the exaggerated stories of a supposed British "double cross" and thus stressed how much the Americans were allowed to see of the Enigma operations at Bletchley.

179   were given a "paper" Enigma: Hastings to Denniston, cable No. 118, December 5, 1941, HW 14/45, PRO.

179   "avoid duplication": Report of Technical Mission to England, A. Sinkov and Leo Rosen, April 11, 1941, "Army and Navy Comint Regs & Papers," NR 4632, HCC.

179   given up on solving the Purple machine: Co-operation with GCCS, "Chronology of Cooperation," NR 2738, HCC, 5.

179   did no more than supply the needed translators: Report of Technical Mission to England, A. Sinkov and Leo Rosen, April 11, 1941, "Army and Navy Comint Regs & Papers," NR 4632, HCC.

6. SUCCESS BREEDS SUCCESS

Page

180   Elmers School took a direct hit: Smith, *Station X,* 54.

180   "B.Q. Party": Memorandum, September 10, 1940, HW 14/7, PRO; letter, January 2, 1941, HW 14/10, PRO; "Plans for GC&CS," HW 14/5, PRO.

181   may result in a complete stoppage: W. G. Welchman, "Notes on Priority," March 11, 1942, HW 14/31, PRO.

181   lacked the experience of the Army operators: Memorandum, August 26, 1940, HW 14/6, PRO.

181   only the most general intelligence: Bennett, *Behind the Battle,* 57–59.

181   Brown, was broken beginning September 2: *BI,* 2: 659.

181   false but still repeated story: DeWeerd, "Churchill, Coventry, and ULTRA," provides a succinct rebuttal of the canard that Churchill "sacrificed" Coventry; see also P. S. Milner-Barry's account in Hinsley and Stripp, eds., *Codebreakers,* 94–95. The story first appeared in *The Ultra Secret* by Frederick Winterbotham, and has been repeated many times since. In fact, once the British learned through ULTRA of the planned German attack they made considerable efforts to disrupt the raid regardless of any risk to ULTRA that might ensue. British bombers were sent to bomb the base from which the Germans were planning to launch their attack, and a patrol of 130 fighters was sent up to intercept the raid. Owing to the relatively poor night-fighting abilities of the RAF at this stage in the war, however, they succeeded in shooting down only one German airplane.

182   was not going to spend a peaceful evening: Gilbert, *Churchill,* 912–14.

182   far surpassed by the Italian marshal's: Churchill, *Second World War,* II: 469.

182   Fliegerkorps X: "North Africa, 1941–43," GC&CS Air & Military, IV: 7; Miller, *War at Sea,* 124.

182   dubbed Light Blue: "North Africa, 1941–43," GC&CS Air & Military, IV: 7, 27; *BI,* 2: 660.

183   Knox had it completely solved: "Naval Section of G.C. and C.S.," HW 3/16, PRO, 29; *BI,* 3(2): 950–52; *BI,* 1: 210.

183   "only intended to deceive": W. F. Clarke, Memorandum to Head of G.C. and C.S., March 24, 1941, HW 14/13, PRO.

183   transmitted the next day: "Admiralty Use of Special Intelligence in Naval Operations," ADM 223/88, PRO, 318; Santoni, *Il vero traditore,* 306–7. One-time pads were later used for transmitting signals intelligence to British naval commanders in the field, but in early 1941 these had not yet been issued; see "Admiralty Use of Special Intelligence in Naval Operations," ADM 223/88, PRO, 17.

184   "blue water" admiral: Bennett, *Behind the Battle,* 77n.

184   "cover plan of my own": Cunningham's full account of the battle is found in Cunningham, *Sailor's Odyssey,* 325–37.

185   "What battleship is that?": Miller, *War at Sea,* 128.

185   "somehow increased *Gloucester*'s speed": Cunningham, *Sailor's Odyssey,* 327.

185   "yellow-livered skunks": Miller, *War at Sea,* 130.

186   "Good Lord . . . we've hit her": Cunningham, *Sailor's Odyssey,* 332.

186   By the grace of God and Dilly: Information from Mavis (Lever) Batey.

186   reflector could be set: Conference III—Enigma Theory—30 March 1945, "Enigma (Conferences, Theory, Related Inf.)," NR 1737, HCC.

186   within a few weeks . . . GC&CS had broken it: *BI,* 2: 668; Hamer, Sullivan, and Weierud, "Enigma Variations," 212–13.

187   "short and cryptic" message: Churchill, *Second World War,* III: 356–58.

188   "Boniface": See, for example, *BI,* 2:4.

188   CX/FJ: Hinsley and Stripp, eds., *Codebreakers,* 23.

188   Typex machines: *BI,* 2: 631.

188   hustled out of the Admiralty map room: Beesly, *Very Special Intelligence,* 99.

188   "C" simply decided otherwise: Bennett, *Behind the Battle,* 58–59.

188   Churchill continually pestered "C": An exasperated reply from Hut 3 at one point observed, "It is generally accepted that few men rival the Minister of Defence [Churchill was both Prime Minister and Minister of Defence] in the mastery of language. And it is certain, in my opinion, that only very rarely would our recipients detect certain of the fine points raised by him if they had both versions"—that is, both the Hut 3 paraphrase and the original decrypt. Memorandum from Hut 3 to C.S.S., January 1, 1943, HW 1/1282, PRO.

189   "a more direct control": "Admiralty Use of Special Intelligence in Naval Operations," ADM 223/88, PRO, 14.

189   first signals intelligence: "Admiralty Use of Special Intelligence in Naval Operations," ADM, 223/88, PRO, 17–18.

190   General Hans Jeschonnek: Miller, *War at Sea,* 162.

190   "Long live the Führer": Hinsley and Stripp, eds., *Codebreakers,* 53.

190   Admiralty prudently decided: "Admiralty Use of Special Intelligence in Naval Operations," ADM 223/88, PRO, 19.

191   I find myself unable to devise: "C" to Prime Minister, June 24, 1941, HW 1/6, PRO.

191   a German submarine sunk earlier: Erskine, "First Naval Enigma Decrypts," advances this speculation; he notes that no German warships were captured around this time and that GC&CS does not appear to have recovered the wiring of rotor VIII by cryptanalytic means.

191   naval Enigma setting for April 28: "History of Hut Eight," NR 4685, HCC, 26.

192   daily Enigma settings for February: Information from the Naval Historical Branch, Ministry of Defence, London.

192   Banburismus was working better: Jack Good, in Hinsley and Stripp, eds., *Codebreakers,* 156–57.

192   proved almost impenetrable: No March traffic was ever broken, and by May 10, Hut 8 had broken only eight days of April traffic; see Erskine, "Naval Enigma: A Missing Link," 497.

193   "dream of all escort vessels": Balme, OH.

193   open fire with every available weapon: Report, Capture of U 110, Commander A. Baker-Cresswell, May 10, 1941, ADM 1/11133, PRO, 10.

193   picking up survivors: *U-110*'s captain, Kapitänleutnant Julius Lemp, drowned. The oft-repeated tale that he was shot by the British boarding party is without foundation and has been categorically denied by Balme himself; see Balme, OH; also a letter to the editor by Balme in the *Daily Telegraph,* February 11, 1988.

193   "finding the results peculiar": Report, Boarding Primrose, Sub Lieutenant D. E. Balme, May 11, 1941, ADM 1/11133, PRO, 13.

194   Allon Bacon, a Royal Navy: Erskine, "Naval Enigma: A Missing Link," 497.

194   arrived at Bletchley that afternoon: Information from the Naval Historical Branch, Ministry of Defence, London.

194   That night at 9:37: Erskine, "Naval Enigma: A Missing Link," 497.

194   could not make it a higher honor: Balme, OH.

194   "Christ, that should be easy": Kahn, *Seizing the Enigma,* 176.

195   arrived on July 4: Information from the Naval Historical Branch, Ministry of Defence, London.

195   signal giving her precise location: *BI,* 2: 163–64.

195   delay of about three days: Erskine, "Naval Enigma: A Missing Link," 499.

195   totaling 818,000 tons: Churchill, *Second World War,* III: 139.

195   By July, U-boat sinkings had dropped: Erskine, "Naval Enigma: A Missing Link," 499.

196   "war with Russia cannot be avoided": Boyd, *Hitler's Japanese Confidant,* 21.

196   added its warnings: Stokesbury, *World War II,* 153.

196   "would be fatal": "C" to Prime Minister, C/6863, June 24, 1941, HW 1/6, PRO.

196   barrage of memoranda: Churchill's opening salvo was to send "C" a copy of an order that had been sent to the British military attaché in Moscow, directing him to make sure that the Russians never referred to any information they were given as having come from a British source (DMI to Military Attaché, Moscow, June 25, 1941, HW 1/8, PRO.) Churchill wrote at the bottom: "Does this satisfy you?" "C" successfully deflected the blow, replying on June 28: "I am satisfied, as all drafts of wires to Moscow based on Most Secret material will be submitted to me," which of course was exactly what Churchill was objecting to—that "C" was exercising unilateral authority to block information as he saw fit. "C" continued to do so.

196   that same day: War Office to British Liaison Staff No. 30, Moscow, HW 1/10, PRO.

197   "last five messages": *BI,* 2: 73.

197   "crush Soviet Russia": Shirer, *Rise and Fall,* 810.

197   as little occupied territory as possible: Breitman, *Official Secrets,* 36; Shirer, *Rise and Fall,* 832.

197   "special tasks": Shirer, *Rise and Fall,* 832.

197   "looters" and "Bolshevik" partisan fighters: Breitman, *Official Secrets,* 43–47, 51, 60.

197   morale and dissent within Germany: "History of the German Police Section," HW 3/155, PRO, para. 1.

197   useful training: *BI,* 2: 670.

198   "Lunch at their mess": "History of the German Police Section," HW 3/155, PRO, para. 1.

198   a spate of reports came in: "The German Police," GC&CS Air & Military, XIII: 234–35.

198   "in competition with each other": German Police Report, ZIP/MSGP.27/ 21.8.41, HW 14/18, PRO.

199   plundering art works: *BI,* 2:671–72.

199   circulated to higher authorities: *BI,* 2: 670.

199   shirty and pedantic memorandum: Memorandum, Lt. Col. Clarke, MI 14, HW 14/19, PRO.

199   intimidation if not of ultimate extermination: Memorandum, German Police, ZIP/MSGP.28/12.9.41, Covering Period 15th–31st Aug 1941, HW 16/6, PRO.

199   "whole districts are being exterminated": Breitman, *Official Secrets,* 92–93.

199   "danger of decipherment": "History of the German Police Section," HW 3/155, PRO, para. 3.

200   not immediately obeyed: "The German Police," GC&CS Air & Military, XIII: 85.

200   at best twice a week: Breitman, *Official Secrets,* 57.

200   did not follow good cipher security practices: Noel Currer-Briggs, in Hinsley and Stripp, eds., *Codebreakers,* 211–15.

201   problem became considerably more difficult: "History of the German Police Section," HW 3/155, PRO, paras. 4, 5.

201   "departures by any means": *BI,* 2: 673.

201   used them as cribs: Appreciation of the 'E' Situation, June to December 1942, HW 14/62, PRO, 5.

201   handful of references to concentration camps: *BI,* 3(2): 736n.

202   for use in a future war crimes tribunal: V. Cavendish Bentinck to Sir Alexander Cadogan, October 8, 1942, HW 14/54, PRO; memorandum, October 16, 1942, HW 14/55, PRO.

202   released in 1996: In 1982, copies of the SS and Police decrypts were turned over to the U.S. Justice Department for use by its Office of Special Investigations, which was hunting down Nazi war criminals who had illegally entered the United States after the war. OSI found nothing in the files of immediate use. In 1996 the National Security Agency declassified the files and sent them to the National Archives ("German WWII Police and SS Traffic," NR 4417, HCC). Breitman implies throughout *Official Secrets* that there was a conspiracy of silence that kept these files classified for so long, and that both justice and history have suffered for it. Yet his own thorough account of what was known about the Nazi atrocities from other sources, both contemporaneously and in the years since, undercuts the strength of that argument. Although the decrypts are a fascinating and chilling window on the German atrocities on the Russian front, they in fact appear to add little if anything to what historians have already thoroughly documented using German official records, letters and scrapbooks of SS and police officers and men, and the testimony of victims. See, for example, Klee, Dressen, and Reiss, eds., *"The Good Old Days,"* which contains many first-person accounts by the perpetrators themselves of executions carried out on the eastern front, particularly horrifying for their nonchalant tone.

202   "pettiness of outlook": Papers Collected by Admiral Godfrey, vol. 22, January 22, 1942, ADM 223/297, PRO.

202   Mess Committee: Memorandum to Commander Saunders, August 8, 1941, HW 14/18, PRO.

202   "No baths at all": Birch to Travis, August 15, 1941, HW 14/18, PRO.

202   only six bombes: Report on Hut 6, July 31, 1941, HW 14/17, PRO.

202   "high grade men": "Squadron Leader Jones' Section," HW 3/164, PRO, 3.

203   shot back a memorandum: August 18, 1941, HW 14/18, PRO.

203   "As a scholar": Letter, Knox to Denniston, HW 14/22, PRO.

203   "If you design a super Rolls-Royce": Letter, Denniston to Knox, November 11, 1941, HW 14/22, PRO.

203   plotting to oust Denniston: Filby, "Bletchley Park and Berkeley Street," 275–76.

204   handed them each a check: Andrew, *Secret Service,* 462.

204   "our wonderful security": Clarke, "Bletchley Park," 92.

204 "very . . . innocent": Milner-Barry, OH.

204 "I think there was a third point": Welchman, *Hut Six Story,* 128.

204 Welchman drafted his personal appeal: The full text appears in *BI,* 2: 655–57.

205 What I do recall: Milner-Barry, "Action This Day."

205 As First Lord: Manchester, *Last Lion,* II: 556.

206 with a staff of about seventy: Filby, "Bletchley Park and Berkeley Street," 277.

206 "dockyard cipher": "History of Hut Eight," NR 4685, HCC, 45. The dockyard cipher and its cryptanalysis is explained more fully by Christopher Morris in Hinsley and Stripp, eds., *Codebreakers,* 233–35, and van der Meulen, "Werftschlüssel."

207 Operation Garden: Hinsley and Stripp, eds., *Codebreakers,* 235; "History of Hut Eight," NR 4685, HCC, 45–46. Early references to "gardening" apparently unassociated with cryptanalysis appear in AIR 14/796, PRO.

207 new system of naming keys: *BI,* 2: 658–68, provides a comprehensive list of keys, though some dates are in error; for the correct date of Shark's appearance, see Erskine, "Naval Enigma: Heimisch and Triton," 168. (Shark, the Atlantic U-boat key, was known to the Germans by the code name "Triton"; Dolphin, the home-waters key, was "Heimisch"; and Porpoise, the Mediterranean key, was "Süd.")

207 lasted a few days at a time: "History of Hut Eight," NR 4685, HCC, 39.

207 Standbys included: "Cryptanalytic Report on Yellow Machine," NR 3175, HCC, 86–88; Bundy, "Some Wartime Experiences," 73; Breaking Possibilities–No. I, HW 14/78, PRO.

207 Welchman forever regretted: "Report on E Operations of the GC & CS at Bletchley Park," NR 3620, HCC, 7.

7. THE MACHINES
Page

208 nine o'clock on Saturday morning: Minutes of Conference, August 16, 1941, "Chronology of Cooperation," NR 2738, HCC.

208 Denniston laid out the situation to "C": Memorandum, Denniston to the Director (Personal), August 5, 1941, HW 14/45, PRO.

208 "still being copied": Washington & E. Traffic, Notes on Correspondence, "Bombe Correspondence," CNSG 5750/441, Crane Files, NACP. In addition, HW 14/45 contains a memo that went to Denniston relaying American requests during this period, and bearing his handwritten instructions. "Request all stekkers [sic] for 1941" is followed in his handwriting with the notation, "Send wheel orders for July only at present."

209 "I am a little uneasy": Memorandum to Commander Denniston, August 5, 1941, HW 14/45, PRO.

209 offered him $10,000 a year: Kahn, *Codebreakers,* 368.

209 Mayer wrote: William Mayer to A.C. of S., G-2, May 10, 1940, Box 100, Herbert O. Yardley Collection, NACP.

209 contract to write up a series of reports: "Yardley Contract," Box 100, Herbert O. Yardley Collection, NACP; Memorandum, January 5, 1942, Ottawa, Canada, "Yardley: The Canadian Experience," Document 200, Herbert O. Yardley Collection, NACP.

209 Friedman . . . reluctantly went along: Alvarez, *Secret Messages,* 67–68.

209 two mathematicians from the Canadian: Wark, "Cryptographic Innocence," 645–46.

210 American response "frigid": Denniston to the Director (Personal), August 5, 1941, HW 14/45, PRO, para. 1.

210 "wholly dependent upon the elimination": "Chronology of Cooperation," NR 2738, HCC, 11.

210 told the Canadians in no uncertain terms: Wark, "Cryptographic Innocence," 650–52.

211 "They make far greater use of these machines": Report, A. G. Denniston, October 31, 1941, HW 14/45, PRO.

211 In many years of service here: "History Machine Branch," NR 3247, HCC, 2.

211 no funds were available: "History Machine Branch," NR 3247, HCC, 3–6.

211 "chagrin was almost unbearable": "The Friedman Lectures on Cryptology," SRH-004, NACP, 170–71.

212 accounts of the Civilian Conservation Corps: "History Machine Branch," NR 3247, HCC, 6–7.

212 fifty machines at $1,000 apiece: Austrian, *Hollerith,* 53, 70.

213 another early user . . . was the War Department: Austrian, *Hollerith,* 51.

213 tabulators with plugged cables: Austrian, *Hollerith,* 243.

213 Marshall Field: Austrian, *Hollerith,* 249.

213 $250 a month: "History Machine Branch," NR 3247, HCC, 4.

213 "language studies": "History Machine Branch," NR 3247, HCC, 10.

214 Samuel Snyder and Lawrence Clark: "History Machine Branch," NR 3247, HCC, 10–11; Snyder, AI; Phillips, AI.

215 scout out the machine room: Rowlett, *Story of MAGIC,* 221–22, 228.

215 Snyder would be dispatched: Snyder, AI; "History Machine Branch," NR 3247, HCC, 14–15.

216 perfectly fit the "S" code's pattern: "History of GYP-1," CNSG 5750/202, Crane Files, NACP, 11–13.

217 one-part code: Jacobsen, AI.

217 Admiral Godfrey radioed: Benson, *U.S. Communications Intelligence,* 20.

217 nearly doubled the number of recoveries: "History of GYP-1," CNSG 5750/202, Crane Files, NACP, 21–22.

217 A few days before the Pearl Harbor attack: "OP-20-GY," CNSG 5750/198, Crane Files, NACP.

217 From December 1941 to June 1942: "History of GYP-1," CNSG 5750/202, Crane Files, NACP, 51.

218 FBI was tailing him: Lewis, AI; Phillips, AI. Kullback, OH, relates a slightly different version of the story, in which the FBI surreptitiously opened the courier's suitcases and photographed the book.

218 immediately gave twenty-five more lines: "History Machine Branch," NR 3247, HCC, 31; Filby, "Floradora and a Unique Break," 411; Kullback, OH; DENN 1/4, A. G. Denniston Papers, CAC, 9. Filby, writing from memory in 1995, states that ten lines were obtained, but both the near-contemporaneous account in "History Machine Branch" and Kullback's recollections agree on the figure of twenty-five. The actual number of lines that the cryptanalysts had to work with was double this number in any case (thus "History Machine Branch" refers to

fifty lines being punched on IBM cards), because the Iceland documents had revealed that the second 5,000 lines in the additive book were all reciprocals of the first 5,000 lines; that is, if line 0001 started with the additive group 56891, then line 5001 started with the group 54219—the two groups when summed together by noncarrying addition yielded 00000. So if lines 0030–0040 were obtained, the cryptanalysts automatically knew lines 5030–5040 as well. According to Denniston, even before the Iceland "pinch," GC&CS had, during the period 1932–39, managed to reconstruct the basic book because it was used unenciphered by the Germans for routine administrative matters.

219   "But we'd get the job done": Lutwiniak, OH.

219   "Dear Alastair": Filby, "Floradora and a Unique Break," 412–13.

219   BERLIN STOP: Filby, "Floradora and a Unique Break," 412.

219   GLEICHLAUTEND AN: Lewis, AI.

220   "We DOOD it": "Exchange Between Kullback and Johnson While Latter at GC and CS," NR 2385, HCC.

220   "Denniston here": Filby, "Floradora and a Unique Break," 415.

220   in about two hours: "History Machine Branch," NR 3247, HCC, 36–37.

220   OHIO: Goodwin, *No Ordinary Time,* 268, Manchester, *Glory and Dream,* 224.

221   "No American will think it wrong": Churchill, *Second World War,* III: 606–8.

221   "Freedom, hope, strength": Goodwin, *No Ordinary Time,* 301.

222   General Headquarters of the new Army: Cline, *Washington Command Post,* 11.

222   Hoover Field: Brinkley, *Washington Goes to War,* 76.

222   incapable of coping: Brinkley, *Washington Goes to War,* 107, 231, 241.

222   thirty dollars a month: McGinnis, AI.

222   D.C. housing officials: Brinkley, *Washington Goes to War,* 243.

223   Marshall had invited the Navy: Parrish, *Roosevelt and Marshall,* 317–19.

223   managers of the new Statler Hotel: Brinkley, *Washington Goes to War,* 118–19.

223   "gracious, Christian womanhood": Bradford, *Elizabeth J. Sommers.*

223   went through the motions: Historical Review of Progress and Accomplishments of OP-20-GA-1; Memorandum for OP-20, November 6, 1942; Memorandum for Commander Wenger, October 21, 1942; all found in "OP-20GA/G-20," CNSG 5750/161, Crane Files, NACP.

225   acquiring the site by condemnation: Letter, November 20, 1942, James Forrestal to Mount Vernon Seminary, "OP-20GA/G-20," CNSG 5750/161, Crane Files, NACP.

225   the 162 students: Information from public affairs office, Naval District Washington; "Mt. Vernon Seminary Taken By Navy for Training School," *Washington Post,* November 25, 1942.

225   $800,000 . . . "vital to the war effort": Brinkley, *Washington Goes to War,* 116–18.

225   On February 7, 1943: "OP-20GM-6/GM-1-C-3/GM-1/GE-1/GY-A-1 Daily War Diary," CNSG 5750/176, Crane Files, NACP.

225   converted to the Navy Exchange: McGinnis, AI.

225   Arlington Hall Junior College: United States Army, "Forty One and Strong."

225   still in the dormitory: Alvarez, *Secret Messages,* 114.

226   only working bathroom: Marston, AI.

226   legions of rodents: Alvarez, *Secret Messages*, 122–23.

226   Preston Corderman: "Notes on History of Signal Intelligence Service," NR 3245, HCC, 85–88; Benson, *U.S. Communications Intelligence*, 80–81.

226   totaled 2,300: Benson, *U.S. Communications Intelligence*, 83.

226   "civilians in uniform" . . . "unmade bed": Phillips, AI.

226   none of them knew how to shoot a weapon: McGinnis, AI.

227   "rigged the whole thing": Lutwiniak, OH.

227   Cecil Phillips: Budiansky, "A Tribute to Cecil Phillips."

228   "If you're so smart": Marston, AI.

228   At the time of America's entry: "Naval Security Group History to World War II," SRH-355, Part I, NACP, 445; "Historical Review of OP-20-G," SRH-152, NACP.

228   right off the boat: McGinnis, AI.

228   the Army protested: Memorandum, October 5, 1944, file 311.5 6-20-42, ABC Files, NACP.

229   "the dislike of Jews": Report, 20 May 1942, HW 14/46, PRO.

229   "every inch a soldier": *Dictionary of National Biography,* s.v. "Tiltman, John Hessell."

229   "must you wear those damned boots" . . . "captured the Park in five minutes": Filby, "Bletchley Park and Berkeley Street."

230   like to learn how to shoot: Hodges, *Turing*, 231–32.

232   had been a bit hasty: Bennett, *Behind the Battle*, 75–76.

232   lobbying to wrest the entire operation away: *BI*, 2: 21–25.

232   complete reorganization: Ralph Bennett, in Hinsley and Stripp, eds., *Code-breakers*, 31.

232   what frequencies to monitor: Notes on Priority, March 11, 1942, HW 14/31, PRO; Notes on a Most Secret Document Dated 7.9.41, September 9, 1941, HW 14/19, PRO.

233   "toughest assignment he had": Benson, *U.S. Communications Intelligence*, 35–36.

233   "take appropriate precautionary measures": Layton, *I Was There*, 123, 140–41.

234   the exclusive domain of OP-20-G: Benson, *U.S. Communications Intelligence*, 43, 47.

234   All twelve bombes were set to work: "History of Hut Eight," NR 4685, HCC, 48–49.

234   *Donner* and *Geier:* Information from the Naval Historical Branch, Ministry of Defence, London.

234   captured document dated January 1941: "History of Hut Eight," NR 4685, HCC, 62. Battleships were actually the first to receive the M4 Enigma, in October 1941, for use on the key known as Neptun (which GC&CS called Barracuda). But due to the limited action these ships saw the traffic carried on these circuits was minimal, and Neptun was never broken during the war. *BI*, 2: 664; information from Ralph Erskine.

234   2,000 rpm: Whitehead, "Cobra and Other Bombes."

235   "four-wheel duds": "History of Hut Eight," NR 4685, HCC, 62; see also Erskine, "Naval Enigma: Heimisch and Triton," 169, and *BI*, 2: 747.

235   In two weeks in January: Miller, *War at Sea*, 292–95; Churchill, *Second World War,* IV: 126.

236   Off Hatteras the tankers sink: Miller, *War at Sea,* 294–98, 302–4.

236   "Hooligan Navy": van der Vat, *Atlantic Campaign,* 347.

237   "bloody incompetence": "The Americans, the Navy Department, and U-boat Tracking," ADM 223/286, PRO. Beesly, *Very Special Intelligence,* 302, has a tamer version of Winn's words.

237   "better information to impart": "The Americans, the Navy Department, and U-boat Tracking," ADM 223/286, PRO; Bath, *Axis Enemy,* 78. Many writers have argued that Admiral King's anglophobia was so great that he willfully refused to learn from British experience and slighted the Atlantic theater in an obsessive pursuit of a "Japan First" strategy. But Cohen and Gooch, *Military Misfortunes,* 59–94, convincingly refute this commonly held view. King certainly disliked the British, though he also disliked just about everyone. He argued strongly in favor of a European invasion in 1943 and appointed many of his best admirals to the Atlantic theater. Cohen and Gooch also make a very persuasive case that the American failures against U-boats in 1942 were the result of a complex interaction of many factors and that the key to British—and ultimately American—success was an integrated intelligence system of the kind Winn helped establish.

237   "Secret Room": "Functions of 'Secret Room' of Cominch Combat Intelligence, Atlantic Section, Anti-Submarine Warfare, WWII," SRMN-038, NACP.

237   "duplicate our work on E": Travis from Tiltman, [18 (?) April 1942], "Bombe Correspondence," CNSG 5750/441, Crane Files, NACP.

238   "Hardly think necessary": For OP-20-G from C.G. & C. S., May 13, 1942, "Bombe Correspondence," CNSG 5750/441, Crane Files, NACP.

238   "exploiting" meant: "A Cryptographic Dictionary," NR 4559, HCC, 36.

238   concealing success: J. N. Wenger, Memorandum for OP-20-GM, Subject: Recent information on "E," August 6, 1942; both in "Bombe Correspondence," CNSG 5750/441, Crane Files, NACP.

238   "it is a gamble": Memorandum for Op-20, Subject: Cryptanalysis of the German (Enigma) Cipher Machine, September 3, 1942; Wenger to G.C. & C.S. for Eachus, September 4, 1942; both in "Bombe Correspondence," CNSG 5750/441, Crane Files, NACP.

238   Jesuitical argument: Memorandum for Director of Naval Communications, "Captain Wenger Memoranda," NR 4419, HCC, 4.

238   only had about thirty bombes: "Squadron Leader Jones' Section," HW 3/164, PRO, 4.

239   overloaded the available machines: Hut 6 Report of July and August 1942, HW 14/51, PRO.

239   Welchman warned Travis: Memorandum, January 5, 1943, HW 14/63, PRO.

239   "full collaboration": Erskine, "Holden Agreement." I am grateful to Ralph Erskine for calling my attention to this important agreement and to its full text, contained in microfiche in "Army and Navy Comint Regs & Papers," NR 4632, HCC.

239   A recent calculation: Davies, "The Bombe," 131.

240   Turing, astonished: Welchman, *Hut Six Story,* 81. Welchman places the date of his discovery in November 1939 but this seems unlikely; see *BI,* 3(2): 955.

240   feel for which relay was tripped: "Squadron leader Jones' Section," HW 3/164, PRO, 2.

240 "machine gun": "Cryptanalytic Report on Yellow Machine," NR 3175, HCC, 34–35.

240 "Consecutive Stecker Knock Out": "Cryptanalytic Report on Yellow Machine," NR 3175, HCC, 39–40.

240 plaguing both designs: Production of High Speed Machines, Hut 6, October 4, 1942, "Bombe Correspondence," CNSG 5750/441, Crane Files, NACP; Progress up to September 29th. 1942, HW 14/53, PRO.

241 Cobra suffered: Whitehead, "Cobra and Other Bombes."

241 open warfare: Memorandum of Meeting with Dr. Radley, May 29, 1943, HW 14/177, PRO.

241 twenty thousand vacuum tubes: Burke, *Information and Secrecy,* 283.

241 shelved that plan: Wenger to G.C. & C.S. For Eachus, September 4, 1942, "Bombe Correspondence," CNSG 5750/441, Crane Files, NACP.

241 crux of the problem: Desch Report, September 15, 1942, "Bombe Correspondence," CNSG 5750/441, Crane Files, NACP.

242 The final design: The Standard #530 Bombe, "Tentative Brief Descriptions of Cryptanalytic Equipment for Enigma Problems," NR 4645, HCC; Brief Resume of OP-20-G and British Activities vis-a-vis German Machine Ciphers, July 14, 1944, "OP-20GY-A/GY-A-1," CNSG 5750/205, Crane Files, NACP, 5.

242 WAVES: S. K. Allyn to Capt. E. S. Stone, April 7, 1943, "Bombe Correspondence," CNSG 5750/441, Crane Files, NACP.

242 "smiled inwardly": Visit to National Cash Register Corporation of Dayton, Ohio, Report of Dr. Turing of G.C. & C.S., December 1942, "Bombe Correspondence," CNSG 5750/441, Crane Files, NACP, 3.

242 ninety-six machines: *BI,* 2:57, incorrectly asserts that an agreement to limit the Navy's production to one hundred machines was a "compromise" reached at GC&CS's insistence in September–October 1942. In fact, the agreement on Enigma cooperation reached during Travis's visit to Washington did not at that time specify the number of bombes to be built. All contemporary documents agree that it was only after Turing's visit in December, and then only on purely technical grounds, that the Navy settled on building ninety-six. Later on the British were actually urging the Americans to increase their bombe production. See Memorandum for Director of Naval Communications, Subj: History of the Bombe Project, May 30, 1944, "Captain Wenger Memoranda," NR 4419, HCC, 7–8; also Erskine, "Holden Agreement."

242 Liberty Ships were coming out: van der Vat, *Atlantic Campaign,* 490; Manchester, *Glory and Dream,* 295.

243 "To American production": Manchester, *Glory and Dream,* 296.

243 Friedman made a blanket arrangement: Marston, AI.

243 Navy also cut a deal: Entries for February 16 and 24, 1942, "File #62 War Diary File on History and Functions of OP-20-G Offices from Dec 42–Oct 45," CNSG 5750/201, Crane Files, NACP.

243 OP-20-G was expanding its machine room: "U.S. Navy Communication Intelligence Organization, Liaison and Collaboration 1941–1945," SRH-197, NACP.

243 Slide Run machine: "History Machine Branch," NR 3247, HCC, Appendix II.

243 "grenades" were principally used: "Tentative Brief Descriptions of Cryptanalytic Equipment for Enigma Problems," NR 4645, HCC; Dudbusting, "Capt. Walter J. Fried Reports/SSA Liaison With GCCS," NR 2612, HCC.

244 Bush suggested that a device: Burke, *Information and Secrecy,* 69–71.

245 gathering dust: "Naval Security Group History to World War II," SRH-355, part I, NACP, 439–40.

245 Other variants of the machine: "Brief Descriptions of RAM Equipment," NR 1494, HCC; "Hypo," NR 1548, HCC.

245 I.C. Machine: "Brief Descriptions of RAM Equipment," NR 1494, HCC, 27–29.

245 "Copperhead": "Brief Descriptions of RAM Equipment," NR 1494, HCC, 24–26. The same principle was incorporated into the design of a later "Brute Force Machine"; see "Request for RAM Equipment," NR 2701, HCC, annex C.

246 "Mike": "Brief Descriptions of RAM Equipment," NR 1494, HCC, 11–13.

246 Eastman Kodak, National Cash Register, and Gray Manufacturing: "RAM Development," NR 2808, HCC; "Brief Descriptions of RAM Equipment," NR 1494, HCC; Development of Rapid Analytical Equipment and Special Devices, February 1, 1944, "OP-20GM-6 etc. War Diary Summaries," CNSG 5750/177, Crane Files, NACP.

246 factor of ten or a hundred: Analysis by the Tetratester, Table 1, "RAM File," NR 3315, HCC.

247 concepts of data processing: "Influence of U.S. Cryptologic Organizations on the Digital Computer Industry," SRH-003, NACP, 4.

8. PARANOIA IS OUR PROFESSION
Page
248 "Accident does not fall on the same side": "German Reactions to Allied Use of Special Intelligence," GC&CS Naval Sigint, VII: 186.

248 respond with only a "yes" or "no": "Battle of the Atlantic," GC&CS Naval, XVIII: 34–35.

248 "out of the question": "German Reactions to Allied Use of Special Intelligence," GC&CS Naval Sigint, VII: 187.

249 *Esso Hamburg* and *Egerland*: "German Reactions to Allied Use of Special Intelligence," GC&CS Naval Sigint, VII: 174.

249 Cape Verde Islands: "German Reactions to Allied Use of Special Intelligence," GC&CS Naval Sigint, VII: 182–83.

249 *Stichwort* would foil any attempt: "History of Hut Eight," NR 4685, HCC, 49.

249 By early autumn of 1940: "The German Navy's Use of Special Intelligence," GC&CS Naval Sigint, VII: 21.

249 Some belated changes: "The German Navy's Use of Special Intelligence," GC&CS Naval Sigint, VII: 16–17.

250 80 percent of the traffic: *BI,* 2: 636.

250 "not enriched by any results of deciphering": "German Reactions to Allied Use of Special Intelligence," GC&CS Naval Sigint, VII: 192.

250 could not possibly possess the cryptologic sophistication: Naval War Staff memorandum, March 18, 1942, quoted in "German Reactions to Allied Use of Special Intelligence," GC&CS Naval Sigint, VII: 189–92.

251 sleeping at "the office": Lieutenant Rudolph J. Fabian to Commander Safford, June 13, 1942, Naval Security Group records, pre-1946, Naval Security Group Command Display.

251 Breakfast was spaghetti: Ballard, *Ultra Active Service,* 148.
252 had a long talk with his deputy: Lieutenant Rudolph J. Fabian to Commander Safford, June 13, 1942, Naval Security Group records, pre-1946, Naval Security Group Command Display.
252 "*We* shall return": Manchester, *American Caesar,* 271.
252 EVACUATE PERSONNEL RADIO INTELLIGENCE UNIT: "Evacuation of USN Comint Personnel from Corregidor," SRH-207, NACP.
253 carrying 111 men: Ballard, *Ultra Active Service,* 148–49.
253 Whitlock immediately faced a predicament: Maneki, *Quiet Heroes,* 3–5.
253 "much to the indignation" . . . wheels of their SIGABA machine: "Evacuation of USN Comint Personnel from Corregidor," SRH-207, NACP.
253 dump it into the ocean: Maneki, *Quiet Heroes,* 78.
254 seven of its ROTC graduates: "The Role of Communications Intelligence in the American–Japanese Naval War," vol. II, SRH-012, NACP, 135.
254 had to "blacklist": A Historical Review of the Progress and Accomplishments of OP-20-GR, 1 January 1941 to 31 December 1943, entry for February 1, 1942, "OP-20-GR," CNSG 5750/183, Crane Files, NACP.
255 NAVY HAD WORD: Frank, "United States Navy," 284.
256 McCollum recalled what happened next: Goren, "Communication Intelligence," 677.
257 "probably a military secret": Goren, "Communications Intelligence," 684.
257 mustachioed Australian adventurer: Frank, "United States Navy," 285–86.
257 befriended the ship's officers: "Compromise of Naval Radio Intelligence from USS Barnett, June 1942," NR 4504, HCC; letter, June 30, 1942, British Admiralty Delegation, Washington, HW 14/47, PRO.
257 dropped the case: Goren, "Communications Intelligence," 669–70.
258 trove of abandoned Japanese code books: Captured Japanese Documents: Their Contribution to OP-20-GYP, "OP-20-GYP History for WWII Era," CNSG 5750/199, Crane Files, NACP, 8.
258 many messages were erroneously enciphered: "History of GYP-1," CNSG 5750/202, Crane Files, NACP, 53.
258 "curse the whole Marine Corps": Rochefort, OH, 118.
258 On August 15: "History of GYP-1," CNSG 5750/202, Crane Files, NACP, 53.
258 Since 1931 the Coast Guard: "Coast Guard War Diary (OP-20-G-70-GU), 1941–43," CNSG 5750/193, Crane Files, NACP.
258 forged a close working relationship: Benson, *U.S. Communications Intelligence,* 48.
259 nationwide roundup of German agents: "Coast Guard War Diary (OP-20-G-70-GU), 1941–43," CNSG 5750/193, Crane Files, NACP, 5–7.
259 "from old newspaper reports": German Clandestine Serial CG2-136, July 18, 1942, SRIC-3226, NACP.
259 meeting on April 2: Benson, *U.S. Communications Intelligence,* 48.
259 the three departments agreed: Report of Conference Appointed to Study Allocation of Cryptanalysis, June 30, 1942, "Army and Navy Comint Regs & Papers," NR 4632, HCC.
259 President signed on July 8: Memorandum for the Director of the Budget, July 8, 1942, "Army and Navy Comint Regs & Papers," NR 4632, HCC.

259   Hoover threatened to begin seizing: Benson, *U.S. Communications Intelligence,* 127.

259   "loyalty, discretion, or intelligence": Benson, *U.S. Communications Intelligence,* 54.

260   "the type of intelligence gathered": "Selected Documents Concerning OSS Operations in Lisbon 5 May–13 Jul 1943," SRH-113, NACP, 17–18.

260   Worse news followed: "Recent Messages Dealing with the Compromise of Japanese Codes in Lisbon," NR 2608, HCC, 1–2.

260   the Joint Chiefs proposed: File 311.5 27 July 43, ABC Files, NACP.

260   play intelligence officer: "Japanese Espionage Activities in the United States, 1941–1943," SRMN-007, NACP, 169–70.

261   order went out from Admiral King: "The Role of Communications Intelligence in the American–Japanese Naval War," vol. II, SRH-012, NACP, 392–94.

261   Rommel announced he was unwell: *BI,* 2: 413.

261   "gossipy" items: Memorandum, September 12, 1942, HW 14/52, PRO; Memorandum to Mr. Cooper, Air Section, September 21, 1942, HW 14/53, PRO.

262   emptying the trash cans: Phillips, AI.

262   WACs dispatched in civilian clothes: Alvarez, *Secret Messages,* 124.

262   with firearms, but with no firearms training: Memorandum for Commander Wenger, October 21, 1942, History of GA-2 (Security Office), "OP-20GA/G-20," CNSG 5750/161, Crane Files, NACP; Alvarez, *Secret Messages,* 123–24.

262   fraternity boys from American University: McGinnis, AI.

262   commander's other headache: McGinnis, AI.

263   Walter Eytan and his brother Ernest: Hinsley and Stripp, eds., *Codebreakers,* 50–51.

263   "We know he's a communist": Patrick Wilkinson, in Hinsley and Stripp, eds., *Codebreakers,* 61.

263   "naval communications": McGinnis, AI.

263   What are you doing now?: "Secrecy," HW, 14/36, PRO.

264   A security officer at Bletchley: James Bellinger to Commander A. G. Denniston, March 23, 1941, HW, 14/13, PRO.

264   "enough sensible people": Walter Eytan, in Hinsley and Stripp, eds., *Codebreakers,* 57.

264   sent a terrifying memo: "Security in G.C. & C.S.," May 11, 1942, HW 14/37, PRO.

265   cautioned against studying Japanese: "History of GYP-1," CNSG 5750/202, Crane Files, NACP, 60.

265   "He is not a suitable person": Memorandum to C.S.S., April 21, 1942, HW 14/35; Memorandum to C.S.S. May 14, 1942, HW 14/37, PRO.

265   special joint U.S.–British war room: "MIS, War Dept., Liaison Activities in the U.K., 1943 to 1945," SRH-153, NACP.

265   Enigma decrypt of September 11: CS/MSS/1391/T1, HW 1/895, PRO.

265   a delay of only one day: "North Africa, 1941–1943," CG&CS Air & Military, IV: 30–31.

266   Breaking Scorpion was made considerably easier: "Appreciation of 'E' Situation, June to December 1942," HW 14/62, PRO, 3.

266   "told them with remarkable assurance": "Brig. Williams and Group Capt. Humphreys Reports Concerning Ultra," NR 4686, HCC, 3.

266   "polite convention": "Brig. Williams and Group Capt. Humphreys Reports Concerning Ultra," NR 4686, HCC, 5–6.

267   We have impeccable evidence: Message, September 11, 1942, HW 14/52.

267   "Some time ago . . . our experts": Brown, "Intelligence"; Kruh, "British–American Cryptanalytic Cooperation," 126.

267   informed Washington that the Cairo code was compromised: Bluebird Incident, "Col. McCormack Trip to London, May–June 1943," NR 3600, HCC, 49–53.

268   failed to reciprocate: *BI,* 2: 58–61.

268   Soviet order of battle: *BI,* 2: 58; an example is an extensive report of Russian Air Force commands, December 5, 1942, HW 14/60, PRO.

268   Oshima toured the Russian front: Boyd, *Hitler's Japanese Confidant,* 65.

268   "relapsed into bad old ways": From Crankshaw, October 16, 1942, HW 14/55, PRO.

268   "contains so little": From Crankshaw, December 2, 1942, HW 14/60, PRO.

269   abruptly shut down: *BI,* 2: 62–63.

269   Germans were reading the Russian codes: Enigma decrypts revealed that Russian Army and Air Force signals were being broken from the very outset of the war with Germany, often within hours of their time of transmission: "Continuous Reading of Russian Army and Air Signals by the Germans," September 6, 1941, HW 14/19, PRO.

269   halt to interception of Russian military traffic: *BI,* 1: 199n; Ref: General Marshall's letter of 23/12/42, "Copies of Letters Between the Field Marshal and General Marshall Etc.," HW 14/60, PRO.

269   "swollen headed": *BI,* 2: 65.

9. THE SHADOW WAR

Page

270   pinch the Panzer Army since July: *BI,* 2: 401.

270   "The convoy sets sail from Naples": *BI,* 2: 424–25.

270   169,000 tons: *BI,* 2: 422.

270   in forty-four of those cases: "North Africa, 1941–1943," CG&CS Air & Military, IV: 209.

271   pro-forma report: "North Africa, 1941–1943," CG&CS Air & Military, IV: 208.

271   three tankers, carrying eight thousand tons: "North Africa, 1941–1943," GC&CS Air & Military, IV: 211–12; *BI,* 2: 442.

271   PANZERARMEE IST ERSCHOEPFT: William Millward, in Hinsley and Stripp, eds., *Codebreakers,* 29; *BI,* 2: 448.

271   twenty-four serviceable tanks . . . Eleven tanks: *BI,* 2: 451, 454.

271   "fierce indignation and dismay": Ralph Bennett, in Hinsley and Stripp, eds., *Codebreakers,* 37.

272   Boniface shows the enemy: *BI,* 2: 459.

272   served under General Edmund Allenby: Bennett, *Behind the Battle,* xxii.

272   overestimating total British strength: *BI,* 5: 43.

274   "foundation of fear": *BI,* 5: 41.

274   "zone of destiny": Churchill, *Second World War,* IV: 112.

274 fed via tame agents: Masterman, *Double-Cross System,* 86.

275 in desperate straits: Masterman, *Double-Cross System,* 109.

275 In late August and September: Erskine, "Bletchley Park Assessment of TORCH."

275 certain British radio activity: *BI,* 2: 482.

275 decrypted a German naval Enigma message: *BI,* 2: 481.

275 early as August 1941: Erskine, "Naval Enigma: An Astonishing Blunder."

275 "worst episode": "History of Hut Eight," NR 4685, HCC, 56–57.

276 "fairly extravagant": "History of Hut Eight," NR 4685, HCC, 61.

276 went over to four wheels: "History of Hut Eight," NR 4685, HCC, 96.

276 June 1, 1944: "History of Hut Eight," NR 4685, HCC, 102.

276 historian Ralph Erskine: Erskine, "Naval Enigma: An Astonishing Blunder."

277 German radio detection unit had arrived: Kozaczuk, *Enigma,* 135–38.

277 showed its own cryptographic imagination: Kozaczuk, *Enigma,* 152.

278 arrived in England at last: Kozaczuk, *Enigma,* 205–9.

278 surrounded by Gestapo agents: Kozaczuk, *Enigma,* 156.

278 They had two questions for him: Kozaczuk, *Geheim-operation WICHER,* 338–41.

278 Bertrand played an even riskier gambit: Kozaczuk, *Enigma,* 211.

279 astonishing broadcast: van der Vat, *Atlantic Campaign,* 407.

279 numbers in the U-boat war: Stokesbury, *World War II,* 131.

279 there was one toilet: Miller, *War at Sea,* 170.

279 Law of Prize . . . unrestricted U-boat warfare: "Battle of the Atlantic," GC&CS Naval, XVIII: 62–67.

280 new and even more savage phase: van der Vat, *Atlantic Campaign,* 413. Dönitz's order was issued in response to a complicated incident in which the *Laconia,* a British liner carrying both Italian POWs and British civilians, including women and children, was torpedoed by *U-156.* The U-boat picked up survivors and was towing four lifeboats when it was attacked by an American Liberator bomber.

280 They shout, even cheer: Lawrence, *Tales of the North Atlantic,* 172.

282 rescue ships had themselves been sunk: Middlebrook, *Convoy,* 112.

282 fatalistic approach to their odds: Stokesbury, *World War II,* 129.

282 "Oh my, how lovely": Middlebrook, *Convoy,* 109.

282 "We were all badly trained": Middlebrook, *Convoy,* 110.

282 "sheer unmitigated hell": Middlebrook, *Convoy,* 105.

282 "a little more attention": *BI,* 2: 548.

283 special indicating system: "History of Hut Eight," NR 4685, HCC, 75.

283 collected and rebroadcast: Erskine, "Kriegsmarine Short Signals," 77.

283 broken by Hut 10: Diary of G. C. McVittie, RLEW, Ronald Lewin Papers, CAC.

283 the matching Norddeich synoptic: "History of Hut Eight," NR 4685, HCC, 74.

284 own indicator system: "History of Hut Eight," NR 4685, HCC, 72.

284 morning of October 29: Kahn, *Seizing the Enigma,* 223–26.

284 reached Bletchley Park on November 24: Information from the Naval Historical Branch, Ministry of Defence, London.

285 three different all-wheel-order runs: "History of Hut Eight," NR 4685, HCC, 77.

285 As a result month of most strenuous endeavor: Admiralty to OPNAV for 20-G (ULTRA), December 13, 1942, "Bombe Correspondence," CNSG 5750/441, Crane Files, NACP.

285   positions of a dozen Atlantic U-boats: Erskine, "Naval Enigma: Heimisch and Triton": 170.

285   just hours too late: "Battle of the Atlantic," GC&CS Naval, XVIII: 269–70.

285   petty wrangle: ADM 1/14256, PRO.

286   discharged for being underage: Erskine, "Kriegsmarine Short Signals," 82.

286   overran and captured a signals unit: *BI,* 2: 298, 404.

286   German POWs and captured documents: Interrogations of German PW Schwartz and Graupe, "German Sigint Activity," NR 3737, HCC; "Disinterred Papers from German Cryptanalysis," NR 3778, HCC; Enemy Cryptanalysis of Converter M-209 Traffic, Major Russell H. Horton, September 26, 1944, "Cover and Deception," NR 3180, HCC.

286   German cryptanalysts in Berlin were employing equipment: "Comparison of Rapid Analytical Machinery," NR 4282, HCC.

287   no detail about allied shipping: "German Navy's Use of Special Intelligence," GC&CS Naval Sigint, VII: 13–15.

287   delay of less than twenty-four hours: "German Navy's Use of Special Intelligence," GC&CS Naval Sigint, VII: 16.

287   determining factor in Dönitz's decision: "German Navy's Use of Special Intelligence," GC&CS Naval Sigint, VII: 68–69.

287   From June to November: "German Navy's Use of Special Intelligence," GC&CS Naval Sigint, VII: 76.

287   doubled the staff: "German Navy's Use of Special Intelligence," GC&CS Naval Sigint, VII: 98.

288   "tardy and conjectural": "Battle of the Atlantic," GC&CS Naval, XVIII: 286–87.

288   unable to break the special "officers-only" naval Enigma settings: Translations of German Intercepts Officer (Officer only) Messages, April 1941–May 1945, Crane Files, NACP.

288   the tried and true "FORT" method: "History of Hut Eight," NR 4685, HCC, 52.

288   went into service in November 1943: Entry for November 16, 1943, "OP-20GM-6/GM-1-C-3/GM-1/GE-1/GY-A-1 Daily War Diary," CNSG 5750/176, Crane Files, NACP; "OP. 20. G. Bombe Policy," November 4, 1943, HW 14/91, PRO.

289   eight "expected convoys": "Battle of the Atlantic," GC&CS Naval, XVIII: 286.

289   convoys were alerted to alter their routes: "Battle of the Atlantic," GC&CS Naval, XVIII: 285.

289   Dönitz's command system: *BI,* 2: 549–50.

289   Dönitz wrote in his log: "German Reactions to Allied Use of Special Intelligence," GC&CS Naval Sigint, VII: 199.

290   convoy from United States to Gibraltar: "German Navy's Use of Special Intelligence," GC&CS Naval Sigint, VII: 103–4.

290   hadn't happened yet: "German Reactions to Allied Use of Special Intelligence," GC&CS Naval Sigint, VII: 200–207.

291   "Now it can be only me or you": van der Vat, *Atlantic Campaign,* 461.

291   *Atlantiksender:* van der Vat, *Atlantic Campaign,* 461–62.

291   "monotonous": "German Reactions to Allied Use of Special Intelligence," GC&CS Naval Sigint, VII: 206.

291   one to three current convoy messages: "German Navy's Use of Special Intelligence," GC&CS Naval Sigint, VII: 107.

291   HX226 . . . ON166: "Battle of the Atlantic," GC&CS Naval, XVII: 305–6.

291   sank fourteen: van der Vat, *Atlantic Campaign,* 450–51.

292   B-Dienst sprang to life: "German Navy's Use of Special Intelligence," GC&CS Naval Sigint, VII: 111.

292   "extending to months": Erskine, "Kriegsmarine Short Signals," 84.

292   TINA: "TINA," NR 901–903, HCC.

292   "We have been successful": Erskine, "Kriegsmarine Short Signals," 84.

292   in Dönitz's hands within hours: Middlebrook, *Convoy,* 137.

292   146,000 tons of shipping: van der Vat, *Atlantic Campaign,* 454–55.

293   signs of cracking morale: *BI,* 2: 567–68.

293   "a weakling and not a true U-boat commander": "Battle of the Atlantic," GC&CS Naval, XVII: 322.

293   three *Offizier* messages: German Naval Communication Intelligence and Compromise of Allied Ciphers, "Battle of the Atlantic," Vol. III, Naval Security Group records, pre-1946, Naval Security Group Command Display, 44–56.

294   series of crushing attacks: Erskine, "Ultra and U.S. Carrier Operations,"; *BI,* 3(1): 213–14; *BI,* 2: 549.

294   By July of the following summer: Erskine, "Ultra and U.S. Carrier Operations," 95–96.

294   "had so much more to lose": Erskine, "Ultra and U.S. Carrier Operations," 96.

294   instructed to remove an essential component: "Technical Intelligence from Allied C.I.," volume IV, "Battle of the Atlantic," Naval Security Group records, pre-1946, Naval Security Group Command Display.

294   infrared detection: McCue, *U-Boats in Biscay,* 28; *BI,* 3(1): 516.

294   *geistige Arbeit:* Bonatz, *Marine Funkaufklärung,* 28, 92. I am grateful to Rebecca Ratcliff for calling my attention to Bonatz's statements and providing a translation.

294   not getting far with their four-wheel machines: "Squadron Leader Jones' Section," HW 3/164, PRO, 8–9.

295   "Chances of breaking Shark": de Grey to Wenger, June 3, 1943, "Bombe Correspondence," CNSG 5750/441, Crane Files, NACP.

295   American bombes had some teething problems: "History of OP-20-G-4e," NR 4640, HCC; entries for May 3, 1943, June 22, 1943, and August 31, 1943, "OP-20GM-6/GM-1-C-3/GM-1/GE-1/GY-A-1 Daily War Diary," CNSG 5750/176, Crane Files, NACP.

295   only really reliable cribs: "History of Hut Eight," NR 4685, HCC, 89.

295   "Hut 6 jobs": "History of Hut Eight," NR 4685, HCC, 90–92; entry for November 14, 1943, "OP-20GM-6/GM-1-C-3/GM-1/GE-1/GY-A-1 Daily War Diary," CNSG 5750/176, Crane Files, NACP.

296   his habitual costume: Phillips, AI.

296   series of exasperated letters: Stevens, letter, July 31, 1942, HW 14/47, PRO; Stevens, letter, August 17, 1942, HW 14/49, PRO; Stevens, letter, September 28, 1942, HW 14/53, PRO.

296   "hopelessly overorganized": Stevens, letter, September 28, 1942, HW 14/53, PRO.

297   Within two weeks: William Friedman, memorandum for Colonel Bullock THRU Colonel Minckler, Subject: Project in the Cryptanalysis of German

Military Traffic in their High-Grade Cipher Machine, September 14, 1942, "Project 68003," NR 3815, HCC.

297   contract with AT&T for $530,000: "Project X68003—Army Bombe," NR 2723, HCC.

297   cost—which would double: Memorandum for Colonel Corderman, "Comparison of Our '003' Type of 'Bombe' with the Rotary Type," NR 2809, HCC, 5.

297   blocking a security clearance: "Copies of Letters between the Field Marshal and General Marshall, etc.," December 2, 1942, HW 14/60, PRO, tells the full story of the dispute over Turing's access to Bell Laboratories.

297   "fundamental principles and details": Frank W. Bullock, Memorandum for File, January 4, 1943, "Project 68003," NR 3815, HCC.

297   "had better get together": Marston, AI.

297   shown the actual prototype: Major G. G. Stevens, Report on Visit to Bell Laboratories, "Project 68003," NR 3815, HCC.

298   memorandum disparaging the American attempt: "Briefly stated the reasons why the British are averse to the Americans exploiting the intercepted German signals encyphered on their machine," May 4, 1943, HW, 14/75, PRO.

298   added to the abuse campaign: Taylor to Clarke, April 5, 1943, "Army and Navy Comint Regs & Papers," NR 4632, HCC, 3.

298   Taylor advised rejecting: Taylor to Clarke, April 5, 1943, "Army and Navy Comint Regs & Papers," NR 4632, HCC, 6–7.

298   "never on God's green earth": Scope of E Operation—other than Personnel, Excerpt from Cable V4772, May 13, 1943, "Col. McCormack Trip to London, May–June 1943," NR 3600, HCC.

298   Welchman urged moderation: Welchman to Travis, "The Americans and 'E,'" HW 14/68, PRO.

299   "BRUSA" agreement: Its full text appears in "Agreement Between British Government Code and Cipher School and U.S. War Department Regarding Special Intelligence," NR 2751, HCC; it has also been published in *Cryptologia* 21 (1997): 30–38.

299   carrier-pigeon handlers: Bundy, "Some Wartime Experiences," 68.

299   grumbling from Hut 3 and Hut 6: P. S. Milner-Barry to Commander Travis, "American Visitors," August 23, 1943, HW 14/86, PRO.

300   celebrated his promotion: Derek Taunt, in Hinsley and Stripp, eds., *Codebreakers,* 109.

300   own bank of ten bombes: 6812th Signal Security Detachment, 1 Feb 1944–7 May 1945, "Decoding German Enigma Messages," NR 3814, HCC.

300   numbered about 290: "History of Special Project Branch, SIS-ETOUSA," NR 4513, HCC.

300   "Have bad cold": Bundy, "Some Wartime Experiences," 71.

301   "lamentable regularity": "History of Hut Eight," NR 4685, HCC, 95.

301   four thousand men and women: Travis to "C," January 8, 1943, HW 14/63, PRO.

301   the watch in Hut 3: The Hut 3 routine is described in detail by Ralph Bennett, in Hinsley and Stripp, eds., *Codebreakers,* 30–40.

302   "We had tremendous talent": Marchant, OH.

302   where dances were regularly held: Hinsley and Stripp, eds., *Codebreakers,* 65, 133.

302   Naval officers at OP-20-G: McGinnis, AI.

302   post-Prohibition liquor laws: Brinkley, *Washington Goes to War,* 228.

303   "Cryptanalytic Bowling Team": Phillips, AI; Budiansky, "A Tribute to Cecil Phillips."

303   "small town enough": Brinkley, *Washington Goes to War,* 245.

303   American nurses: Litoff and Smith, *We're in this War, Too,* 12, 29–30; Treadwell, *Women's Army Corps,* 16–17.

303   paying women two-thirds: Nicholson, *Kiss the Boys Good-bye.*

303   "Think of the humiliation!": Treadwell, *Women's Army Corps,* 24–25.

304   "a goddamned mess!": Brinkley, *Washington Goes to War,* 238–39.

304   received 35, applications: Wilcox, *Sharing the Burden.*

304   beset with wild rumors: Treadwell, *Women's Army Corps,* 191–214.

304   pregnancy rates were lower: Great Britain, *Report of Committee on Amenities,* 50.

304   staff of OP-20-G: "Historical Review of OP-20-G," SHR-152, NACP.

304   staff of the U.S. Army's SIS: Phillips, AI; "Achievements of the Signal Security Agency in World War II," SRH-349, NACP, 3; Treadwell, *Women's Army Corps,* 316.

304   "experiment" . . . 1,676 Wrens: "Squadron Leader Jones' Section," HW 3/164, 3, 14.

305   "still dreaming": Diana Payne, in Hinsley and Stripp, eds., *Codebreakers,* 132–33.

305   say they were "writers": Diana Payne, in Hinsley and Stripp, eds., *Codebreakers,* 135.

305   lied about her age: Joan (Clarke) Clift, AI.

305   "routine but detailed work": Treadwell, *Women's Army Corps,* 316.

10. COMMAND OF THE ETHER

Page

306   William Weisband: Preface to United States, *VENONA.*

306   "That sounds interesting": Phillips, AI; Budiansky, "A Tribute to Cecil Phillips."

307   "throw a steam-roller": Appendix dated August 1, 1942, Stevens to London, July 31, 1942, HW 14/47, PRO.

307   request was never forwarded: Memorandum, September 8, 1943, "U.S. British Agreements on Comint Effort 1942–1943," NR 4013, HCC. The words "not sent" are written across this memorandum in pencil.

308   "frequently gets the impression": Corderman from Fried, November 1, 1944, "Clark Files," NR 4566, HCC.

308   ceased sending Berkeley Street: IB 32164, "Clark Files," NR 4566, HCC, 15.

309   so was Copperhead: War Diary, OP-20-G-4-D-2, Summary for Period 26 November 1944 to 25 December 1944, 8 January 1945, "OP-20G Section War Diaries (Various)." CNSG 5750/159, Crane Files, NACP. Two thousand messages of Washington–Moscow and Moscow–Washington traffic were searched for common differences in cipher groups using Copperhead. The five Russian systems were designated ZZA through ZZE by Arlington Hall; ZZE was the system on which Phillips made his original break, which turned out to be used by the KBG (Benson, AI). The designations changed several times; later the trade code apparently was referred to as ZET and the KGB system was ZDJ.

Russian traffic in general was referred to as "Blue" by Arlington Hall; see "Signal Security Agency General Cryptanalytic Branch—Annual Report FY 1945," NR 4360, HCC, 5–7.

310   in the British Embassy in Washington: "Development of the 'G'—'Homer' ['Gomer'] Case," October 11, 1952, Public Release Copies of Records Relating to Project VENONA, NACP.

310   chronicled day by day: Benson, *VENONA Historical Monographs;* preface to United States, *VENONA.*

310   Weisband refused either to admit or deny: Report of Washington Field Office, FBI, November 27, 1953, "William Woolf Weisband," in United States, *VENONA,* 170.

311   John Cairncross: Hinsley and Stripp, eds., *Codebreakers,* 26, 207–8.

311   code name was BARON: West, "VENONA."

311   3,651 pages: "Cipher Work Carried out at S.S.A. on GEE," NR 970, HCC.

312   rotated in a complex pattern: "One-Time Pad," NR 1440, HCC.

312   110 people were working: "Signal Security Agency General Cryptanalytic Branch—Annual Report FY 1945," NR 4360, HCC, 23–24.

313   Tiltman set to work on it: Hinsley and Stripp, eds., *Codebreakers,* 161; Fox, "Colossal Adventures"; Tutte, "FISH and I." In binary noncarrying arithmetic, addition is carried out bit by bit, with the rules $1 + 1 = 0$, and $1 + 0 = 1$. This has the interesting property that addition is exactly the same as subtraction, so encipherment and decipherment can be performed through the exact same machine.

313   favored by Major Ralph Tester: Memorandum on non-Morse Situation, September 18, 1943, HW 14/88, PRO. The Swedish government, with the aid of the brilliant mathematician Arne Beurling, also achieved a break in the German teleprinter traffic using hand methods. On the morning after the German invasion of Denmark and Norway on April 9, 1940, the German government demanded Swedish permission to use the Swedish West Coast cable for communications between Berlin and Oslo. Stockholm hemmed and hawed so as not to arouse suspicions, then gave its seemingly reluctant approval on April 14. A few days later Swedish telephone technicians tapped into the line and Beurling, exploiting the depth of messages that had been sent in the same key, broke the first traffic only a few weeks later. From then until 1944, when the messages began to defy hand methods of cryptanalysis, the Swedes read three hundred thousand German teletype messages. See Försvarets Radioanstalt, *A Swedish Success.*

314   arrived at the "Newmanry" in May 1943: *BI,* 3(1): 479.

314   thirty miles per hour: Jack Good, in Hinsley and Stripp, eds., *Codebreakers,* 163. A full technical description of Robinson and Colossus appears in "Report of British Attack on Fish," NR 1596, HCC; also Flowers, "Design of Colossus."

315   catch fire: Good, "Early Work," 73–74.

315   versus about eighty thousand Enigma messages: *BI,* 2: 29.

315   did not even know the name: *BI,* 4: 12.

315   depending on who you talked to: The original name seems to have been Illicit Series and spawned its other forms over time, probably in part for the reasons that William Friedman, who spent several months at Bletchley Park in 1943,

identified in his report on the ISOS and ISK sections on November 25, 1943: "The term 'Illicit Series' is perhaps a poor one and perhaps arose from the fact that in the early days the traffic emanated from secret transmitters operated by espionage agents located in enemy and neutral countries. At the present time, however, most of the traffic at least so far as the stations located in Europe are concerned, is handled on an official or quasi-official basis, so that the term 'Illicit' or 'Clandestine' is perhaps not properly applicable." See: System GEQ (Orange), "Ger ASA Memoranda on GEQ (Orange Machine) Dip Traffic," NR 4318, HCC, 3.

315   provided Dilly Knox's "ISK": See Nigel West's introduction in Masterman, *Double-Cross System,* xv.

316   breaking the Abwehr Enigma traffic: *BI,* 4: 108; Deavours, "Lobsters, Crabs, and Abwehr Enigma"; Hamer, Sullivan, and Weierud, "Enigma Variations," 219–20.

316   five of those attempts ISOS disclosed: *BI,* 4: 95.

316   "dimly, very dimly": Masterman, *Double-Cross System,* 58–59.

316   had become a certainty: *BI,* 4: 108–11.

317   went to Lisbon: Masterman, *Double-Cross System,* 115–16; *BI,* 4: 112.

317   convoy from Liverpool to Malta: Masterman, *Double-Cross System,* 116; *BI,* 4: 114.

317   "sorry they arrived late": *BI,* 5: 63.

318   invasion was imminent: Bennett, *Behind the Battle,* 261.

318   three days later: Masterman, *Double-Cross System,* 156–58.

318   awarding him the Iron Cross: Masterman, *Double-Cross System,* 173–75.

318   "run the war for twenty minutes": Boyd, *Hitler's Japanese Confidant,* 118.

318   Oshima's reports continued to provide reassurance: Boyd, *Hitler's Japanese Confidant,* 127–28.

318   supplied subtly skewed data: Masterman, *Double-Cross System,* 178–82.

319   each day at 9:00 A.M.: Layton, *I Was There,* 473.

319   carried about 70 percent: Hugh Denham, in Hinsley and Stripp, eds., *Codebreakers,* 277.

319   "CINC COMBINED FLEET WILL VISIT": Midway and Yamamoto: Properly Revisited, "Army and Navy Comint Regs & Papers," NR 4632, HCC.

319   Layton and Nimitz: Layton, *I Was There,* 473–74.

319   Lanphier fired: Miller, *War at Sea,* 371.

320   eleven people: "Japanese Army Codes Solution Section," NR 2418, HCC.

320   the main system had been split: "Technical History, Japanese Army Problem, Part II Historical Development," NR 2718, HCC, 4–10.

320   one to as much as eight months: "History of Cryptanalysis of Japanese Army Codes," NR 3072, HCC, 4–5.

320   only two dozen: "Japanese Army Codes Solution Section," NR 2418, HCC.

320   three to four million IBM cards: "History of Machine Branch," NR 3247, HCC, 40–41; "History of Cryptanalysis of Japanese Army Codes," NR 3072, HCC, 36–38.

321   Statistical calculations: "History of Cryptanalysis of Japanese Army Codes," NR 3072, HCC, 104.

321   BPPS RCRC: "Technical History, Japanese Army Problem, Part II Historical Development," NR 2718, HCC, 4–6.

321 captured message forms: "History of Military Cryptanalytic Branch (to 30 June 1944)," NR 2719, HCC, II/IV/D/1; Lewis, AI.

321 10 × 10 conversion square: "SSA—A New Course in Japanese Army Systems," NR 2836, HCC, 33.

322 CBB sent a cable: Maneki, *Quiet Heroes,* 35–36; "History of Military Cryptanalytic Branch (to 30 June 1944)," NR 2719, HCC, II/IV/C/1.

322 tripled, to 270: "Japanese Army Codes Solution Section," NR 2418, HCC; "History of Military Cryptanalytic Branch (to 30 June 1944)," NR 2719, HCC, II/IV/C/2.

322 more complex encipherment system: "Japanese Army Codes Solution Section," NR 2418, HCC.

322 the word *sentō:* Michael Loewe, in Hinsley and Stripp, eds., *Codebreakers,* 259.

323 Tiltman once spent weeks: Some Reminiscences by Brigadier John H. Tiltman, "Army and Navy Comint Regs & Papers," NR 4632, HCC, 6–7.

323 provided a reliable entry point: "History of Cryptanalysis of Japanese Army Codes," NR 3072, HCC, 138–40; "JAT Write-Up—Selections from JMA Traffic," NR 3225, HCC.

323 By early June 1943: "Achievements of the Signal Security Agency in World War II," SRH-349, NACP, 24–28.

323 unload at Wewak: "Achievements of the Signal Security Agency in World War II," SRH-349, NACP, 19–20.

324 dubbed CAMEL: "History of Cryptanalysis of Japanese Army Codes," NR 3072, HCC, 245.

324 "entire cryptographic library": Maneki, *Quiet Heroes,* 40.

324 hanging up the sopping wet pages: Drea, *MacArthur's Ultra,* 92–93.

324 thirty-six thousand messages: Drea, *MacArthur's Ultra,* 93.

324 a fully automatic deciphering system: "History of Military Cryptanalytic Branch (to 30 June 1944)," NR 2719, HCC, II/IV/D/18; "History of Cryptanalysis of Japanese Army Codes," NR 3072, HCC, 13.

325 MacArthur thwarted it: Drea, *MacArthur's Ultra,* 28–30.

325 distributed ULTRA far more freely: Memorandum for Director of Naval Communications, March 9, 1943, "OP-20-G File on Army/Navy Collaboration, 1941 to 1945," SRH-200, NACP.

326 penetrated most of the German ruses: A very complete analysis of all ULTRA signals relating to the Ardennes offensive was produced by GC&CS on December 28, 1944: "Indications of the German Offensive of December 1944," HW 14/118, PRO.

327 "fire brigade": Bennett, *Behind the Battle,* 279.

327 cost Hitler two hundred thousand men: Stokesbury, *World War II,* 355.

328 "for the final solution": Walter Eytan, in Hinsley and Stripp, eds., *Codebreakers,* 60.

328 Hitler is dead: Alec Dakin, in Hinsley and Stripp, eds., *Codebreakers,* 56.

328 "demonstrations of any sort": Signal 1433/5/D70/L37, May 5, 1945, "Chronological Files, 30 April 1945–9 July 1945," Crane Files, NACP.

328 Two hours before: "Achievements of the Signal Security Agency in World War II," SRH-349, NACP, 19.

EPILOGUE: LEGACY

Page

330   even President Truman was kept in the dark: Moynihan, *Secrecy,* 70–72.

331   innovations in computing: "Influence of U.S. Cryptologic Organizations on the Digital Computer Industry," SRH-003, NACP, 5–8; Burke, "Pendergass Report."

331   were political liberals: Phillips, AI.

331   cranking out cryptic crossword puzzles: Lewis, AI.

331   Turing naively told them: Hodges, *Turing,* 496–57.

332   offered a salary of £5,000: Hodges, *Turing,* 496–97.

334   "we were lucky": Welchman, *Hut Six Story,* 169.

334   for six months refused: Carter W. Clarke, Memorandum for General Bissell, Subject: Army–Navy Agreement regarding Ultra, March 4, 1944, "Army and Navy Comint Regs & Papers," NR 4632, HCC.

335   Hans-Joachim Frowein: "European Axis Signal Intelligence in World War II as Revealed by 'TICOM' Investigations and by Other Prisoner of War Inter-rogations and Captured Material, Principally German, Volume II—Notes on German High Level Cryptography and Cryptanalysis," Army Security Agency, May 1, 1946, National Cryptologic Museum.

335   pluggable reflector: "Operation D," July 25, 1944, HW, 14/108, PRO; German Army and Air Force Enigma, 27 March 1945, "Enigma (Conferences, Theory, Related Inf.)," NR 1737, HCC; memorandum of meeting, March 29, 1944, Frank B. Rowlett, "File—Administrative Records of Army Security Agency, General Cryptanalytic Branch," NR 3070, HCC.

335   without warning some Luftwaffe keys: "Enigma Uhr," Hut 6, July 17, 1944, HW 14/108, PRO.

336   evidence that Japanese commanders paid it much heed: Barnhart, "Japanese Intelligence," 426–28.

336   Japanese cryptanalysts did succeed: "The Japanese Intelligence System," SRH-254, NACP; "Japanese Success in Reading Air Movement Crypto-graphic Systems," NR 2727, HCC; "Appreciation of Japanese Signal Intelli-gence," NR 4547, HCC.

APPENDIXES

Page

345   German Navy on May 1, 1937: "History of Hut Eight," NR 4685, HCC, 13.

345   groups from a printed list: "Enigma General Procedure," NR 1679, HCC, 10–11, 15; see also Erskine, "Naval Enigma: A Missing Link," 494–96.

346   known as "Baby": "Turing's Treatise on the Enigma," NR 964, HCC, 141; Hinsley and Stripp, eds., *Codebreakers,* 114; "Squadron Leader Jones' Sec-tion," HW 3/164, PRO, 1.

346   possible to recover the Grund: Memorandum, J. H. Howard to Commander Engstrom, August 21, 1942, Subject: Location of Basic Setting After Success-ful Bombe Solution, "Grenades," NR 2338, HCC.

347   constructing the cipher alphabet: A complete, though not terribly clear, de-scription of the theory behind Banburismus is found in "History of Hut Eight," NR 4685, HCC, 16–20.

349   Royal Flags Wave Kings Above: Hinsley and Stripp, eds., *Codebreakers,* 158.
351   frequency of letters in Purple cipher text: Deavours and Kruh, *Machine Cryptography,* 235, incorrectly state that the "sixes" have a higher frequency in cipher text because each is replaced by only six substitutes, whereas the "twenties" each have twenty possible substitutes. But that effect is exactly canceled out by the fact that there are twenty of the "twenties" versus only six of the "sixes"—in other words, although each of the "twenties" is replaced by twenty different substitutes, there are more of them to begin with. In fact what determines the frequency with which the letters in each group appear in cipher text is solely the average plain-text frequency of the letters that make up each group.
352   in romanized Japanese text: Deavours and Kruh, *Machine Cryptography,* 236–37.
352   Rosen's chance discovery: Rowlett, *Story of MAGIC,* 148–49.
355   fast, medium, or slow cycle: Japanese Machines, March 29, 1945, "Enigma (Conferences, Theory, Related Inf.)," NR 1737, HCC.
357   OP-20-G was operating about 445 receivers: Benson, *U.S. Communications Intelligence,* 67.
357   Army network at this time: Benson, *U.S. Communications Intelligence,* 85.
357   far-flung network in the United Kingdom: "Col. McCormack Trip to London, May–June 1943," NR 3600, HCC, 40.
359   RAM: "Brief Descriptions of RAM Equipment," NR 1494, HCC; "Rapid Analytic Machinery Needed for Research," NR 2803, HCC; "Development of RAM," NR 2808, HCC; "Hypo," NR 1548, HCC; "Request for RAM Equipment," NR 2701, HCC; "RAM File," NR 3315, HCC; "The 5202," NR 2748, HCC.
359   Bombes: "Squadron Leader Jones' Section," HW 3/164, PRO; "Cryptanalytic Report on Yellow Machine," NR 3175, HCC; "Bombe History," NR 1736, HCC; "Information on 'Bombes' Taken from Mr Fletchers Private Files," HW 3/93, PRO; Crawford, "Autoscritcher and Superscritcher"; "Tentative Brief Descriptions of Cryptanalytic Equipment for Enigma Problems," NR 4645, HCC; Whitehead, "Cobra and Other Bombes."
361   Colossi: Hinsley and Stripp, eds., *Codebreakers,* 139–92; Randell, "Colossus"; Flowers, "Design of Colossus."

Glossary and Abbreviations

Abwehr German espionage service

additive a number added to a code group during encipherment, to disguise its identity

B-Dienst Beobachtungsdienst, German Navy radio "observation service"

bigram a combination of two letters

book breaking the process of recovering the linguistic meanings of code groups

"C" also CSS; chief of British secret service

CBB Central Bureau Brisbane

cipher a letter-for-letter (as opposed to word-for-word) encryption system

code an encryption system in which each word, phrase, punctuation mark, number, etc., is assigned a multidigit number (or sometimes, a multiletter alphabetic string) and listed in a code book supplied to users

code group a group of letters or numerals that stands for an entire word or phrase, as listed in a code book

crash the occurrence of the same letter at the same position in an Enigma message and crib that have been placed in a trial alignment, indicating an impossible position

crib a phrase of probable plain text matched against an encrypted message as an aid to its solution

depth the alignment of two or more code messages such that code groups in corresponding positions of each have been enciphered with the same additive; in a cipher, the alignment of messages such that letters in corresponding position have been enciphered with the same key

D/F direction finding

DNI Director of Naval Intelligence

double hit the occurrence of two pairs of identical enciphered code groups at equal distances apart in two messages

dud an Enigma message in which the stecker, wheel order, and ring setting are known but the initial rotor position is not

enciphered code a code in which the code groups are disguised by a further process of encipherment before transmission, usually by summing them in turn to a sequence of additives drawn from an additive book, though also sometimes by transposition or substitution

FCC Federal Communications Commission

Fish Allied code name for German non-Morse, i.e. teleprinter, ciphers

GC&CS Government Code and Cypher School

GCHQ Government Communications Headquarters

indicator a code group transmitted within a message, often disguised by its own encipherment system, which tells the recipient the setting or additive starting point that was used to encipher the message text

ISK Illicit Series Knox (also "Intelligence Series" and other variants); the decrypts produced by Dilly Knox's section at Bletchley Park responsible for nonsteckered Enigma traffic, notably that used by the Abwehr and its agents (hence "illicit" signals)

ISOS Illicit Series Oliver Strachey; decrypts of the Abwehr's hand cipher systems

key the sequence of alphabetic substitutions or additives used to encipher a given message; also, the settings or starting points used for a given message or for a given group of stations for a certain fixed period of time

MAGIC American code name for intelligence derived from high-level Japanese diplomatic ciphers

MI5 British counterintelligence

MI6 British Secret Intelligence Service

NSA National Security Agency

ONI Office of Naval Intelligence (U.S. Navy)

OP-20-G U.S. Navy radio intelligence bureau

OSS Office of Strategic Services

P/L plain language; an unencrypted message

plain text the text of a message before encryption

Purple American code name for the cipher machine used for high-level Japanese diplomatic traffic

RAF Royal Air Force

RAM Rapid Analytical Machinery

ring setting on the Enigma machine, the position of the outer "tire" relative to the inner "wheel" on each rotor; the ring carries the letters A to Z on its circumference as well as the turnover notches that trigger the adjacent rotor to advance

SIGABA U.S. Army cipher machine; in the U.S. Navy, known as the Electric Cipher Machine

SIS Signal Intelligence Service, the name of the U.S. Army's code breaking bureau in the 1930s and early 1940s; also Secret Intelligence Service, the British agency also known as MI6, responsible for running agents behind enemy lines

stecker the plug board on an Enigma machine that acts to scramble the identity of letters before and after passing through the rotors

stripping the process of identifying and subtracting out the additive sequence used to encipher a message in an enciphered code, thereby laying bare the basic code groups

ULTRA Allied code name for intelligence derived from high-level Axis communications

wheel order the left-to-right sequence of numbered rotors placed in an Enigma machine for a given setting

W/T wireless telegraphy; i.e., radio messages transmitted by Morse Code

Y the interception of enemy radio signals and the decryption, usually in the field, of low- and medium-level codes and ciphers

Bibliography

UNPUBLISHED SOURCES

National Archives at College Park, College Park, Md. (NACP)
ABC Files, Record Group 165
Crane Files, Record Group 38
Herbert O. Yardley Collection, Record Group 457
Historic Cryptographic Collection, pre–World War I Through World War II, ca. 1891–ca. 1981, Record Group 457
Public Release Copies of Records Relating to Project VENONA, Record Group 457
SRH series: Studies on Cryptology, 1917–1977, Record Group 457
SRIC series: Messages of German Intelligence/Clandestine Agents, 1942–1945, Record Group 457
SRMA series: United States Army Records Relating to Cryptology, 1927–1952, Record Group 457
SRMN series: United States Navy Records Relating to Cryptology, 1941–1945, Record Group 457

Public Record Office, Kew, U.K. (PRO)
ADM 1, Admiralty and Secretariat Papers
ADM 53, Ships' Logs
ADM 199, War History Cases and Papers
ADM 223, Naval Intelligence Papers 1921–1961
AIR 14, Files of Headquarters Bomber Command
HW 1, Signals Intelligence Passed to the Prime Minister, Messages and Correspondence
HW 3, Government Code and Cypher School and Predecessors: Personal Papers, Unofficial Histories, Foreign Office X Files and Miscellaneous Records
HW 14, Government Code and Cypher School: Directorate, Second World War Policy Papers
HW 16, Government Code and Cypher School: German Police Section, Decrypts of German Police Communications During Second World War
PREM 3, Prime Minister's Office: Operations Papers

National Cryptologic Museum, Fort George G. Meade, Md. (NCM)
European Axis Signal Intelligence in World War II as Revealed by "TICOM" Investigations
Government Code & Cypher School Official Histories of World War II (copies of the Naval and Air & Military volumes in this series are also in HW 11 at the U.K. Public Record Office)

Churchill Archives Centre, Churchill College, Cambridge University, Cambridge, U.K. (CAC)
A. G. Denniston Papers, DENN
Ronald Lewin Papers, RLEW

Naval Security Group Command Display, Pensacola, Fla.
Naval Security Group records, pre-1946

ORAL HISTORIES

Dennis William Babbage, Imperial War Museum Sound Archive, London (IWM)
David Edward Balme, IWM
Joan Bindon, IWM
Peter John Ambrose Calvocoressi, IWM
Earle F. Cook, National Security Agency Oral History, National Cryptologic Museum, Fort George G. Meade, Md. (NCM)
Prescott Currier, NCM
Tommy Flowers, IWM
Solomon Kullback, NCM
William Lutwiniak, NCM
Herbert Stanley Marchant, IWM
Philip Stuart Milner-Barry, IWM
Joseph J. Rochefort, U.S. Naval Institute, Annapolis, Md.
Frank Rowlett, NCM
Arthur Egerton Watts, IWM

AUTHOR'S INTERVIEWS AND CORRESPONDENCE

David Alvarez, historian, St. Mary's College
Ralph Bennett, duty officer, Hut 3
Robert Louis Benson, National Security Agency official
Joan (Clarke) Clift, intercept operator, Chicksands
Graham Ellsbury, computer scientist, developer of bombe simulator
Ralph Erskine, historian of naval Enigma
David Hatch, director, Center for Cryptologic History, National Security Agency
Phillip Jacobsen, intercept operator, OP-20-G
David Kahn, historian of cryptology
Frank Lewis, cryptanalyst, Arlington Hall
John Lundstrom, historian, Milwaukee Public Museum
George McGinnis, radio engineer, OP-20-G

Dale Marston, chief, machine branch, Arlington Hall
William Newman, son of Max Newman, Bletchley Park cryptanalyst
David O'Keefe, signals intelligence historian, Canadian Department of National
Defence
Cecil Phillips, cryptanalyst, Arlington Hall
John Prestwich, military adviser, Hut 3
Rebecca Ratcliff, historian in residence, 1998–99, National Security Agency
Samuel Snyder, cryptanalyst, Arlington Hall
Dennis Underwood, member of British "Y" service
Forrest E. Webb, IBM operator, Station Hypo

BIBLIOGRAPHIES

Cantwell, John D. *The Second World War: A Guide to Documents in the Public Record Office.* London: H.M. Stationery Office, 1992.
Sexton, Donal J., Jr. *Signals Intelligence in World War II: A Research Guide.* Bibliographies of Battles and Leaders, No. 18. Westport, Conn.: Greenwood Press, 1996.
United States. National Security Agency. Center for Cryptologic History. *Sources on Cryptology.* Fort George G. Meade, Md.

BOOKS AND PUBLISHED REPORTS

Alvarez, David. *Secret Messages: Codebreaking and American Diplomacy, 1930–1945.* Lawrence, Kans.: University Press of Kansas, 2000.
Andrew, Christopher. *Her Majesty's Secret Service: The Making of the British Intelligence Community.* New York: Penguin, 1987.
Austrian, Geoffrey D. *Herman Hollerith, Forgotten Giant of Information Processing.* New York: Columbia University Press, 1982.
Ballard, Geoffrey St. Vincent. *On ULTRA Active Service: The Story of Australia's Signal Intelligence Operations in World War II.* Privately printed, 1991. Geoffrey Ballard, 22 Prospect Road, Rosanna, Victoria 3084, Australia.
Barnhart, Michael A. "Japanese Intelligence before the Second World War: 'Best Case' Analysis." In *Knowing One's Enemies: Intelligence Assessment Before the Two World Wars,* edited by Ernest R. May. Princeton: Princeton University Press, 1984.
Bath, Alan Harris. *Tracking the Axis Enemy: The Triumph of Anglo-American Naval Intelligence.* Lawrence, Kans.: University Press of Kansas, 1998.
Beesly, Patrick. *Very Special Intelligence: The Story of the Admiralty's Operational Intelligence Centre, 1939–1945.* London: Hamilton, 1977.
———. *Room 40: British Naval Intelligence, 1914–1918.* New York: Harcourt Brace Jovanovich, 1982.
Bennett, Ralph. *Behind the Battle: Intelligence in the War with Germany, 1939–45.* Revised edition. London: Pimlico, 1999.
Benson, Robert Louis. *A History of U.S. Communications Intelligence during World War II: Policy and Administration.* United States Cryptologic History, series IV, World War II, Vol. 8. Fort George G. Meade, Md.: Center for Cryptologic History, National Security Agency, 1997.

———. *VENONA Historical Monographs.* Nos. 1–5. Fort George G. Meade, Md.: Center for Cryptologic History, National Security Agency.

Bertrand, Gustave. *Enigma, ou la plus grande enigme de la guerre 1939–1945.* Paris: Plon, 1973.

Bonatz, Heinz. *Die deutsche Marine Funkaufklärung.* Darmstadt: Wehr & Wissen, 1970.

Boyd, Carl. *Hitler's Japanese Confidant: General Oshima Hiroshi and Magic Intelligence, 1939–1945.* Lawrence, Kans.: University Press of Kansas, 1993.

Bradford, Faith. *Elizabeth J. Somers.* Norwood, Mass.: Plimpton Press, 1937.

Breitman, Richard. *Official Secrets: What the Nazis Planned, What the British and Americans Knew.* New York: Hill and Wang, 1998.

Brinkley, David. *Washington Goes to War.* New York: Knopf, 1988.

Burke, Colin. *Information and Secrecy: Vannevar Bush, Ultra, and the Other Memex.* Lanham, Md.: Scarecrow Press, 1994.

Burns, James McGregor. *Roosevelt: The Lion and the Fox.* New York: Harcourt, Brace, 1956.

———. *Roosevelt: The Soldier of Freedom.* New York: Harcourt Brace Jovanovich, 1970.

Calvocoressi, Peter. *Top Secret Ultra.* New York: Pantheon, 1980.

Churchill, Winston. *The World Crisis, 1911–1918.* 1923–29. 4 vols. Reprint. London: Hamlyn, 1974.

———. *The Second World War.* 6 vols. Boston: Houghton Mifflin, 1948–53.

Clayton, Aileen. *The Enemy Is Listening.* New York: Ballantine, 1980.

Cline, Ray S. *Washington Command Post: The Operations Division.* Washington, D.C.: Office of the Chief of Military History, Dept. of the Army, 1951.

Cohen, Eliot A., and John Gooch. *Military Misfortunes: The Anatomy of Failure in War.* New York: Free Press, 1990.

Cunningham, Viscount. *A Sailor's Odyssey.* London: Hutchinson, 1951.

Deavours, Cipher A., and Louis Kruh. *Machine Cryptography and Modern Cryptoanalysis.* Dedham, Mass.: Artech House, 1985.

Dinnerstein, Leonard. *Antisemitism in America.* New York: Oxford University Press, 1994.

Dorwart, Jeffrey. *Conflict of Duty: The U.S. Navy's Intelligence Dilemma, 1919–1945.* Annapolis, Md.: Naval Institute Press, 1983.

Drea, Edward J. *MacArthur's ULTRA: Codebreaking and the War against Japan, 1942–1945.* Lawrence, Kans.: University Press of Kansas, 1992.

Eiduson, Bernice T. *Scientists: Their Psychological World.* New York: Basic Books, 1962.

Eisenhower, Dwight D. *At Ease: Stories I Tell to Friends.* New York: Doubleday, 1967.

Erskine, Ralph. "Breaking Naval Enigma." In *Enigma Symposium 1995,* edited by Hugh Skillen. Privately printed, 1995. Hugh Skillen, 56 St Thomas Drive, Pinner HA5 4SS, England.

Fitzgerald, Penelope. *The Knox Brothers.* New York: Coward, McCann & Geoghegan, 1977.

Försvarets Radioanstalt. *A Swedish Success: Breaking the German Geheimschreiber in WW2.* Stockholm: Försvarets Radioanstalt, 1997.

Geyer, Michael. "National Socialist Germany: The Politics of Information." In *Knowing One's Enemies: Intelligence Assessment Before the Two World Wars,* edited by Ernest R. May. Princeton: Princeton University Press, 1984.

Gilbert, Martin. *Winston S. Churchill: Finest Hour, 1939–1941.* London: Heinemann, 1983.

Goodwin, Doris Kearns. *No Ordinary Time: Franklin and Eleanor Roosevelt: The Home Front in World War II.* New York: Simon and Schuster, 1994.

Graves, Robert. *Good-bye to All That.* 1929. Reprint. New York: Doubleday, 1957.

Great Britain. *Report of the Committee on Amenities and Welfare Conditions in the Three Women's Services.* London: H.M. Stationery Office, 1942.

Hinsley, F. H., et al. *British Intelligence in the Second World War.* 5 vols. New York: Cambridge University Press, 1979–90.

Hinsley, F. H., and Alan Stripp, eds. *Codebreakers: The Inside Story of Bletchley Park.* Paperback. Oxford: Oxford University Press, 1994.

Hodges, Andrew. *Alan Turing: The Enigma.* New York: Simon & Schuster, 1983.

Jones, R. V. *The Wizard War: British Scientific Intelligence, 1939–1945.* New York: Coward, McCann & Geoghegan, 1978.

Kahn, David. "United States Views Germany and Japan in 1941." In *Knowing One's Enemies: Intelligence Assessment Before the Two World Wars,* edited by Ernest R. May. Princeton: Princeton University Press, 1984.

———. *Seizing the Enigma: The Race to Break the German U-Boat Codes, 1939–1943.* Boston: Houghton Mifflin, 1991.

———. *The Codebreakers.* Revised ed. New York: Scribner, 1996.

Klee, Ernst, Willi Dressen, and Volker Riess, eds. *"The Good Old Days": The Holocaust as Seen by Its Perpetrators and Bystanders.* New York: Free Press, 1988.

Kozaczuk, Wladyslaw. *Enigma: How the German Machine Cipher Was Broken and How it Was Read by the Allies in World War Two.* Frederick, Md.: University Publications of America, 1984.

———. *Geheim-operation WICHER: Polnische Mathematiker knacken den deutschen Funkschlüssel.* Konbenz: Bernard & Graef Verlag, 1989.

Lawrence, Hal. *Tales of the North Atlantic.* Toronto: McClelland and Stewart, 1985.

Layton, Edwin T., Roger Pineau, and John Costello. *"And I Was There": Pearl Harbor and Midway—Breaking the Secrets.* New York: Morrow, 1985.

Lewin, Ronald. *Ultra Goes to War.* London: Hutchinson, 1978.

Litoff, Judy Barrett, and David C. Smith. *We're in This War, Too: World War II Letters from American Women in Uniform.* New York: Oxford University Press, 1994.

McCue, Brian. *U-Boats in the Bay of Biscay: An Essay in Operational Analysis.* Washington, D.C.: National Defense University Press, 1990.

Manchester, William. *The Glory and the Dream.* New York: Bantam, 1975.

———. *American Caesar: Douglas MacArthur, 1880–1964.* Boston: Little, Brown, 1978.

———. *The Last Lion, Winston Spencer Churchill.* 2 vols. New York: Little, Brown, 1988.

Maneki, Sharon A. *The Quiet Heroes of the Southwest Pacific Theater: An Oral History of the Men and Women of CBB and FRUMEL.* United States Cryptologic History, series IV, World War II, Vol. 7. Fort George G. Meade, Md.: Center for Cryptologic History, National Security Agency, 1996.

Marks, Leo. *Between Silk and Cyanide: A Codemaker's War 1941–1945*. New York: Free Press, 1999.

Masterman, John C. *The Double-Cross System in the War of 1939–1945*. New Haven: Yale University Press, 1972.

Middlebrook, Martin. *Convoy: The Battle for Convoys SC122 and HX229*. London: Allen Lane, 1976.

Miller, A. Ray. *The Cryptographic Mathematics of Enigma*. Fort George G. Meade, Md.: Center for Cryptologic History, National Security Agency.

Miller, Nathan. *War at Sea: A Naval History of World War II*. New York: Scribner, 1995.

Millett, Allan R., and Peter Maslowski. *For the Common Defense: A Military History of the United States of America*. New York: Free Press, 1984.

Milner-Barry, P. S. "C. H. O'D. Alexander—A Personal Memoir." In *The Best Games of C. H. O'D. Alexander*, edited by Harry Golombek and Bill Harston. London: Oxford University Press, 1976.

Morison, Samuel Eliot. *The History of U.S. Naval Operations During the Second World War*. 15 vols. Boston: Little, Brown, 1947–62.

Moynihan, Daniel Patrick. *Secrecy*. New Haven: Yale University Press, 1998.

Newman, James R. *The World of Mathematics*. New York: Simon and Schuster, 1956.

Nicholson, Jenny. *Kiss the Boys Goodbye*. London: Hutchinson, 1944.

Overy, R. J. *The Air War: 1939–1945*. New York: Stein and Day, 1980.

Parker, Frederick D. *A Priceless Advantage: U.S. Navy Communications Intelligence and the Battles of Coral Sea, Midway, and the Aleutians*. United States Cryptologic History, series IV, World War II, Vol. 5, Fort George G. Meade, Md.: Center for Cryptologic History, National Security Agency, 1993.

—— *Pearl Harbor Revisited: United States Navy Communications Intelligence, 1924–1941*. United States Cryptologic History, series IV, World War II, Vol. 6. Fort George G. Meade, Md.: Center for Cryptologic History, National Security Agency, 1994.

Parrish, Thomas. *Roosevelt and Marshall*. New York: William Morrow, 1989.

Prados, John. *Combined Fleet Decoded: The Secret History of American Intelligence and the Japanese Navy in World War II*. New York: Random House, 1995.

Prange, Gordon W., Donald M. Goldstein, and Katharine V. Dillon. *Miracle at Midway*. New York: McGraw-Hill, 1981.

Randell, Brian. "The Colossus." In *A History of Computing in the Twentieth Century*, edited by N. Metropolis et al. New York: Academic Press, 1980.

Rohwer, Jürgen. *The Critical Convoy Battles of March 1943: The Battle for HX299/SC122*. Translated by Derek Masters. London: I. Allan, 1977.

Rowlett, Frank B. *The Story of MAGIC*. Laguna Hills, Calif.: Aegean Park Press, 1998.

Santoni, Alberto. *Il vero traditore: il ruolo documentato di ULTRA nella guerra del Mediterraneo*. Milan: Mursia, 1981.

Sassoon, Siegfried. *Memoirs of a Fox-Hunting Man*. 1928. Reprint. London: Faber and Faber, 1960.

Shirer, William L. *The Rise and Fall of the Third Reich*. New York: Simon and Schuster, 1960.

Smith, Michael. *Station X: The Codebreakers of Bletchley Park*. London: Channel Four Books, 1998.

Stokesbury, James L. *A Short History of World War II.* New York: Morrow, 1980.

Thompson, Julian. *The Imperial War Museum Book of The War at Sea.* London: Sidgwick & Jackson, 1996.

Thurber, James. "Exhibit X." In *The Beast in Me and Other Animals.* New York: Harcourt Brace Jovanovich, 1973.

Treadwell, Mattie E. *The Women's Army Corps.* Washington, D.C.: Office of the Chief of Military History, Dept. of the Army, 1954.

United States. Congress. Joint Committee on the Investigation of the Pearl Harbor Attack. *Investigation of the Pearl Harbor Attack.* Washington, D.C., 1946.

United States. Congress. Senate. Committee on Naval Affairs. *Women's Auxiliary Naval Reserve.* Washington, D.C., 1942.

United States. National Security Agency. Center for Cryptologic History. *Pioneers in U.S. Cryptology.* Fort George M. Meade, Md.

United States. National Security Agency and Central Intelligence Agency. *VENONA: Soviet Espionage and the American Response, 1939–1957.* Washington, D.C., 1996.

United States. Navy Department. *The Story of You in Navy Blue.* Washington, D.C., 1944.

van der Vat, Dan. *The Atlantic Campaign: The Great Struggle at Sea 1939–1945.* London: Grafton, 1990.

Watt, Donald Cameron. "British Intelligence and the Coming of the Second World War in Europe." In *Knowing One's Enemies: Intelligence Assessment Before the Two World Wars,* edited by Ernest R. May. Princeton: Princeton University Press, 1984.

Wedemeyer, Albert C. *Wedemeyer on War and Peace.* Stanford, Calif.: Hoover Institution Press, 1987.

Welchman, Gordon. *The Hut Six Story: Breaking the Enigma Codes.* Revised ed. Cleobury Mortimer, U.K.: M & M Baldwin, 1997.

Whitehead, David. "The U-Boat Ciphers and the 4-Wheel Bombes." In *The Enigma Symposium 1995,* edited by Hugh Skillen. Privately printed, 1995. Hugh Skillen, 56 St Thomas Drive, Pinner HA5 4SS, England.

Wilcox, Jennifer. *Sharing the Burden: Women in Cryptology during World War II.* Fort George M. Meade, Md.: National Cryptologic Museum.

Woolf, Leonard. *Downhill All the Way.* New York: Harcourt Brace Jovanovich, 1967.

———. *The Journey Not the Arrival Matters.* New York: Harcourt Brace Jovanovich, 1969.

Yardley, Herbert O. *The American Black Chamber.* Indianapolis: Bobbs-Merrill, 1931.

ARTICLES

Atha, Robert I. "Bombe! 'I Could Hardly Believe It!'" *Cryptologia* 9 (1985): 332–36.

Beesly, Patrick. "Who Was the Third Man at Pyry?" *Cryptologia* 11 (1987): 78–80.

Bennett, Ralph. "Ultra and Some Command Decisions." *Journal of Contemporary History* 16 (1981): 131–51.

Bloch, Gilbert. "Enigma Before Ultra: Polish Work and the French Connection." *Cryptologia* 11 (1987): 142–55.

———. "Enigma Before Ultra: The Polish Success and Check." *Cryptologia* 11 (1987): 227–34.

———. "Enigma Before Ultra." *Cryptologia* 12 (1988): 178–84.

Bloch, Gilbert, and Ralph Erskine. "Enigma: The Dropping of Double Encipherment." *Cryptologia* 10 (1986): 134–41.

Boyd, Carl. "Anguish Under Siege: High-Grade Japanese Signal Intelligence and the Fall of Berlin." *Cryptologia* 13 (1989): 193–209.

Brown, Kathryn. "Intelligence and the Decision to Collect It: Churchill's Wartime Diplomatic Signals Intelligence." *Intelligence and National Security* 10 (1995): 449–67.

Budiansky, Stephen. "A Tribute to Cecil Phillips—And Arlington Hall's 'Meritocracy.'" *Cryptologia* 23 (1999): 97–107.

———. "Too Late for Pearl Harbor." *Naval Institute Proceedings,* December 1999, 47–51.

Bundy, William P. "Some of My Wartime Experiences." *Cryptologia* 11 (1987): 65–77.

Burke, Colin. "An Introduction to an Historic Computer Document: The 1946 Pendergass Report Cryptanalysis and the Digital Computer." *Cryptologia* 17 (1993): 113–23.

Chandler, W. W. "The Installation and Maintenance of Colossus." *Annals of the History of Computing* 5 (1983): 260–62.

Chapman, J. W. M. "No Final Solution: A Survey of the Cryptanalytical Capabilities of German Military Agencies, 1926–35." *Intelligence and National Security* 1 (1986): 13–47.

Clarke, William F. "Government Code and Cypher School: Its Foundation and Development with Special Reference to Its Naval Side." *Cryptologia* 11 (1987): 219–26.

———. "The Years Between." *Cryptologia* 12 (1988): 52–58.

———. "Bletchley Park 1941–1945." *Cryptologia* 12 (1988): 90–97.

Coombs, Allen W. M. "The Making of Colossus." *Annals of the History of Computing* 5 (1983): 253–59.

Crawford, David J. "The Autoscritcher and Superscritcher: Aids to Cryptanalysis of the German Enigma Cipher Machine, 1944–1946." *IEEE Annals of the History of Computing* 14 (1992): 9–22.

Currier, Prescott. "My 'Purple' Trip to England in 1941." *Cryptologia* 20 (1996): 193–201.

Davies, Donald. "The Bombe—A Remarkable Logic Machine." *Cryptologia* 23 (1999): 108–38.

Deavours, C. A. "How the British Broke Enigma." *Cryptologia* 4 (1980): 129–32.

———. "La Methode des Batons." *Cryptologia* 4 (1980): 240–47.

———. "The Autoscritcher." *Cryptologia* 19 (1995): 137–48.

———. "Lobsters, Crabs, and the Abwehr Enigma." *Cryptologia* 21 (1997): 193–99.

Deavours, C. A., and Louis Kruh. "The Turing Bombe: Was It Enough?" *Cryptologia* 14 (1990): 331–49.

Denniston, A. G. "The Government Code and Cypher School Between the Wars." *Intelligence and National Security* 1 (1986): 48–70.

Dewdney, A. K. "On Making and Breaking Codes: Part I." *Scientific American,* October 1988, 144–47.

DeWeerd, Harvey A. "Churchill, Coventry, and ULTRA." *Aerospace Historian,* December 1980, 227–29.

Drea, Edward J. "Reading Each Other's Mail: Japanese Communication Intelligence, 1920–1941." *Journal of Military History* 55 (April 1991): 185–205.

————. "Were the Japanese Army Codes Secure?" *Cryptologia* 19 (1995): 113–36.

Ellsbury, Graham. "Breaking the Enigma: Two Emulators for Cryptological Machines." Project Report. Department of Computer Science, Birkbeck College, University of London, 1998.

Erskine, Ralph. "Enigma and the Polish Contribution." Review of *Enigma: How the German Machine Cipher Was Broken and How it Was Read by the Allies in World War Two,* by Wladyslaw Kozaczuk. *Cryptologia* 9 (1985): 316–23.

————. "GC and CS Mobilizes 'Men of the Professor Type.'" *Cryptologia* 10 (1986): 50–59.

————. "Naval Enigma: The Breaking of Heimisch and Triton." *Intelligence and National Security* 3 (1988): 162–82.

————. "A Bletchley Park Assessment of German Intelligence of TORCH." *Cryptologia* 13 (1989): 135–42.

————. "Naval Enigma: A Missing Link." *International Journal of Intelligence and Counterintelligence* 3 (1989): 493–508.

————. "The Soviets and Naval Enigma: Some Comments." *Intelligence and National Security* 4 (1989): 503–11.

————. "The German Naval Grid in World War II." *Cryptologia* 16 (1992): 39–51.

————. "Ultra and Some U.S. Navy Carrier Operations." *Cryptologia* 19 (1995): 81–96.

————. "Naval Enigma: An Astonishing Blunder." *Intelligence and National Security* 11 (1996): 468–73.

————. "Churchill and the Start of the Ultra–Magic Deals." *International Journal of Intelligence and Counterintelligence* 10 (1996): 57–74.

————. "Eavesdropping on 'Bodden': ISOS v. the Abwehr in the Straits of Gibraltar." *Intelligence and National Security* 12 (1997): 110–29.

————. "When a Purple Machine Went Missing: How Japan Nearly Discovered America's Greatest Secret." *Intelligence and National Security* 12 (1997): 185–89.

————. "The First Naval Enigma Decrypts of World War II." *Cryptologia* 21 (1997): 42–46.

————. "Joan E. L. Murray, MBE." *Intelligence and National Security* 13 (1998): 213–14.

————. "Kriegsmarine Short Signal Systems—and How Bletchley Park Exploited Them." *Cryptologia* 23 (1999): 65–92.

————. "The Holden Agreement on Naval Sigint: The First BRUSA?" *Intelligence and National Security* 14 (1999): 187–97.

Erskine, Ralph, and Frode Weierud. "Naval Enigma: M4 and its Rotors." *Cryptologia* 11 (1987): 235–44.

Ewing, Alfred. "Some Special War Work—Part I." *Cryptologia* 4 (1980): 193–203.

————. "Some Special War Work—Part II." *Cryptologia* 5 (1981): 33–39.

Filby, P. William. "Bletchley Park and Berkeley Street." *Intelligence and National Security* 3 (1988): 272–84.

————. "Floradora and a Unique Break into One-Time Pad Ciphers." *Intelligence and National Security* 10 (1995): 408–22.

Flowers, Thomas A. "The Design of Colossus." *Annals of the History of Computing* 5 (1983): 239–52.

Fox, Barry. "Colossal Adventures." *New Scientist,* May 10, 1997.

Frank, Larry J. "The United States Navy v. the *Chicago Tribune*." *The Historian* 42 (1980): 284–303.

Gillogly, James J. "Ciphertext-Only Cryptanalysis of Enigma." *Cryptologia* 19 (1995): 405–13.

Gish, Donald M. "A New Pearl Harbor Villain: Churchill. A Cryptologic Analysis." *International Journal of Intelligence and Counterintelligence* 6 (1993): 369–88.

Godson, Susan H. "The Waves in World War II." *Naval Institute Proceedings,* December 1981, 46–51.

Good, I. J. "Early Work on Computers at Bletchley." *Cryptologia* 3 (1979): 65–77.

Goren, Dina. "Communications Intelligence and the Freedom of the Press. The *Chicago Tribune*'s Battle of Midway Dispatch and the Breaking of the Japanese Naval Code." *Journal of Contemporary History* 16 (1981): 663–90.

Hamer, David H. "ENIGMA: Actions Involved in the 'Double Stepping' of the Middle Rotor." *Cryptologia* 21 (1997): 47–50.

Hamer, David H., Geoff Sullivan, and Frode Weierud. "Enigma Variations: An Extended Family of Machines." *Cryptologia* 22 (1998): 211–29.

Hinsley, Harry. "The Counterfactual History of No Ultra." *Cryptologia* 20 (1996): 308–24.

Jukes, Geoff. "The Soviets and Ultra." *Intelligence and National Security* 3 (1988): 233–47.

Kahn, David. "Why Germany Lost the Code War." *Cryptologia* 6 (1982): 26–31.

———. "Churchill Pleads for the Intercepts." *Cryptologia* 6 (1982): 47–49.

———. "Roosevelt, Magic, and Ultra." *Cryptologia* 6 (1992): 289–319.

———. "An Enigma Chronology." *Cryptologia* 17 (1993): 237–46.

Kasparek, Christopher, and Richard Woytak. "In Memoriam Marian Rejewski." *Cryptologia* 6 (1982): 19–25.

Kozaczuk, Wldyslaw. "Enigma Solved." *Cryptologia* 6 (1982): 32–33.

Kruh, Louis. "How to Use the German Enigma Cipher Machine: A Photographic Essay." *Cryptologia* 7 (1983): 291–96.

———. "Stimson, the Black Chamber, and the 'Gentlemen's Mail' Quote." *Cryptologia* 12 (1988): 65–89.

———. "British–American Cryptanalytic Cooperation and an Unprecedented Admission by Winston Churchill." *Cryptologia* 13 (1989): 123–34.

———. "Why Was Safford Pessimistic About Breaking the German Enigma Cipher Machine in 1942?" *Cryptologia* 14 (1990): 253–57.

Layton, Edwin T. "America Deciphered Our Code." *Naval Institute Proceedings,* June 1979, 98–100.

Lewis, Frank W. "The Day of the Dodo." *Cryptologia* 14 (1990): 11–27.

Lewis, Graydon A. "Setting the Record Straight on Midway." *Cryptologia* 22 (1998): 99–101.

Lundstrom, John E. "A Failure of Radio Intelligence: An Episode in the Battle of the Coral Sea." *Cryptologia* 7 (1983): 97–118.

Mache, Wolfgang. "Geheimschreiber." *Cryptologia* 10 (1986): 230–247.

Mead, David. "The Breaking of the Japanese Army Administrative Code." *Cryptologia* 18 (1994): 193–203.

Miller, A. Ray. "The Cryptographic Mathematics of Enigma." *Cryptologia* 19 (1995): 65–80.

Milner-Barry, P. S. "Action This Day: The Letter from Bletchley Park Cryptanalysts to the Prime Minister, 21 October 1941." *Intelligence and National Security* 1 (1986): 272–76.

———. "The Soviets and Ultra: A Comment on Jukes' Hypothesis." *Intelligence and National Security* 3 (1988): 248–50.

Pendergass, J. T. "Cryptanalytic Use of High-Speed Digital Computing Machines." *Cryptologia* 17 (1993): 124–47.

Rejewski, Marian. "Mathematical Solution of the Enigma Cipher." *Cryptologia* 6 (1982): 1–18.

———. "Remarks on Appendix 1 to *British Intelligence in the Second World War* by F. H. Hinsley." *Cryptologia* 6 (1982): 74–83.

Smith, Thomas T. "The Bodden Line: A Case-study of Wartime Technology." *Intelligence and National Security* 6 (1991): 447–57.

Tutte, W. T. "Fish and I." Lecture, June 19, 1998, Centre for Applied Cryptographic Research, University of Waterloo.

United States. Army. Intelligence and Security Command. "Forty One and Strong: Arlington Hall Station," *Cryptologia* 9 (1985): 306–10.

United States. National Security Agency. National Cryptologic Museum. "In Memoriam—Solomon Kullback." *Cryptologia* 19 (1995): 149–50.

van der Meulen, Michael. "Werftschlüssel: A German Navy Hand Cipher System—Part I." *Cryptologia* 19 (1995): 349–64.

———. "Werftschlüssel: A German Navy Hand Cipher System—Part II." *Cryptologia* 20 (1996): 37–54.

Wark, Wesley K. "Cryptographic Innocence: The Origins of Signals Intelligence in Canada in the Second World War." *Journal of Contemporary History* 22 (1987): 639–65.

Welchman, Gordon. "From Polish Bomba to British Bombe: The Birth of Ultra." *Intelligence and National Security* 1 (1986): 71–110.

West, Nigel. "VENONA: The London Connection." Paper presented at the Cryptologic History Symposium, October 27–29, 1999, National Security Agency, Fort George G. Meade, Md.

Whitehead, David. "Cobra and Other Bombes." *Cryptologia* 20 (1996): 289–307.

Woytak, Richard. "A Conversation with Marian Rejewski." *Cryptologia* 6 (1982): 50–60.

Acknowledgments

I would like to express my deep appreciation to the many historians, librarians, archivists, curators, and cryptologic experts and enthusiasts who helped me in my research for this book. I owe a special debt to Ralph Erskine, the preeminent historian of naval Enigma, who probably feels he wrote this book himself after having answered scores of e-mailed inquiries from me on everything from how Hut 8 recovered the ring settings on the naval Enigma to how the name of an obscure German naval vessel was spelt. Besides being a veritable human tidal wave of historical knowledge of all things naval and cryptanalytical, Ralph was an unfailing source of encouragement and sage advice throughout, and I am especially grateful for the many leads to important archival records that he freely shared with me. In the few places where our interpretations of historical events differ I know he will forgive me.

My special thanks also go to Ralph Bennett, distinguished historian and Hut 3 veteran; John Taylor, national security archivist at the National Archives at College Park; Rebecca Ratcliff, National Security Agency historian in residence, 1998–99; Robert Louis Benson, NSA official and author of groundbreaking studies in cryptologic history; Graham Ellsbury, who not only gave me dinner in London but also explained to me how the Turing bombe worked so I could actually understand it; David Alvarez, professor of politics at St. Mary's College, California; David Hatch, director of the NSA's Center for Cryptologic History; David O'Keefe, signals intelligence historian at the Canadian Department of National Defence; George McGinnis, Naval Cryptologic Veterans Association; and David Kahn, historian of cryptology. My discussions and correspondence with all of them were a source of pleasure and invaluable information throughout the project.

I am also very grateful to many others who helped: the veterans of Bletchley Park, Arlington Hall, and OP-20-G whom I interviewed or corresponded with (their names appear in the Bibliography); Robert S. Scherr, archivist, Naval Security Group Headquarters; Tom Johnson, NSA Center for Cryptologic History; the volunteers and staff of the National Cryptologic Museum, Fort George G. Meade, Maryland; the reading-room staffs of the National Archives at College Park and the U.K. Public Record Office; Frode Weierud, European Center for Nuclear Research (CERN); Petty Officer AnTuan Guerry, Naval District Washington; Mark Dunton, reader information services department, Public Record Office; Brian Randell, University of Newcastle, England; Jeffrey Goldberg, Cranfield University, England; Bernard Cavalcante, Operational Archives Branch, Naval Historical Center, Washington; John Harper and Tony Sale, bombe rebuild project, Bletchley Park Trust; Kathryn McKee, technical services librarian, St. John's College Library, Cambridge University; Jenny Mountain, Churchill Archives Centre, Cambridge University; Kate Johnson, sound archivist, Imperial War Museum, London; Lee Gladwin, archivist, National Archives at College Park; Colin Burke, University of Maryland Baltimore County; Evelyn Cherpak, curator, Naval Historical Collection, Naval War College, Newport, Rhode Island.

Dave Merrill produced the maps with his ever-impressive combination of artistry and intelligence.

Photographs from the collection of the Imperial War Museum are reproduced by permission of the Trustees of the Imperial War Museum, London. I thank Christine Large of the Bletchley Park Trust for permission to reproduce the photograph of the mansion at Bletchley Park, and the National Security Agency and the U.S. Navy for providing copies of photographs from their collections.

Index